NONLINEAR ASPECTS OF TELECOMMUNICATIONS

Discrete Volterra Series and Nonlinear Echo Cancellation

ELECTRONIC ENGINEERING SYSTEMS SERIES

Series Editor: J. K. FIDLER, *University of York*

Asssociate Series Editor: PHIL MARS, *University of Durham*

Published and Forthcoming Titles

KNOWLEDGE-BASED SYSTEMS FOR ENGINEERS AND SCIENTISTS
Adrian A. Hopgood, The Open University

OPTIMAL AND ADAPTIVE SIGNAL PROCESSING
Peter M. Clarkson, Illinois Institute of Technology

CIRCUIT SIMULATION METHODS AND ALGORITHMS
Jan Ogrodzki, Warsaw University of Technology

THE ART OF SIMULATION USING PSPICE - ANALOG AND DIGITAL
Bashir Al-Hashimi, Staffordshire University

PRINCIPLES AND TECHNIQUES OF ELECTROMAGNETIC COMPATIBILITY
Christos Christopoulos, University of Nottingham

LEARNING ALGORITHMS: THEORY AND APPLICATIONS IN SIGNAL PROCESSING, CONTROL AND COMMUNICATIONS
Phil Mars, J. R. Chen, and Raghu Nambiar
University of Durham

INTRODUCTION TO INSTRUMENTATION AND MEASUREMENTS
Robert B. Northrop, University of Connecticut

FUNDAMENTALS OF NONLINEAR DIGITAL FILTERING
Jaakko Astola and Pauli Kuosmanen, Tampere University of Technology

WIDEBAND CIRCUIT DESIGN
Herbert J. Carlin, Cornell University and Pier Paolo Civalleri, Turin Polytechnic

SYNTHETIC APERTURE RADAR PROCESSING
Giorgio Franceschetti and Riccardo Lanari, IRECE, Naples

CONTINUOUS-TIME ACTIVE FILTER DESIGN
Theodore Deliyannis, University of Patras,
Yichuang Sun, University of Hertfordshire,
and J. K. Fidler, University of York

NONLINEAR ASPECTS OF TELECOMMUNICATIONS
Andrzej Borys, University of Technology and Agriculture, Bydgoszcz, Poland

DESIGN AUTOMATION OF INTEGRATED CIRCUITS
Ken Nichols, University of Southampton

NONLINEAR ASPECTS OF TELECOMMUNICATIONS

Discrete Volterra Series and Nonlinear Echo Cancellation

Andrzej Borys

CRC Press
Boca Raton London New York Washington, D.C.

Library of Congress Cataloging-in-Publication Data

Borys, Andrzej
Nonlinear aspects of telecommunications : discrete volterra series and nonlinear echo cancellation / Andrzej Borys.
p. cm. -- (Electronic engineering systems series)
Includes bibliographical references and index.
ISBN 0-8493-2571-4
1. Echo suppression (Telecommunication) 2. Volterra equations. I. Title. II. Series

TK5102.98 .B67 2000
621.382--dc21

00-057180
CIP

Direct all inquiries to CRC Press LLC, 2000 N.W. Corporate Blvd., Boca Raton, Florida 33431.

International Standard Book Number 0-8493-2571-4
Library of Congress Card Number 00-057180
Printed in the United States of America 1 2 3 4 5 6 7 8 9 0
Printed on acid-free paper

Preface

The area covered by this book is the discrete Volterra series and its application to cancellation of nonlinear echo in digital telecommunication systems. This is the first volume of Nonlinear Aspects of Telecommunications, which, as planned, will contain altogether three volumes.

This volume consists of two chapters. The first chapter is devoted to explanation of basics of the discrete Volterra series. The basic definitions and notions regarding the discrete Volterra series, conditions for convergence and stability, and matrix representation for multiple-input and multiple-output nonlinear digital systems are presented. A very important problem of approximating a nonlinear digital system with the use of the discrete Volterra series is dealt with on many pages. The newest achievements in this area, not available in textbooks, are presented here. Finally, other possible approximations for nonlinear digital systems are discussed.

The second chapter deals with the problem of nonlinear echo cancellation. After introductory material regarding basics of adaptive cancellers, structures for nonlinear echo cancellers using nonlinear transversal filters for baseband transmission are analyzed. Nonlinear echo cancellers for voiceband transmission, as well as interleaved structures, are presented in the last section of this chapter.

The book can serve as a text for graduate students of telecommunications. It can be also helpful for engineers involved in the design of telecommunication systems.

At the end of this short preface, I would like to express my sincere thanks to Zbyszek Zakrzewski, who typed the manuscript and did all the drawings. I am also indebted to Navin Sullivan and Felicia Shapiro, editors of the CRC Press, for their support and kind cooperation. My special thanks goes to Zdzistaw Drzycimski, the Head of the Institute for Telecommunications of ATR Bydgoszcz, for his continuous encouragement.

Author

Dr. Andrzej Borys studied electrical engineering at the Technical University of Gdańsk, Poland from 1969 to 1974, where he received his M.S.E.E. degree. He earned a Ph.D. in electrical engineering from the Technical University of Poznań, Poland in 1980. Since 1974, Dr. Borys has been with the Institute for Telecommunications, University of Technology and Agriculture (ATR) Bydgoszcz, Poland, first as a research and teaching assistant and then as an associate professor. From 1983 to 1993, Dr. Borys visited and did research at the Institute of Telecommunications, ETH Zürich, Switzerland; the Institute of Telecommunications of the University of Kaiserslautern, Germany; and the Institute of Telecommunications and Computing Research Center of the Technical University Hamburg–Harburg, Germany. During his stay at the University of Kaiserslautern, he was a research fellow of the Alexander von Humboldt Foundation, Bonn, Germany. Dr. Borys has published more than 50 research papers in electronics and telecommunications. At present, his research interests lie in the areas of telecommunications and computer networks. Dr. Borys is a member of the Association of Polish Electrical Engineers, the Scientific Society of Bydgoszcz, and the Societas Humboltiana Polonorum.

Dedication

To my youngest brother,
Zbyszek

Table of Contents

1 Basics of Discrete Volterra Series

1.0 INTRODUCTION

The discrete Volterra series is one of the powerful tools used in analysis of nonlinear digital systems. In this book, we present its application to nonlinear echo cancellation. Presentation of solutions of nonlinear echo cancellers is, however, preceded by theoretical material in this chapter. This material is necessary for understanding of an analysis presented in the next chapter. The first chapter is long and seems maybe to be partly very advanced because of the usage of functional analysis. The range of functional analysis applied here is, however, needed for explanation and solving of some problems related with the discrete Volterra series, such as its convergence, approximation of a digital nonlinear system by the discrete Volterra series, and so on. All the fundamental notions of functional analysis used are fully explained and illustrated by examples. It is assumed to do this at the place of their first occurrence in the text. No special appendices are devoted to them. The functional analysis allows us to present many known results in a more general framework, as well as to present the recent achievements in the field, which are inherently related to the use of this analysis.

This chapter is devoted to basics of the discrete Volterra series. Thus, after making the introductory remarks, we continue with Section 1.1, presenting the basic definitions regarding the discrete Volterra series. Then, the basic notions related to nonlinear sampled-data or, using another name for such systems, nonlinear digital systems, are explained. Section 1.3 is devoted to the multidimensional Z transform as introductory material for the next section, which deals with the discrete Volterra series representation in the Z domain. Further relations of this representation with the Laplace transform and the discrete Fourier transform representations are discussed as well. The very important topic of convergence and stability of the discrete Volterra series, and of its components, is treated in Section 1.5. The next section presents the matrix representation for multiple-input and multiple-output nonlinear digital systems, which have the representation in the form of a discrete Volterra series. Here, for the first time in this book, the first notion of the functional analysis is introduced, that is, of the norm of a vector, and of a vector-valued sequence. In Section 1.7, the notion of fading memory of a nonlinear discrete-time system described by a nonlinear operator is introduced. Most of the material presented there is, however, devoted to the explanation of basic notions of functional analysis used in the following sections. These notions are mapping, function, operator, transformation, functional, linear space of which elements are vectors, or scalar- or vector-valued sequences, normed space, metric space, convergence and continuity with respect to a given metric, and, very important in this book, the l^∞ space. The topic of fading memory is continued in the next section, where the definitions of the related notions of decaying memory and of approximately-finite memory are introduced, and discussed as well. The relations existing between the definitions of memories mentioned above are explained. In Section 1.9, the approximation of nonlinear discrete-time systems, possessing fading memory, or equivalently approximately-finite memory with an additional property regarding continuity of a certain functional obtained from the operator describing a system by the discrete Volterra series is considered. Some recent results due to Boyd and Chua, [1] and due to Sandberg, [2, 3, 4] are presented. Section 1.10 concerns a specialized version of the discrete Volterra series for binary signals. Here, realization structures of the Volterra series approximator dependent on a defined measure of the memory length of a system considered, and dependent on the strength measure of system nonlinearities, are presented. The main task of Section 1.11 is consideration of the so-called associated, or extended, expansions and their relations to the original representations. The last section of this chapter is devoted to presentation of other approximations different from the discrete Volterra series, such as the approximation with the use of lattice mapping, the approximation using sigmoidal functions, and the approximation exploiting radial functions.

1.1 BASIC DEFINITIONS

The purpose of this section is to provide the reader with some basic definitions regarding the discrete Volterra series. In the forthcoming sections, we will extend these definitions in many directions, which are useful in applications.

To define a discrete Volterra series, let us imagine such a nonlinear discrete system of which response to any input signal (from an allowable set of input signals) can be described in the form of a sum of partial responses. In other words, we postulate the output signal (output sequence), $y(k)$ as a response to the input signal (input sequence) $x(k)$, where k means a discrete time, to be expressed in the form

$$y(k) = y^{(0)}(k) + y^{(1)}(k) + y^{(2)}(k) + y^{(3)}(k) + \ldots = \sum_{n=0}^{\infty} y^{(n)}(k) \qquad (1.1)$$

Note that $y^{(0)}(k)$, $y^{(1)}(k)$, $y^{(2)}(k)$, $y^{(3)}(k)$ in Equation 1.1 mean the zero-order, first-order, second-order, third-order, respectively, partial responses, and so on. Figure 1.1 illustrates Equation 1.1. In this figure, we assume that the components of the zero-order, first-order, second-order, third-order, and so on are related in some way with (or correspond to) respective nonlinearity orders. So the zero-order component is an independent-of-input-signal component, the first-order component is a linear component, the second-order component is a quadratic one, and the third-order component is a cubic one.

When the partial responses in Equation 1.1 have the form

$$y^{(0)}(k) = h^{(0)}(k) \tag{1.2a}$$

$$y^{(1)}(k) = \sum_{i=-\infty}^{\infty} h^{(1)}(k,i)x(i) \tag{1.2b}$$

$$y^{(2)}(k) = \sum_{i_1=-\infty}^{\infty} \sum_{i_2=-\infty}^{\infty} h^{(2)}(k, i_1, i_2)x(i_1)x(i_2) \tag{1.2c}$$

and in general,

$$y^{(n)}(k) = \sum_{i_1=-\infty}^{\infty} \sum_{i_2=-\infty}^{\infty} \cdots \sum_{i_n=-\infty}^{\infty} h^{(n)}(k, i_1, i_2, \ldots, i_n)x(i_1)x(i_2)\ldots x(i_n) \tag{1.2d}$$

where $n \in \mathbb{N}$, $\mathbb{N} = \{1, 2, 3, \ldots\}$ meaning the set of positive integers, then Equation 1.1 (with Equations 1.2) describes a discrete Volterra series for time-dependent systems.

It is worthwhile to make at this point a terminological remark regarding the order of occurrence of indices in the multiple sums, as, for example, in Equation 1.2d. In Equation 1.2d, the order of occurrence of the indices $i_1, \ldots, i_n$ is the following: $i_1, i_2, i_3, \ldots, i_n$. That order is used by many authors. However, the others use reverse order, that is, $i_n, \ldots, i_3, i_2, i_1$. Note that, using the reverse order of indexing, Equation 1.2d can be rewritten as

$$y^{(n)}(k) = \sum_{i_n=-\infty}^{\infty} \cdots \sum_{i_2=-\infty}^{\infty} \sum_{i_1=-\infty}^{\infty} h^{(n)}(k, i_1, i_2, \ldots, i_n)x(i_1)x(i_2)\ldots\ x(i_n) \tag{1.2d}$$

In this book, we prefer to use the first means of writing out the indices under the symbols of multiple sums.

Note that $y^{(0)}(k)$ in Equation 1.2a does not depend upon the input signal $x(k)$, but eventually upon time, and equals $h^{(0)}(k)$. Here $h^{(0)}(k)$ is the zero-order impulse response. We assume that $|h^{(0)}(k)| \neq \infty$ for every $k \in \mathbb{Z}$, where $\mathbb{Z}$ means the set of integers, i.e., $\mathbb{Z} = \{\ldots, -2, -1, 0, 1, 2, \ldots\}$. Furthermore, $h^{(1)}(k, i)$ is a standard impulse response $h(k, i)$ known from the theory of linear systems. Note that, in this case, it depends upon time k. In our terminology, it is of course the first-order impulse response. $h^{(n)}(k, i_1, i_2, \ldots, i_n)$, $n = 2, \ldots$, mean impulse responses of orders greater that 1, related with higher order terms ($n \geq 2$) in Equation 1.1.

These responses depend upon time. Note also that it is assumed $i_1 = i$ in Equation 1.2b for simplicity of notation.

For systems independent of time, the partial responses will look like

$$y^{(0)}(k) = h^{(0)} \tag{1.3a}$$

$$y^{(1)}(k) = \sum_{i=-\infty}^{\infty} h^{(1)}(i)x(k-i) \tag{1.3b}$$

$$y^{(2)}(k) = \sum_{i_1=-\infty}^{\infty} \sum_{i_2=-\infty}^{\infty} h^{(2)}(i_1, i_2)x(k-i_1)x(k-i_2) \tag{1.3c}$$

and in general,

$$y^{(n)}(k) = \sum_{i_1=-\infty}^{\infty} \sum_{i_2=-\infty}^{\infty} \cdots \sum_{i_n=-\infty}^{\infty} h^{(n)}(i_1, i_2, \ldots, i_n)x(k-i_1)x(k-i_2)\ldots x(k-i_n) \tag{1.3d}$$

(a) A nonlinear discrete system as a black box

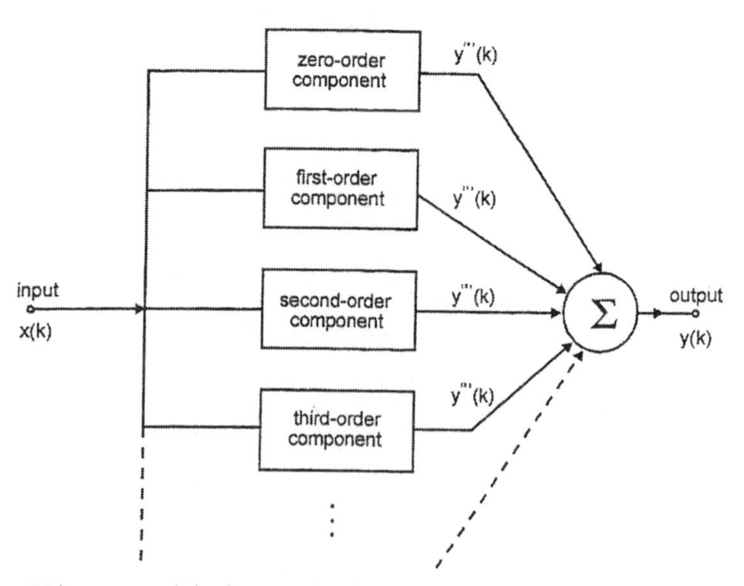

(b) Input-output relation for a system in (a) as described by Equation 1.1

FIGURE 1.1 A nonlinear discrete system described by (1.1).

As before, in Equation 1.2a, $h^{(0)}$ is the zero-order impulse response. Note that this response is independent of time k, so it can be seen as a dc (direct current) component at the output of a system. We assume $|h^{(0)} \neq \infty$. Moreover, $h^{(1)}(i)$ is a standard impulse response $h(i)$ known from the theory of linear systems. Note that in this case, it does not depend upon time k. In our terminology, it is, of course, the first-order impulse response. $h^{(n)}(i_1, i_2, \ldots, i_n)$, $n = 2, \ldots$, mean impulse responses of orders higher than 1, related with higher order terms ($n \geq 2$) in Equation 1.1. In this case, these responses do not depend upon time. Note finally that it is assumed $i_1 = i$ in Equation 1.3b for simplicity of notation.

We assume that all the series both in Equations 1.2 as well as in Equations 1.3 are absolutely convergent.

A notion of stationarity is very useful in the theory of systems. Regarding nonlinear systems having Volterra series representations, we say that they are stationary, according to the definition of Rugh, [5] when their impulse responses fulfill the following equations:

$$h^{(0)}(k) = h^{(0)}(0) \stackrel{df}{=} g^{(0)} = const \tag{1.4a}$$

$$h^{(1)}(k, i_1) = h^{(1)}(0, i_1 - k) \stackrel{df}{=} g^{(1)}(k - i_1) \tag{1.4b}$$

$$h^{(2)}(k, i_1, i_2) = h^{(2)}(0, i_1 - k, i_2 - k) \stackrel{df}{=} g^{(2)}(k - i_1, k - i_2) \tag{1.4c}$$

and in general,

$$h^{(n)}(k, i_1, i_2, \ldots, i_n) = h^{(n)}(0, i_1 - k, i_2 - k, \ldots, i_{-k}) \stackrel{df}{=} g^{(n)}(k - i_1, k - i_2, \ldots, k - i_n) \tag{1.4d}$$

for every $k \in \mathbb{Z}$. In Equation 1.4d, $n \in \mathbb{N}$.

Let us rewrite Equation 1.4b assuming $i_l = i$ for notational simplicity; then, we obtain

$$h^{(1)}(k, i) = h^{(1)}(0, i - k) = h^{(1)}(0, -m) = g^{(1)}(m) \tag{1.5}$$

where $m = k - i$.

Having Equation 1.5 for the linear case, we shall illustrate now the property of stationarity on this example (see Figure 1.2). We assume for purpose of illustration that arguments i, k, m, of the functions in Figure 1.2 are real numbers. That is, i, k, $m \in \mathbb{R}$ in Figure 1.2, where $\mathbb{R}$ is the set of real numbers. This convention of plotting a function with integer arguments as if these arguments would be real ones will be also used in the forthcoming sections of the book. Moreover, note that the values of the functions in Figure 1.2 are also values belonging to the set of real numbers $\mathbb{R}$.

We start with Figure 1.2a, where the function $h^{(1)}(k, i)$ for the time instant $k = 0$ is shown; it has a typical form for real systems of a decaying curve for $i \rightarrow \infty$.

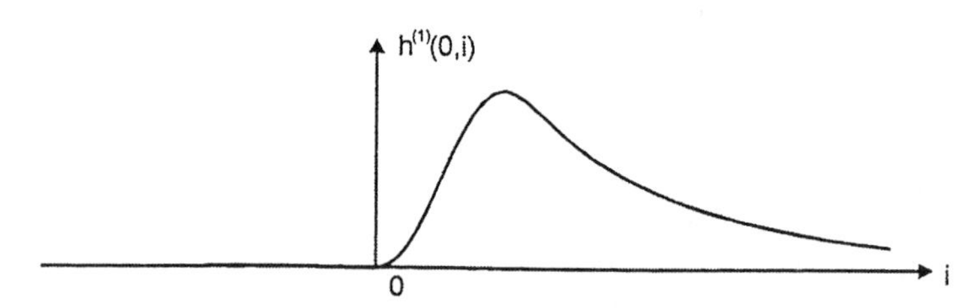

(a) Assumed characteristic of $h^{(1)}(k,i)$ for the time instant k=0

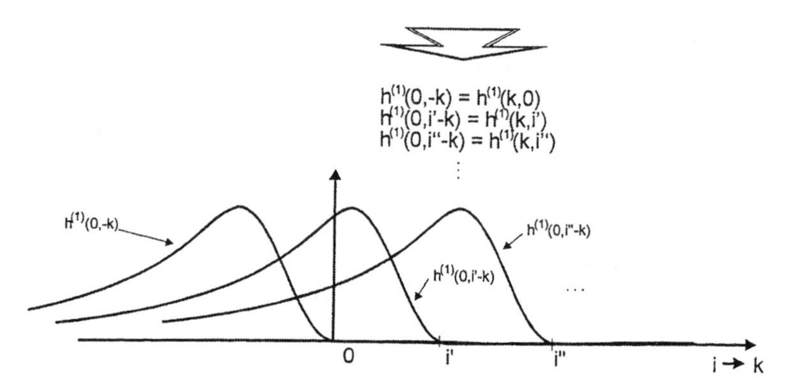

(b) Application of stationarity property to function $h^{(1)}(k,i)$

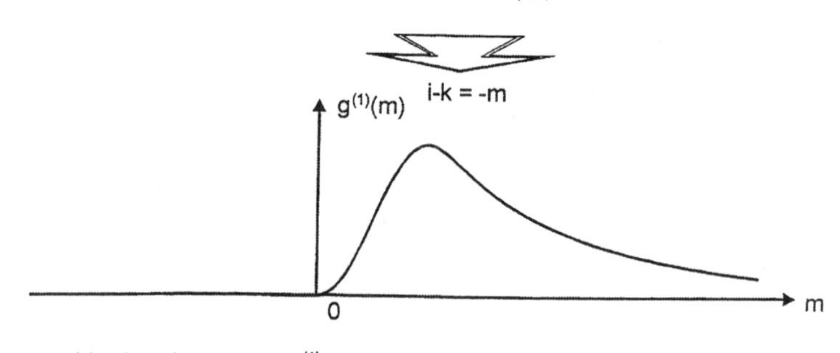

(c) Resulting impulse response $g^{(1)}(m)$

FIGURE 1.2 Illustration of stationarity property in the linear case.

Note that this function is repeated in all curves shown in Figure 1.2b, where the curves are presented after applying the stationarity property for the times $i = 0$, $i = i'$, and $i = i''$. These curves are mirror versions of that shown in Figure 1.2a, and are shifted by 0, i', and i'', respectively. Furthermore, note that the operation illustrated can be interpreted as reducing one variable in the two-variable function $h^{(1)}(k, i)$, which brings a one-variable function as a result, with the shifting parameter equal to 0, i', i'', and so on. Finally, after introducing a variable $m = k - i$, the result presented in Figure 1.2b can be summarized in a single one-variable function $g^{(1)}(m)$, as shown in Figure 1.2c.

Similarly, we can interpret Equations 1.4b, 1.4c, and, in general Equation 1.4d. The stationarity property applied in these equations can be seen as reducing one variable from them, making the corresponding impulse responses independent of a moment of applying an input signal.

Now we will show that the description by the Volterra series given by 1.1 and 1.2 reduces to that given by 1.1 and 1.3, when the stationarity property is applied. To show this, we begin with Equation 1.4a; using this equation in 1.2a gives

$$y^{(0)}(k) = g^{(0)} \tag{1.6}$$

Similarly, substituting 1.4d in general expression for the system partial response of the nth order given by 1.2d, we get

$$y^{(n)}(k) = \sum_{i_1=-\infty}^{\infty} \sum_{i_2=-\infty}^{\infty} \cdots \sum_{i_n=-\infty}^{\infty} g^{(n)}(k-i_1, k-i_2, \ldots, k-i_n)x(i_1)x(i_2)\cdots x(i_n) \tag{1.7}$$

Changing then the variables $k - i_p = m_p$, $p = 1, \ldots, n$, in 1.7 leads to

$$y^{(n)}(k) = \sum_{m_1=+\infty}^{-\infty} \sum_{m_2=+\infty}^{-\infty} \cdots \sum_{m_n=+\infty}^{-\infty} g^{(n)}(m_1, m_2, \ldots, m_n)x(k-m_1)x(k-m_2)\cdots x(k-m_n) \tag{1.8}$$

Finally, note that summation order in (1.8) is of no importance, so we can rewrite (1.8) in the form

$$y^{(n)}(k) = \sum_{m_1=-\infty}^{\infty} \sum_{m_2=-\infty}^{\infty} \cdots \sum_{m_n=-\infty}^{\infty} g^{(n)}(m_1, m_2, \ldots, m_n)x(k-m_1)x(k-m_2)\cdots x(k-m_n) \tag{1.9}$$

In conclusion, we see that Equations 1.6 and 1.9 represent system description in the form of a Volterra series with partial responses as given by Equation 1.3 for time-independent systems, with the only difference being that impulse responses are now denoted by $g^{(n)}(m_1, m_2, \ldots, m_n)$, instead of $h^{(n)}(i_1, i_2, \ldots, i_n)$.

The next important notion related to the Volterra series is the notion of homogeneity. Let us define the partial response given by 1.2a or 1.3a as being homogeneous of zero degree. Furthermore, observe that for the remaining partial responses, we can write

$$y(k;\ \text{with input signal}\ \alpha x) = \alpha^n \cdot y^{(\ )n}(k;\ \text{with input signal x}) \tag{1.10}$$

where α is some nonzero real number and $n = 1, 2, \ldots$. So taking 1.10 into account, we can say that partial response $y^{(n)}(k)$ is homogeneous of degree n.

Note that using the homogeneity property just explained, we are now able to define the notion of the order of a partial response as introduced intuitively in the beginning of this section. So a given partial response is of the order n, $n = 0, 1, 2, \ldots$, if and only if it is homogeneous of degree n. Moreover, note that the property of homogeneity can be also used to define the order of an impulse response. Similarly as before, we consider the impulse response being of the nth order if and only if it is related with partial response, which is homogeneous of degree n.

Consider now impulse responses $h^{(n)}(i_1, i_2, \ldots, i_n)$ in Equations 1.3 possessing such a property that they become equal to zero whenever one of the arguments i_1, $i_2, \ldots, i_n$ becomes negative. In other words, for $h^{(n)}(i_1, i_2, \ldots, i_n)$ to be equal to zero, it is sufficient that one of the arguments $i_1, i_2, \ldots, i_n$ is negative. Note that systems having such impulse responses are called causal.[5]

Both causality and noncausality of impulse responses are illustrated in Figure 1.3 using examples of the zero, first, and second order impulse responses.

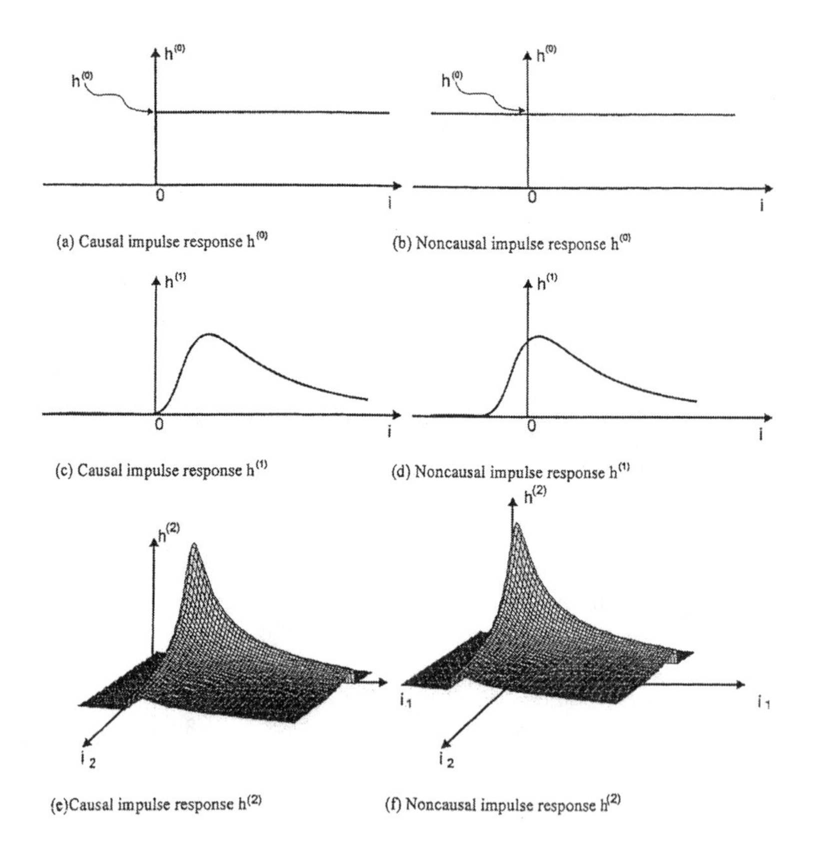

FIGURE 1.3 Illustration of causality and noncausality of impulse responses.

Note that by the use of causality property in Equations 1.1 and 1.3, the Volterra series description for time-independent systems simplifies to

$$\begin{aligned} y(k) = {} & h^{(0)} + \sum_{i=0}^{\infty} h^{(1)}(i)x(k-i) \\ & + \sum_{i_1=0}^{\infty} \sum_{i_2=0}^{\infty} h^{(2)}(i_1, i_2)x(k-i_1)x(k-i_2) \\ & + \sum_{i_1=0}^{\infty} \sum_{i_2=0}^{\infty} \sum_{i_3=0}^{\infty} h^{(3)}(i_1, i_2, i_3)x(k-i_1)x(k-i_2)x(k-i_3)\ldots \end{aligned} \tag{1.11}$$

where lower summation limits are now equal to zero, and $h^{(0)}$ is meant as the causal impulse response of the zero-order, as shown in Figure 1.3a. Further simplification of 1.11 is possible for input signals that are equal to zero for times $k < 0$, that is, for signals applied to a system at time $k = 0$. Then, we get

$$\begin{aligned} y(k) = {} & h^{(0)} + \sum_{i=0}^{k} h^{(1)}(i)x(k-i) \\ & + \sum_{i_1=0}^{k} \sum_{i_2=0}^{k} h^{(2)}(i_1, i_2)x(k-i_1)x(k-i_2) \\ & - \sum_{i_1=0}^{k} \sum_{i_2=0}^{k} \sum_{i_3=0}^{k} h^{(3)}(i_1, i_2, i_3)x(k-i_1)x(k-i_2)x(k-i_3).. \end{aligned} \tag{1.12}$$

where upper summation limits are also finite. These limits are now equal to k, which means that each of the partial responses in Equation 1.12 is a sum over a finite number of elements.

At the end of this section, we make a remark regarding the question of when the Volterra series describes a linear system, and a nonlinear one. In this context, note that a system is a linear one if and only if it fulfills the equation

$$y(\alpha x_1 + \beta x_2) = \alpha y(x_1) + \beta y(x_2) \tag{1.13}$$

where y means a system response, x_1 and x_2 are some input signals, and α and β are real numbers.

Taking into account Equation 1.1 with Equations 1.2 or 1.3, observe that Equation 1.13 is fulfilled only when the Volterra series is reduced to

$$y(k) = y^{(1)}(k) = \sum_{i=-\infty}^{\infty} h^{(1)}(k, i)x(i) \tag{1.14a}$$

for time-dependent systems, or to

$$y(k) = y^{(1)}(k) = \sum_{i=-\infty}^{\infty} h^{(1)}(i)x(k-i) \tag{1.14b}$$

for time-independent systems.

One thing more needs some explanation in this section, namely, a notion of symmetric, triangular, and regular nonlinear impulse responses (note that here, nonlinear stands for all of them: zero-order, first-order, second-order, and so on). Incidentally, note also that the adjective "nonlinear" is used in the literature before the words "impulse response," when one wants to underline consideration of a nonlinear system described by the Volterra series.

We refer now to the stationary case as described by Equation 1.11, and define nonlinear symmetric impulse responses in the following way:

$$h_{sym}^{(n)}(i_1, i_2, \ldots, i_n) \stackrel{df}{=} \frac{1}{n!} \sum_{\substack{\text{over all n!} \\ \text{permutations} \\ \text{of indices} \\ i_1, i_2, \ldots, i_n}} h^{(n)}(i_1, i_2, \ldots, i_n) \tag{1.15}$$

Explanation 1.1

Note that it follows from Equation 1.15 that the first four nonlinear symmetric impulse responses look like

$$\begin{aligned}
h_{sym}^{(0)} &= \frac{1}{0!}h^{(0)} = h^{(0)} \\
h_{sym}^{(1)}(i) &= \frac{1}{1!}h^{(1)}(i) = h^{(1)}(i) \\
h_{sym}^{(2)}(i_1, i_2) &= \frac{1}{2!}[h^{(2)}(i_1, i_2) + h^{(2)}(i_2, i_1)] = \frac{1}{2}[h^{(2)}(i_1, i_2) + h^{(2)}(i_2, i_1)] \\
h_{sym}^{(3)}(i_1, i_2, i_3) &= \frac{1}{3!}[h^{(3)}(i_1, i_2, i_3) + h^{(3)}(i_1, i_3, i_2) + h^{(3)}(i_2, i_1, i_3) + h^{(3)}(i_2, i_3, i_1) \\
&\quad + h^{(3)}(i_3, i_1, i_2) + h^{(3)}(i_3, i_2, i_1)]
\end{aligned}$$

All the symmetric impulse responses of higher orders are constructed similarly.

Explanation 1.2

Observe that

$$y^{(2)}(k) = \sum_{i_1=0}^{\infty}\sum_{i_2=0}^{\infty} h^{(2)}(i_1, i_2)x(k-i_1)x(k-i_2)$$
$$= \sum_{i_1=0}^{\infty}\sum_{i_2=0}^{\infty} h^{(2)}(i_2, i_1)x(k-i_1)x(k-i_2)$$

and

$$y^{(3)}(k) = \sum_{i_1=0}^{\infty}\sum_{i_2=0}^{\infty}\sum_{i_3=0}^{\infty} h^{(3)}(i_1, i_2, i_3)x(k-i_1)x(k-i_2)x(k-i_3)$$
$$= \sum_{i_1=0}^{\infty}\sum_{i_2=0}^{\infty}\sum_{i_3=0}^{\infty} h^{(3)}(i_1, i_3, i_2)x(k-i_1)x(k-i_2)x(k-i_3)$$
$$= \sum_{i_1=0}^{\infty}\sum_{i_2=0}^{\infty}\sum_{i_3=0}^{\infty} h^{(3)}(i_2, i_1, i_3)x(k-i_1)x(k-i_2)x(k-i_3)$$
$$= \sum_{i_1=0}^{\infty}\sum_{i_2=0}^{\infty}\sum_{i_3=0}^{\infty} h^{(3)}(i_2, i_3, i_1)x(k-i_1)x(k-i_2)x(k-i_3) = \ldots$$

Note that we shall have similar relations for partial response of higher orders.

Taking now into account the results presented in Explanation 1.1 and Explanation 1.2, we see that Equation 1.11 can be rewritten in the form

$$\begin{aligned} y(k) = {} & h_{sym}^{(0)} + \sum_{i=0}^{\infty} h_{sym}^{(1)}(i)\,x(k-i) + \sum_{i_1=0}^{\infty}\sum_{i_2=0}^{\infty} h_{sym}^{(2)}(i_1, i_2)\,x(k-i_1)x(k-i_2) \\ & + \sum_{i_1=0}^{\infty}\sum_{i_2=0}^{\infty}\sum_{i_3=0}^{\infty} h_{sym}^{(3)}(i_1, i_2, i_3)x(k-i_1)x(k-i_2)x(k-i_3) + \ldots \end{aligned} \tag{1.16}$$

Comparing Equation 1.11 with Equation 1.16, we see that they have the same form. Moreover, the corresponding components in these equations are equal to each other. So it does not matter whether the impulse responses are meant in Equation 1.11 as the ordinary or symmetric ones. Note that the advantage of symmetrization of the nonlinear impulse responses in Equation 1.11 lies in the fact that, after performing this operation, there exist many groups of elements having the same values. For each of these groups, we take only one element in descriptions using the triangular and regular impulse responses. Thus a reduction of elements, which are summed for a given nonlinearity order n, is performed to arrive at a partial response of that order, $y^{(n)}(k)$.

To find an alternative description of Equation 1.11, which uses triangular impulse responses, we shall introduce a special multivariable step function defined by[5]

$$\bar{\varepsilon}(i_1, i_2, \ldots, i_{n-1}) = \begin{cases} 0, \text{ if any } i_z < 0 \\ 1, i_1 = i_2 = \cdots = i_{n-1} = 0 \\ \vdots \\ \dfrac{n!}{m_1!\ldots m_z!}, i_{s_1} = \cdots = i_{s_1+m_1-1} = 0, \ldots, \\ \qquad i_{s_z} = \cdots = i_{s_z+m_z-1} = 0 \\ \vdots \\ n!, i_1, i_2, \cdots, i_{n-1} > 0, s_1, \cdots, s_z \in \{1, 2, \cdots, n-1\}, n > 1 \end{cases} \tag{1.17}$$

Numbers m_1, ..., m_z in Equation 1.17 mean numbers of zeros in groups of zeros occurring one after another in the sequence of arguments i_1, ..., i_{n-1}, with such a property that the numbers between the zeros are positive numbers.

Before going further, let us illustrate now the function given by Equation 1.17.

Explanation 1.3

Note from Equation 1.17 that we have

- for $n = 2$

$$\bar{\varepsilon}(i_1) = \begin{cases} 0, i_1 < 0 \\ 1, i_1 = 0 \\ 2, i_1 > 0 \end{cases}$$

- for $n = 3$

$$\bar{\varepsilon}(i_1, i_2) = \begin{cases} 0, \text{ if any of } i_1, i_2 < 0 \\ 1, i_1 = i_2 = 0 \\ 6, \text{ if } i_1 = 0, i_2 > 0 \text{ or } i_1 > 0, i_2 = 0 \\ 6, i_1, i_2 > 0 \end{cases}$$

- for $n = 4$

$$\bar{\varepsilon}(i_1, i_2, i_3) = \begin{cases} 0, \text{ if any of } i_1, i_2, i_3 < 0 \\ 1, i_1 = i_2 = i_3 = 0 \\ 12, \text{ if } i_1 = 0, i_2 = 0, i_3 > 0 \text{ or} \\ \qquad i_1 = 0, i_2 > 0, i_3 = 0 \text{ or} \\ \qquad i_1 > 0, i_2 = 0, i_3 = 0 \\ 24, \text{ if } i_1 = 0, i_2 > 0, i_3 > 0 \text{ or} \\ \qquad i_1 > 0, i_2 = 0, i_3 > 0 \text{ or} \\ \qquad i_1 > 0, i_2 > 0, i_3 = 0 \\ 24, i_1, i_2, i_3 > 0 \end{cases}$$

and so on.

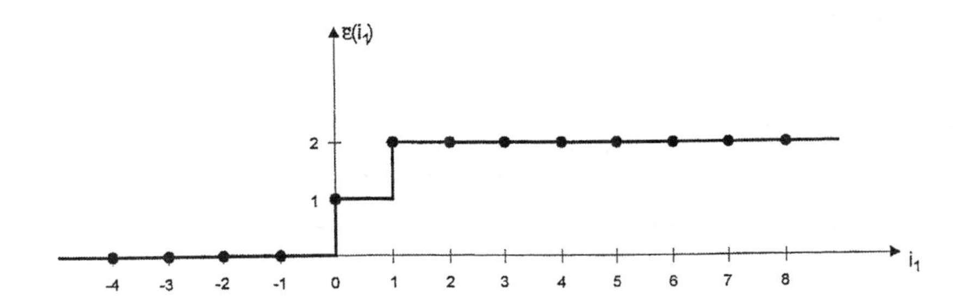

(a) Plot of function ε(i1)

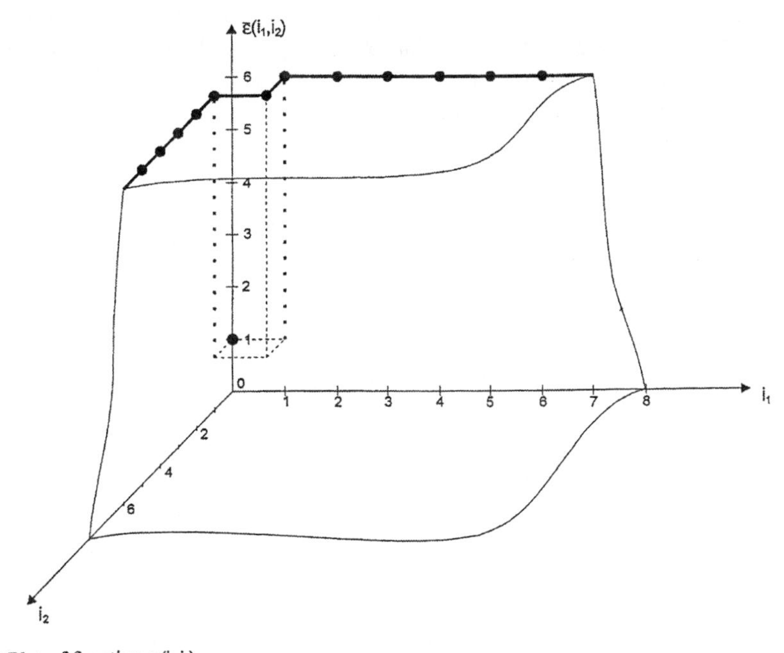

(b) Plot of function ε(i1,i2)

FIGURE 1.4 Plots of functions $\bar{\varepsilon}(i_1)$ and $\bar{\varepsilon}(i_1, i_2)$.

The functions $\bar{\varepsilon}(i_1)$ and $\bar{\varepsilon}(i_1, i_2)$ are illustrated in Figure 1.4.

It is interesting to note that the function $\bar{\varepsilon}(i_1)$, as illustrated in Figure 1.4a, is not identical to the standard step function defined by

$$(k) = \begin{cases} 1, k = 0, 1, 2, 3, \cdots \\ 0, k < 0 \end{cases} \tag{1.18}$$

The function $\varepsilon(k)$ is plotted in Figure 1.5.

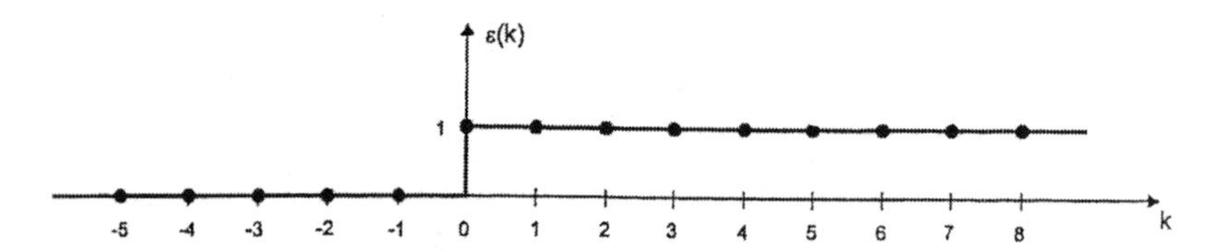

FIGURE 1.5 Standard step function $\varepsilon(k)$.

Incidentally, note that having the function $\varepsilon(k)$, the impulse response $h^{(0)}$ in Equation 1.11 can be expressed as $h^{(0)}\varepsilon(k)$ with $h^{(0)}$ meaning only some constant. In contrast to this, see in Figure 1.3a that $h^{(0)}$ there means both the impulse response of the zero-order and a constant.

It has been shown[5] that the triangular impulse responses can be obtained using the following relations:

$$h_{tri}^{(0)} \stackrel{df}{=} h_{sym}^{(0)} = h^{(0)}, \quad n = 0 \tag{1.19a}$$

$$h_{tri}^{(1)}(i) \stackrel{df}{=} h^{(1)}(i), n = 1 \tag{1.19b}$$

$$h_{tri}^{(n)}(i_1, i_2, \cdots, i_n) \stackrel{df}{=} h_{sym}^{(n)}(i_1, i_2, \cdots, i_n)\bar{\varepsilon}(i_1 - i_2, i_2 - i_3, \ldots, i_{n-1} - i_n), n > 1 \tag{1.19c}$$

We now illustrate Equations 1.19.

Explanation 1.4

Note from Equation 1.19c that we have

- for $n = 2$

$$h_{tri}^{(2)}(i_1, i_2) = h_{sym}^{(2)}(i_1, i_2)\bar{\varepsilon}(i_1 - i_2) = \begin{cases} 0, i_1 < i_2 \\ h_{sym}^{(2)}(i_1, i_2), i_1 = i_2 \\ 2h_{sym}^{(2)}(i_1, i_2), i_1 > i_2 \end{cases}$$

- for $n = 3$

$$h_{tri}^{(3)}(i_1, i_2, i_3) = h_{sym}^{(3)}(i_1, i_2, i_3)\bar{\varepsilon}(i_1 - i_2, i_2 - i_3) = \begin{cases} 0, \text{if any of } i_1 < i_2, i_2 < i_3 \\ h_{sym}^{(3)}(i_1, i_2, i_3), i_1 = i_2 = i_3 \\ 6h_{sym}^{(3)}(i_1, i_2, i_3), \text{if } i_1 = i_2, i_2 > i_3 \text{ or} \\ \qquad i_1 > i_2, i_2 = i_3 \\ 6h_{sym}^{(3)}(i_1, i_2, i_3), i_1 > i_2, i_2 > i_3 \end{cases}$$

and so on.

To arrive at the Volterra series containing triangular impulse responses, we note first that the functions $h^{(n)}(i_1, i_2, \ldots, i_n)$ on the right-hand side of Equation 1.15 can be also meant as triangular impulse responses. Using this in Equation 1.16, we get, finally

$$\begin{aligned} y(k) &= h_{tri}^{(0)} + \sum_{i=0}^{\infty} h_{tri}^{(1)}(i)x(k-i) + \sum_{i_1=0}^{\infty}\sum_{i_2=0}^{\infty} h_{tri}^{(2)}(i_1, i_2)x(k-i_1)x(k-i_2) \\ &+ \sum_{i_1=0}^{\infty}\sum_{i_2=0}^{\infty}\sum_{i_3=0}^{\infty} h_{tri}^{(3)}(i_1, i_2, i_3)x(k-i_1)x(k-i_2)x(k-i_3) + \cdots \end{aligned} \tag{1.20}$$

Changing the variables ($i_1 = i_2 + m_1$, $i_2 = i_3 + m_2$, ..., $i_{n-1} = i_n + m_{n-1}$) in the expressions under the summation symbols in Equation 1.20, successively for $n = 2, 3, \ldots$, we obtain

$$\begin{aligned} y(k) &= h^{(0)} + \sum_{i=0}^{\infty} h_{tri}^{(1)}(i)x(k-i) + \sum_{m_1=0}^{\infty}\sum_{i_2=0}^{\infty} h_{tri}^{(2)}(m_1+i_2, i_2)x(k-m_1-i_2)x(k-i_2) \\ &+ \sum_{m_1=0}^{\infty}\sum_{m_2=0}^{\infty}\sum_{i_3=0}^{\infty} h_{tri}^{(3)}(m_1+m_2+i_3, m_2+i_3, i_3)x(k-m_1-m_2-i_3)x(k-m_2 i_3)x(k-i_3) + \cdots \end{aligned} \tag{1.21}$$

Note that the range of the variables m_i in Equation 1.21 is the same as that of the variables i_i.

Let us now replace quite formally the variables m_i by the variables i_l, $i = 1, 2, \ldots$, in Equation 1.21, and introduce then the regular impulse responses defined as[5]

$$h_{reg}^{(0)} = h_{tri}^{(0)} = h_{sym}^{(0)} = h^{(0)}, n = 0 \tag{1.22a}$$

$$h_{reg}^{(1)}(i) = h_{tri}^{(1)}(i) = h_{sym}^{(1)}(i) = h^{(1)}(i), n = 1 \tag{1.22b}$$

$$\begin{aligned} h_{reg}^{(n)}(i_1, i_2, \ldots, i_n) &= h_{tri}^{(n)}(i_1+i_2+\cdots+i_n, i_2+\cdots+i_n, \ldots, i_n) \\ &= h_{sym}^{(n)}(i_1+i_2+\cdots+i_n, i_2+\cdots+i_n, \ldots, i_n)\bar{\varepsilon}(i_1, i_2, \ldots, i_{n-1}), n > 1 \end{aligned} \tag{1.22c}$$

Performing the above operations leads to

$$\begin{aligned} y(k) &= h^{(0)} + \sum_{i=0}^{\infty} h_{reg}^{(1)}(i)x(k-i) + \sum_{i_1=0}^{\infty}\sum_{i_2=0}^{\infty} h_{reg}^{(2)}(i_1, i_2)x(k-i_1-i_2)x(k-i_2) \\ &+ \sum_{i_1=0}^{\infty}\sum_{i_2=0}^{\infty}\sum_{i_3=0}^{\infty} h_{reg}^{(3)}(i_1, i_2, i_3)x(k-i_1-i_2-i_3)x(k-i_2-i_3)x(k-i_3) + \cdots \end{aligned} \tag{1.23}$$

All the four kinds of nonlinear impulse responses are illustrated by an example of an impulse response of the second order.

Example 1.1

Consider an impulse response of the second order $h^{(2)}(i_1, i_2) \neq 0$, $0 \leq i_1 \leq 8$, $0 \leq i_2 \leq 8$.

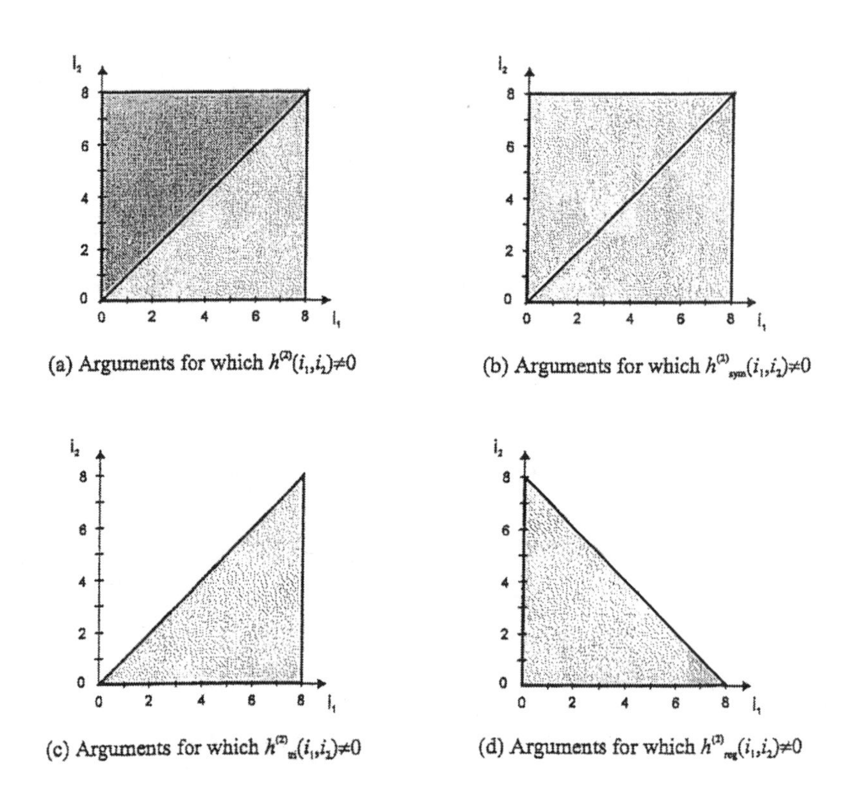

(a) Arguments for which $h^{(2)}(i_1,i_2)\neq 0$

(b) Arguments for which $h^{(2)}_{sym}(i_1,i_2)\neq 0$

(c) Arguments for which $h^{(2)}_{tri}(i_1,i_2)\neq 0$

(d) Arguments for which $h^{(2)}_{reg}(i_1,i_2)\neq 0$

FIGURE 1.6 Illustration of notions of symmetric, triangular, and regular nonlinear impulse responses.

We assume here that $h^{(2)}(i_1, i_2)$ has nonzero values in the range $0 \leq i_1 \leq 8$, $0 \leq i_2 \leq 8$. Figure 1.6a shows schematically the arguments i_1 and i_2 for which the function $h^{(2)}(i_1, i_2)$ is nonzero. Note that it is assumed in Figure 1.6a that the values in the lower triangle shown are different from those occurring in the upper triangle. In Figure 1.6b, the effect of performing the symmetrization operation is seen: there is full symmetry between nonzero values in the upper and lower triangle. Figure 1.6c shows that the two triangles of Figure 1.6b are reduced to only one, with the function values adjusted, respectively, to preserve the values of the function $y(k)$. Finally, note that the triangle of Figure 1.6d is placed the other way compared to that of Figure 1.6c.

1.2 NONLINEAR SAMPLED-DATA SYSTEMS

Generally, we can say that electrical signals are continuous in analog systems and discrete in digital ones. However, because we have two variables, time and amplitude, which can be made discrete in processes illustrated schematically in Figures 1.7 and 1.8, four situations are possible. These situations are shown in Figure 1.9.

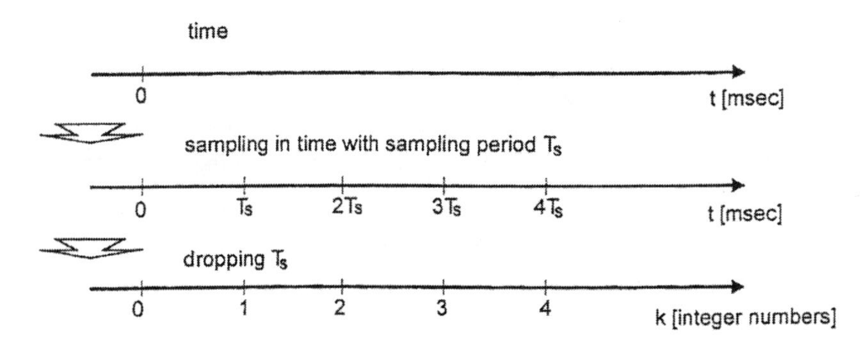

FIGURE 1.7 Discretization (digitalization) in time shown schematically.

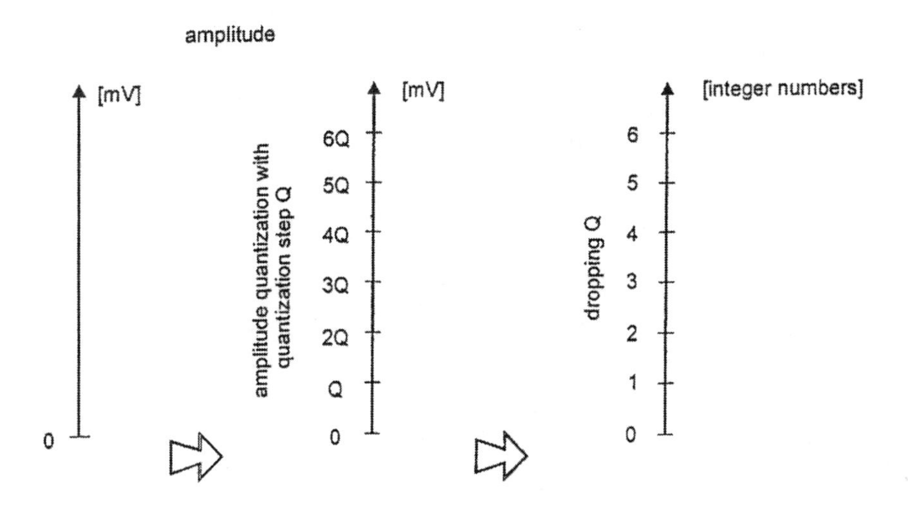

FIGURE 1.8 Discretization (digitalization) of amplitude shown schematically.

In Figure 1.9, $x(t)$ denotes a continuous signal; that is, signal that is continuous both in time and amplitude. This fully continuous signal is also called an analog one. Furthermore, when the amplitude of a signal can take on only discrete values from a prescribed set of numbers, then we have a signal denoted as $x_A(t)$ in Figure

1.9. Here the subscript A underlines the fact that this is a signal with a discrete amplitude. On the other hand, when the signal $x(t)$ is sampled in time, as shown on the right-hand side of Figure 1.9, we arrive at a discrete signal $x(k)$. Finally, the signal $x(k)$ can take on only discrete values both in time and amplitude, as illustrated by means of a signal $x_A(k)$ in Figure 1.9. The signal $x_A(k)$ is a fully digital one. In the literature, the signals $x(k)$ and $x_A(k)$ are generally referred to as sampled-data signals because of sampling in time. Moreover, note that in most considerations, no distinction is made between them. So we also do not distinguish between these signals in this book unless a situation forces us to show the difference. Hence, when we write $x(k)$, this stands for both the signals: discrete only in time, and discrete in both time and amplitude. Furthermore, note that, in the literature, the signals $x(k)$ and $x_A(k)$ are both frequently referred to simply as discrete signals. We follow this convention in this book.

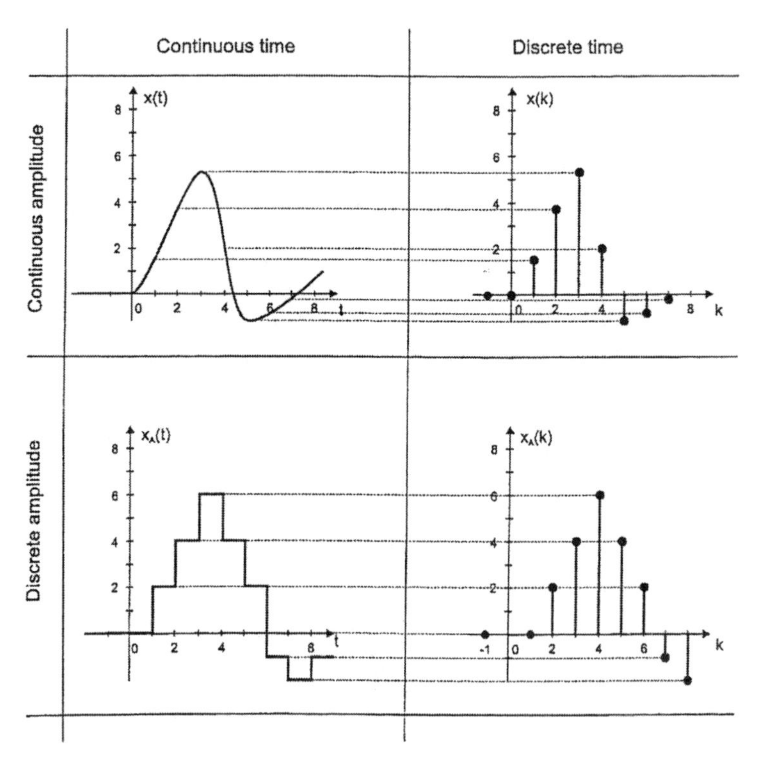

FIGURE 1.9 Four possible situations when making time and/or amplitude of an electrical signal discrete.

Now we can give the definition of sampled-data systems. Sampled-data systems are systems that process sampled-data signals or discrete signals. Furthermore, nonlinear sampled-data systems are sampled-data systems that additionally possess some nonlinearities. Finally, we mention that in this book we will use equivalently the terms nonlinear digital systems and nonlinear discrete systems when referring to nonlinear sampled-data systems.

We present now two examples of nonlinear sampled-data systems. Note that the first example presented comes from the automatic control area, and is a classical one of a linear plant with a nonlinear feedback. The second example comes from the telecommunications area, and describes a digital transmission channel with a quadratic nonlinearity involved.

Example 1.2

Figure 1.10 shows a linear plant with a nonlinear feedback.

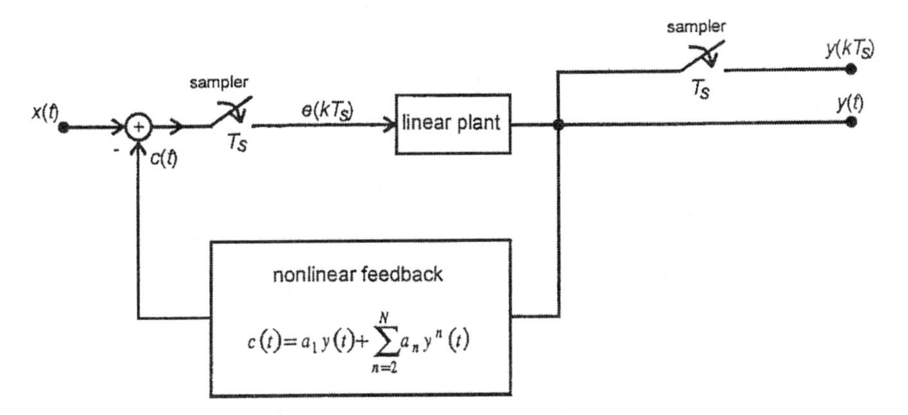

FIGURE 1.10 An example of a linear plant with nonlinear polynomial-type feedback.

Observe from Figure 1.10 that the presented system is not a purely digital one. We see here both analog signals such as $x(t)$, $y(t)$, and $c(t)$, and discrete ones, such as $e(kT_s)$ and $y(kT_s)$. Note that the signals $e(kT_s)$ and $y(kT_s)$ have kT_s, that is, k times the sampling period T_s as an argument. When we drop T_s, which is a common practice in the signal processing literature, we obtain $e(k)$ and $y(k)$. We are then consistent with the notation assumed in Figure 1.9.

Example 1.3

Figure 1.11 shows a telecommunication tract consisting of a transmitter, a nonlinear channel with quadratic nonlinearity and corrupted by an additive noise, and a receiver.

Note that the system in Figure 1.11 is not purely digital, similar to that shown in the previous figure. This is because of the occurrence of analog signals $x_c(t)$ and $y_c(t)$, besides the digital ones $x(k)$ and $y(k)$. However, when the telecommunication tract in Figure 1.11 is considered as a black box between the points, where the signals $x(k)$ and $y(k)$ are written, then such a box can be assumed to represent a digital channel. Note that the transmitter in Figure 1.11, among other things, converts

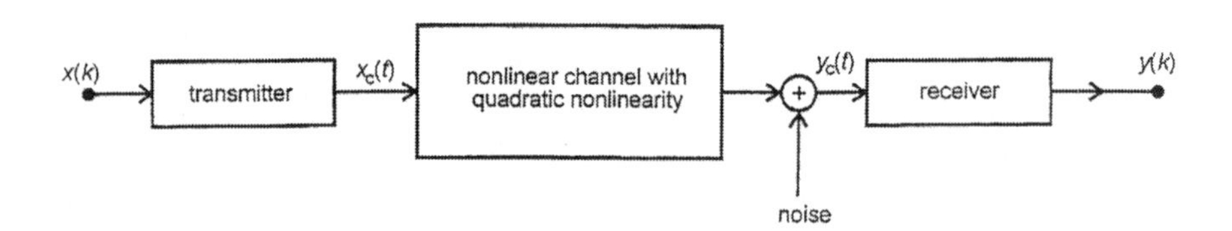

FIGURE 1.11 An example of a nonlinear telecommunication channel possessing quadratic nonlinearity and corrupted by additive noise.

the digital signal $x(k)$ into the analog one $x_c(t)$. Similarly, the analog signal $y_c(t)$ is converted by the receiver into the digital one $y(k)$.

Note that for some parts of the system, or the whole system, as the relation between $y(kT_s)$ and $e(kT_s)$ in Figure 1.10 and the relation between $e(k)$ and $y(k)$ in Figure 1.11 can be described by the discrete Volterra series. Other examples of nonlinear sampled-data systems will follow in the course of this book, some of which will be discussed in great detail.

1.3 MULTIDIMENSIONAL Z TRANSFORM

The Z transform is a basic tool in analysis and synthesis of linear discrete systems. It can be viewed as the other form of the discrete Fourier transform or as a special representation of the Laplace transform for discrete signals.

Let us begin with standard definitions of the Z transform for linear systems and/or one-dimensional signals. We define two-sided Z transform of a sequence of numbers $x(k) = \{..., x(-2), x(-1), x(0), x(1), x(2), ...\}$ as

$$X(z) \overset{df}{=} Z\{x(k)\} = \sum_{k=-\infty}^{\infty} x(k)z^{-k} \tag{1.24}$$

Note that the set of values of elements of the sequence is denoted above by $x(k)$. Such a shortened notation of sequences will be used in this book. The actual meaning of $x(k)$, that is, whether it means only an element $x(k)$ of the sequence $\{x(k)\}$, or the whole, will follow from the context.

Similarly, one-sided Z transform for causal sequences; that is, for sequences of the form $\{..., 0, 0, 0, x(0), x(1), x(2), ...\}$, is given by

$$X(z) = Z\{x(k)\} = \sum_{k=0}^{\infty} x(k)z^{-k} \tag{1.25}$$

In Equations 1.24 and 1.25, the variable z means the complex variable $z = a + jb$, where a and b are real numbers, and $j = \sqrt{-1}$. Moreover, note that the two-sided Z transform applies to sequences with discrete time going from $k = -\infty$ to k

$= +\infty$, whereas the one-sided Z transform applies to sequences with discrete time changing also from $k = -\infty$ to $k = +\infty$, but with all elements identically zeros for $k<0$.

It is worth noting that, when the sequence $\{x(k)\}$ has identically zero values for negative discrete time instants, i.e., $x(k) = 0$ for $k = -1, -2, -3, \ldots$, then the two-sided Z transform reduces to the one-sided transform for this sequence. This is the reason we do not distinguish between the one- and two-sided Z transforms in notation in this book. In each case, it will be clear from the context which of the transforms is actually used.

To find a relation between the so-called discrete-time Fourier transform and the Z transform, recall now the definitions of the forward and inverse discrete-time Fourier transforms. These definitions have forms:

$$X(e^{j\Omega}) \stackrel{df}{=} F\{x(k)\} = \sum_{k=-\infty}^{\infty} x(k)e^{-jk\Omega} \tag{1.26a}$$

and

$$x(k) \stackrel{df}{=} F^{-1}\{X(e^{j\Omega})\} = \frac{1}{2\pi}\int_{-\pi}^{\pi} X(e^{j\Omega})e^{jk\Omega}d\Omega \tag{1.26b}$$

where the normalized frequency Ω is given by

$$\Omega = \omega T_s \tag{1.26c}$$

with ω being an angular frequency and T_s meaning the sampling period.

Taking into account the fact that the complex variable z in Equation 1.24 can be rewritten in the polar form; that is, as

$$z = \sqrt{a^2 + b^2}e^{j\arctan\left(\frac{b}{a}\right)} = |z|e^{j\varphi} \tag{1.27}$$

we obtain then, from Equation 1.24

$$X(z) = \sum_{k=-\infty}^{\infty} (x(k)|z|^{-k})e^{-jk\varphi} \tag{1.28}$$

Comparing Equation 1.28 with Equation 1.26a and identifying φ with Ω, we see that the Z transform of the sequence $\{x(k)\}$ is equal to the discrete-time Fourier transform of the sequence $\{x(k)|z|^{-k}\}$.

To proceed further, we define now a Dirac-impulse sequence as

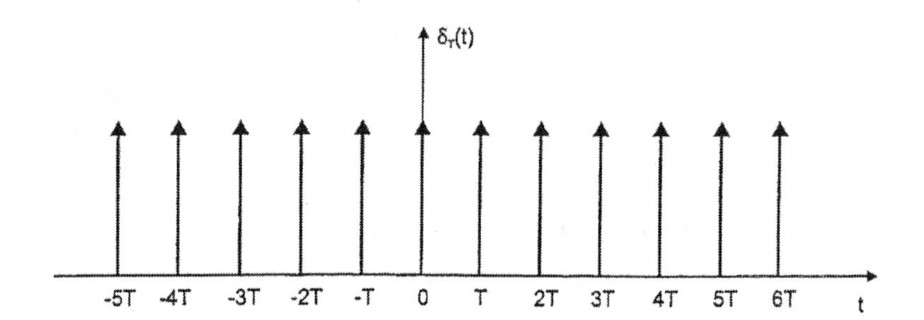

FIGURE 1.12 Dirac-impulse sequence as a function of time.

$$\delta_T(t) = \sum_{k=-\infty}^{\infty} \delta(t-kT) \tag{1.29a}$$

where δ means a Dirac impulse (distribution) having the following property

$$\int_{-\infty}^{\infty} \delta(t)dt = 1 \tag{1.29b}$$

The function given by Equation 1.29a is illustrated in Figure 1.12.

Note from Figure 1.12 that the Dirac-impulse sequence consists of an infinite number of Dirac impulses lying equidistantly. The Dirac impulses are denoted by arrows in the figure, and the period T is a distance between the nearest two impulses.

It can be shown that a sampled ideally, continuous signal can be expressed mathematically as a multiplication of the original signal by the Dirac-impulse sequence with $T = T_s$, where T_s is a sampling period. Hence, denoting such a signal by $x_s(t)$, we can describe it as

$$x_s(t) = x(t)\cdot\delta_T(t) = x(t)\sum_{k=-\infty}^{\infty} \delta(t-kT_s) \tag{1.30a}$$

or, using the selectivity property of the Dirac-impulse ($x(t)\delta(t-t_0) = x(t_0)\delta(t-t_0)$) as

$$x_s(t) = \sum_{k=-\infty}^{\infty} x(kT_s)\delta(t-kT_s) = \sum_{k=-\infty}^{\infty} x(k)\delta(t-kT_s) \tag{1.30b}$$

Figure 1.13 illustrates the relationship between the original continuous signal $x(t)$, sampled signal $x_s(t)$ with continuous time variable t, and resulting sequence consisting of discrete values. Note now that, because the function $x_s(t)$ given by

Equation 1.30b has a continuous time variable t, we can calculate its Fourier transform. For this purpose we recall, however, first the definition of the Fourier integral (Fourier transform) for continuous signals (in the two-sided version here). It is given by

$$X(j\omega) \overset{df}{=} \int_{-\infty}^{\infty} x(t)\exp(-j\omega t)dt \qquad (1.31)$$

for the forward transform, and by

$$x(t) \overset{df}{=} \frac{1}{2\pi}\int_{-\infty}^{\infty} X(j\omega)\exp(j\omega t)d\omega \qquad (1.31b)$$

for the inverse transform. Applying then Equation1.31a to Equation 1.30b, we get

$$X_s(j\omega) = \int_{-\infty}^{\infty}\left(\sum_{k=-\infty}^{\infty} x(k)\delta(t-kT_s)\right)\exp(-j\omega t)dt = \sum_{k=-\infty}^{\infty} x(k)\exp(-j\omega kT_s) \qquad (1.32a)$$

Note the use of the selectivity property of the Dirac-impulse and the property given by Equation 1.29b in Equation 1.32a, which give

$$\int_{-\infty}^{\infty} \exp(-j\omega t)\delta(t-kT_s)dt = \exp(-j\omega kT_s) \qquad (1.32b)$$

Substituting now, quite formally, $s = j\omega$ in Equation1.32a we obtain

$$X_s(s) = \sum_{k=-\infty}^{\infty} x(k)\exp(-skT_s) \qquad (1.33)$$

Assuming afterwards

$$z = \exp(sT_s) \qquad (1.34a)$$

in Equation 1.33 leads to

$$X_s(s) = \sum_{k=-\infty}^{\infty} x(k)z^{-k} \qquad (1.34b)$$

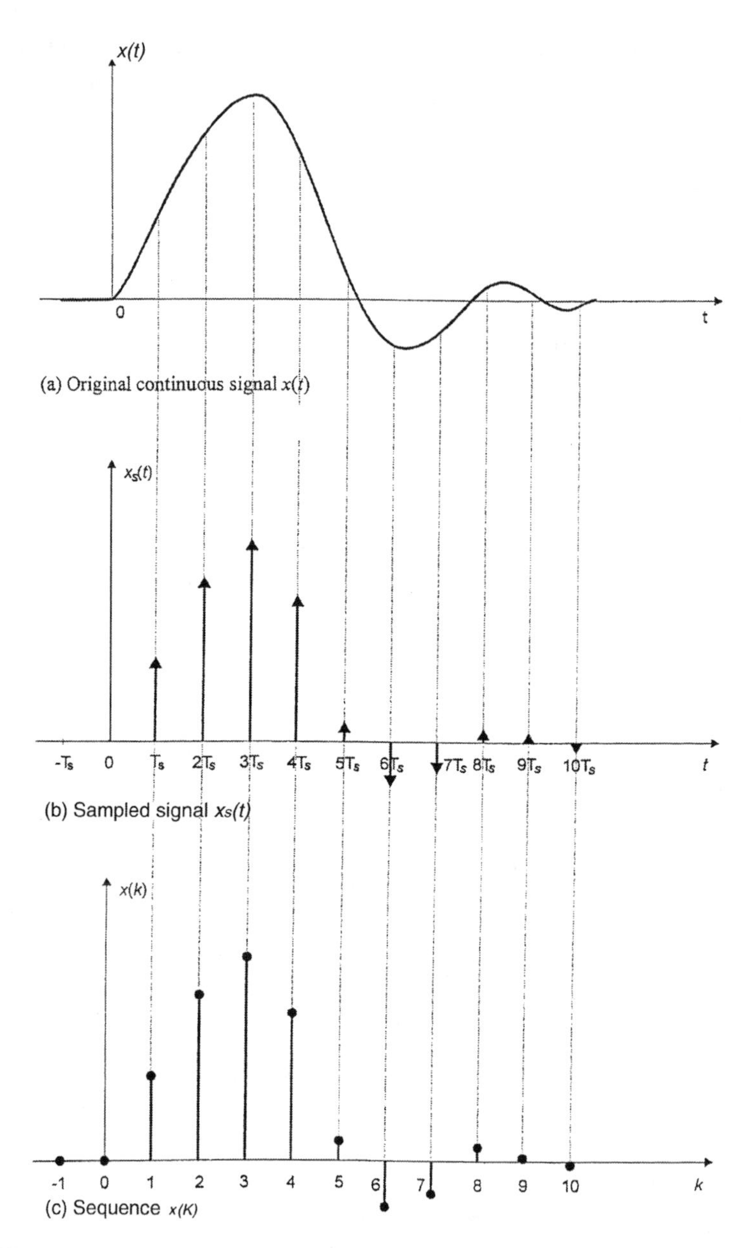

FIGURE 1.13 Illustration of differences between original continuous signal $x(t)$, ideally sampled signal $x_s(t)$, and sequence $x(k)$.

The expression on the right-hand side of Equation 1.34b, according to Equation 1.24, is nothing other than the two-sided Z transform of the sequence $x(k)$. On the other hand, note that the left-hand side of Equation 1.34b is the two-sided Laplace transform of the function $x_s(t)$. We recall here that the two-sided Laplace transform is defined by the following equations:

$$X(s) \overset{df}{=} \int_{-\infty}^{\infty} x(t)\exp(-st)dt \tag{1.35a}$$

that is of the forward transform, and

$$x(t) \overset{df}{=} \frac{1}{2\pi j} \int_{\sigma - j\infty}^{\sigma + j\infty} X(s)\exp(st)ds \tag{1.35b}$$

being the definition of the inverse transform, where the complex variable s is given by

$$s \overset{df}{=} \sigma + j\omega \tag{1.35c}$$

Comparing Equation 1.31a with Equation 1.35a, we see that both definitions are identical when $s = j\omega$ can be assumed. Note that such an assumption was made to arrive at Equation 1.33 from 1.32a.

Furthermore, comparison of Equation 1.34b with 1.24 leads to the conclusion that

$$X(z = \exp(sT_s)) = X_s(s) \tag{1.36}$$

Equation 1.36 means that the Z transform of the sequence $x(k)$ is equal to the Laplace transform of the signal $x_s(t)$, being the continuous in time representation of the original signal $x(t)$. The variables z and s are then related by Equation 1.34a.

Finally, it is also worth noting that

$$X_s\left(s = j\frac{\Omega}{T_s}\right) = X(e^{j\Omega}) \tag{1.37}$$

follows from Equations 1.26a, 1.26c, and 1.34a. Equation 1.37 shows that the Fourier transform of the signal $x_s(t)$ is equal to the discrete-time Fourier transform of the sequence $x(k)$.

Let us now recapitulate the discussion about the relations of the Z transform with the Fourier and Laplace transforms. For this purpose, we present Figure 1.14, where all the relationships just described are illustrated.

To complete the expressions 1.24 and 1.25 regarding the forward Z transform, recall that the definition of the inverse Z transform is given by

$$x(k) \overset{df}{=} \frac{1}{2\pi j} \oint_c X(z)z^{k-1}dz \tag{1.38}$$

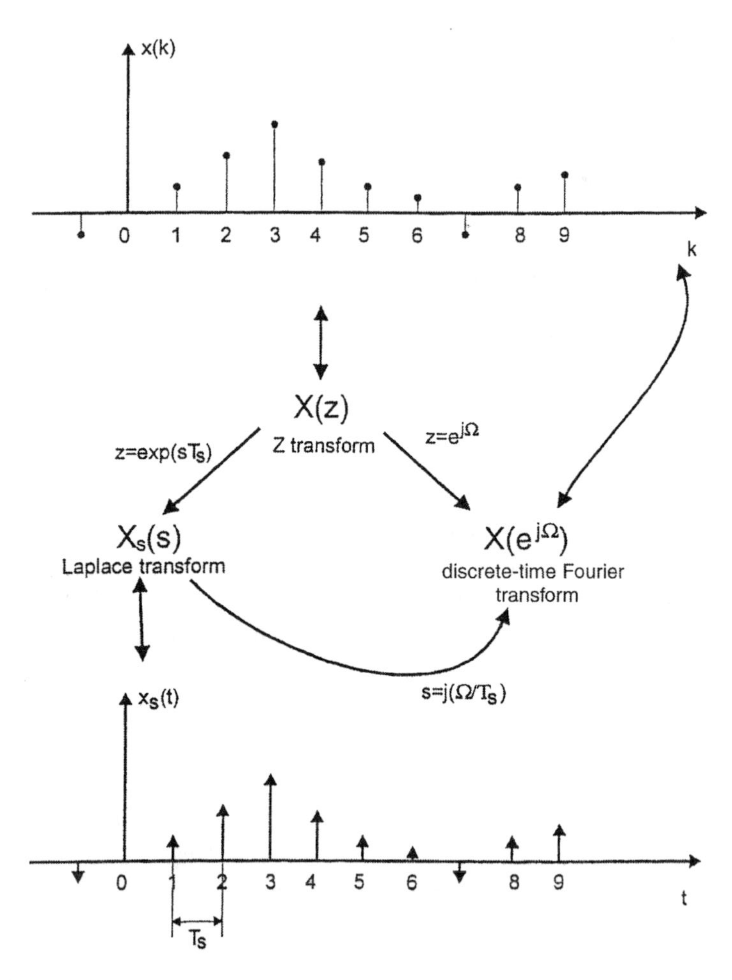

FIGURE 1.14 Illustration of the relationship of the Z transform with the Fourier and Laplace transforms.

where C is such a contour in the complex plane z, that lies in the area where $X(z)$ is stable (in other words, the series 1.24 or 1.25 converges for all values of z in this area).

Having made this short review of the one-dimensional Z transform and its relationships with other transforms, we can go further to extend the definitions 1.24, 1.25, and 1.38 to the multidimensional case. So, having a sequence of which elements depend upon $n>1$ arguments $k_1, k_2, \ldots, k_n$, we define two-sided n-dimensional Z transform as

$$X(z_1, z_2, \ldots, z_n) \stackrel{df}{=} Z\{x(k_1, k_2, \ldots, k_n)\} = \sum_{k_1=-\infty}^{\infty} \sum_{k_2=-\infty}^{\infty} \cdots \sum_{k_n=-\infty}^{\infty} x(k_1, k_2, \ldots, k_n) z_1^{-k_1} z_2^{-k_2} \ldots z_n^{-k_n} \tag{1.39}$$

Similarly, the one-sided n-dimensional Z transform for sequences, which have nonzero values only if the arguments $k_1 \geq 0, k_2 \geq 0, \ldots, k_n \geq 0$, is expressed by

$$X(z_1, z_2, \ldots, z_n) \stackrel{df}{=} Z\{x(k_1, k_2, \ldots, k_n)\} = \sum_{k_1=0}^{\infty}\sum_{k_2=0}^{\infty}\ldots\sum_{k_n=0}^{\infty} x(k_1, k_2 \ldots, k_n) z_1^{-k_1} z_2^{-k_2} \ldots z_n^{-k_n} \tag{1.40}$$

Note that we do not distinguish in our notation between the one- and two-sided n-dimensional Z transforms, as in the one-dimensional case. Which of the transforms is considered at the moment will follow from the context.

Regarding the inverse n-dimensional Z transform, Equation 1.38 extends in this case to the following form:

$$x(k_1, k_2, \ldots, k_n) \stackrel{df}{=} \frac{1}{(2\pi j)^n} \oint_{C_1}\oint_{C_2}\ldots\oint_{C_n} X(z_1, z_2, \ldots, z_n) \cdot z_1^{k_1-1} z_2^{k_2-1} \ldots z_n^{k_n-1} dz_1 dz_2 \ldots dz_n \tag{1.41}$$

where $C_1, C_2, \ldots, C_n$, are appropriate contours in the complex planes $z_1, z_2, \ldots, z_n$, respectively.

Let us now illustrate calculation of the forward multidimensional Z transform by an example.

Example 1.4

Consider the following sequence:

$$x(k_1, k_2) = \begin{cases} 0, \text{ if any of } k_1, k_2 < 0 \text{ or both are negative} \\ k_1 - \sin(k_2\Omega), k_1, k_2 \geq 0 \end{cases}$$

where Ω is a normalized frequency equal to $\omega T_s = 2\pi f / f_s$. Note that this sequence depends upon two arguments k_1 and k_2.

To calculate the Z transform of the above sequence, we use the definition 1.40). Then, we get

$$\begin{aligned} X(z_1, z_2) &= \sum_{k_1=0}^{\infty}\sum_{k_2=0}^{\infty} (k_1 - \sin(k_2\Omega)) z_1^{-k_1} z_2^{-k_2} \\ &= \sum_{k_1=0}^{\infty}\sum_{k_2=0}^{\infty} k_1 z_1^{-k_1} z_2^{-k_2} - \sum_{k_1=0}^{\infty}\sum_{k_2=0}^{\infty} \sin(k_2\Omega) z_1^{-k_1} z_2^{-k_2} \\ &= \sum_{k_1=0}^{\infty} k_1 z_1^{-k_1} \cdot \sum_{k_2=0}^{\infty} z_2^{-k_2} - \sum_{k_1=0}^{\infty} z_1^{-k_1} \cdot \sum_{k_2=0}^{\infty} \sin(k_2\Omega) z_2^{-k_2} \end{aligned}$$

Now we shall use the one-dimensional Z transforms, available in almost every book on the conventional Z transform and its applications. What we need here is the knowledge of the following Z transform pairs:

$$1 \leftrightarrow \frac{z}{z-1}$$

$$k \leftrightarrow \frac{z}{(z-1)^2}$$

$$\sin(k\Omega) \leftrightarrow \frac{z\sin(\Omega)}{z^2 - 2z\cos(\Omega) + 1}$$

Applying these transforms in the expression above, we get finally

$$X(z_1, z_2) = \frac{z_1}{(z_1-1)^2} \cdot \frac{z_2}{z_2-1} - \frac{z_1}{z_1-1} \cdot \frac{z_2 \sin(\Omega)}{z_2^2 - 2z_2\cos(\Omega) + 1}$$

Now take a look at Example 1.4 once again, and observe that the linearity property of the two-dimensional Z transform is fulfilled here. Furthermore, using Equation1.40, we show quite generally that this property holds for every n-dimensional Z transform. For our purpose, we assume the sequence $x(k)$ of having the form of a linear combination of two sequences as

$$x(k_1, \ldots, k_n) = \begin{cases} 0, \text{ if any of } k_1, \ldots, k_n < 0 \\ \alpha x_1(k_1, \ldots, k_n) + \beta x_2(k_1, \ldots, k_n) \end{cases} \tag{1.42}$$

where $k_1, \ldots, k_n \epsilon \{0, 1, 2, \ldots\}$. Note that the sequences x, x_1, and x_2 in Equation 1.42 depend upon n arguments. Moreover, α and β in 1.42 are some constants.

Observe that, in principle, we use here the same definition of linearity as that given by Equation 1.13 regarding linearity of systems. Hence, substituting Equation 1.42 into Equation 1.40 gives

$$X(z_1, \ldots, z_n) = \sum_{k_1=0}^{\infty} \cdots \sum_{k_n=0}^{\infty} [\alpha x_1(k_1, \ldots, k_n) + \beta x_2(k_1, \ldots, k_n)] z_1^{-k_1} \ldots z_n^{-k_n} \tag{1.43}$$

Equation 1.43 can be rewritten in the form

$$X(z_1, \ldots, z_n) = \alpha \sum_{k_1=0}^{\infty} \cdots \sum_{k_n=0}^{\infty} x_1(k_1, \ldots, k_n) z_1^{-k_1} \ldots z_n^{-k_n} + \beta \sum_{k_1=0}^{\infty} \cdots \sum_{k_n=0}^{\infty} x_2(k_1, \ldots, k_n) z_1^{-k_1} \ldots z_n^{-k_n} \tag{1.44a}$$

or equivalently

$$X(z_1, \ldots, z_n) = \alpha X_1(z_1, \ldots, z_n) + \beta X_2(z_1, \ldots, z_n) \tag{1.44b}$$

The result, Equation 1.44b, proves the linearity of the multidimensional one-sided Z transform. Moreover, observe that the proof of linearity expressed by Equations 1.42, 1.43, and 1.44 does not change for the sequences with nonzero values for negative values of the arguments k_i, $i = 1, \ldots, n$, and the two-sided Z transforms.

Now we prove two more properties known very well from the theory of one-dimensional Z transform and its applications, namely, representations for right-shifting and convolution summation in the z domain. We shall start with the first one, and with causal sequences; that is, with sequences that have identically zero values for negative k_i, $i = 1, \ldots, n$. Take into account one of such sequences, which is delayed m_1 discrete time units in the argument k_1, delayed m_2 discrete time units in the argument k_2, and so on. We denote it by

$$u(k_1, k_2, \ldots, k_n) = x(k_1 - m_1, k_2 - m_2, \ldots, k_n - m_n) \quad (1.45a)$$

where $m_i \in \{0, 1, 2, \ldots\}$, $i = 1, \ldots, n$, and apply the definition 1.40. This gives

$$U(z_1, z_2, \ldots, z_n) = \sum_{k_1=0}^{\infty} \sum_{k_2=0}^{\infty} \cdots \sum_{k_n=0}^{\infty} x(k_1 - m_1, k_2 - m_2, \ldots, k_n - m_n) z_1^{-k_1} z_2^{-k_2} \cdots z_n^{-k_n} \quad (1.45b)$$

Substituting then the variables $k_1 - m_1 = r_1$, $k_2 - m_2 = r_2$, $\ldots$, $k_n - m_n = r_n$ in Equation 1.45b leads to

$$U(z_1, z_2, \ldots, z_n) = \sum_{r_1=-m_1}^{-1} \sum_{r_2=-m_2}^{-1} \cdots \sum_{r_n=-m_n}^{-1} x(r_1, r_2, \ldots, r_n) z_1^{-(r_1+m_1)} z_2^{-(r_2+m_2)} \cdots z_n^{-(r_n+m_n)}$$
$$+ \sum_{r_1=0}^{\infty} \sum_{r_2=0}^{\infty} \cdots \sum_{r_n=0}^{\infty} x(r_1, r_2, \ldots, r_n) z_1^{-(r_1+m_1)} z_2^{-(r_2+m_2)} \cdots z_n^{-(r_n+m_n)} \quad (1.46)$$

We use now in Equation 1.46 the fact that the sequence $x(r_1, r_2, \ldots, r_n)$ is causal. So we conclude that all the sums in Equation 1.46 with the lower and upper summation limits $-m_i$, $i = 1, \ldots, n$, and -1, respectively, equal zero. In consequence, we obtain

$$U(z_1, z_2, \ldots, z_n) = z_1^{-m_1} z_2^{-m_2} \cdots z_n^{-m_n} \sum_{r_1=0}^{\infty} \sum_{r_2=0}^{\infty} \cdots \sum_{r_n=0}^{\infty} x(r_1, r_2, \ldots, r_n) z_1^{-r_1} z_2^{-r_2} \cdots z_n^{-r_n} \quad (1.47a)$$

and finally,

$$U(z_1, z_2, \ldots, z_n) = z_1^{-m_1} z_2^{-m_2} \ldots z_n^{-m_n} X(z_1, z_2, \ldots, z_n) \quad (1.47b)$$

It can be shown, using the same kind of argument, that the relation 1.47b holds also for noncausal sequences with the transforms $U(z_1, z_2, \ldots, z_n)$ and $X(z_1, z_2, \ldots, z_n)$ understood as the two-sided ones.

We show now what the n-dimensional convolution-summation property for causal impulse responses and causal signals looks like in the Z domain. Note that this property is an extension of the very well-known one-dimensional (linear) convolution-summation property to the n dimensions. It can be expressed as

$$u(k_1, k_2, \ldots, k_n) \overset{df}{=} \sum_{i_1=0}^{\infty}\sum_{i_2=0}^{\infty}\cdots\sum_{i_n=0}^{\infty} h^{(n)}(i_1, i_2, \ldots, i_n)g(k_1-i_1, k_2-i_2, \ldots, k_n-i_n) \tag{1.48}$$

where $h^{(n)}(i_1, i_2, \ldots, i_n)$ is a nth order causal impulse response, and $g(k_1-i_1, k_2-i_2, \ldots, k_n-i_n)$ plays the role of a causal n-dimensional right-shifted signal.

Applying the one-sided n-dimensional Z transform to Equation 1.48 gives

$$U(z_1, z_2, \ldots, z_n) = \sum_{k_1=0}^{\infty}\sum_{k_2=0}^{\infty}\cdots\sum_{k_n=0}^{\infty}\sum_{i_1=0}^{\infty}\sum_{i_2=0}^{\infty}\cdots\sum_{i_n=0}^{\infty} h^{(n)}(i_1, i_2, \ldots, i_n) \tag{1.49}$$
$$g(k_1-i_1, k_2-i_2, \ldots, k_n-i_n)z_1^{-k_1}z_2^{-k_2}\cdots z_n^{-k_n}$$

Introducing then new variables $k_1-i_1 = r_1$, $k_2-i_2 = r_2$, ..., $k_n-i_n = r_n$, in Equation 1.49 leads to

$$U(z_1, z_2, \ldots, z_n) = \sum_{k_1=0}^{\infty}\sum_{k_2=0}^{\infty}\cdots\sum_{k_n=0}^{\infty}\left(\sum_{i_1=0}^{\infty}\sum_{i_2=0}^{\infty}\cdots\sum_{i_n=0}^{\infty} h^{(n)}(i_1, i_2, \ldots, i_n)\cdot z_1^{-i_1}z_2^{-i_2}\ldots z_n^{-i_n}\right)$$
$$g(r_1, r_2, \cdots, r_n)z_1^{-r_1}z_2^{-r_2}\cdots z_n^{-r_n} \tag{1.50}$$

Observing that the expression in parentheses in Equation 1.50 is the n-dimensional Z transform of $h^{(n)}(i_1, i_2, \ldots, i_n)$, and that the variables $r_1, r_2, \ldots, r_n$ change from $-\infty$ to $+\infty$, we can rewrite Equation 1.50 in the following form:

$$U(z_1, z_2, \ldots, z_n) = H^{(n)}(z_1, z_2, \ldots, z_n)\cdot\sum_{r_1=-\infty}^{\infty}\sum_{r_2=-\infty}^{\infty}\cdots\sum_{r_n=-\infty}^{\infty} g(r_1, r_2, \ldots, r_n)z_1^{-r_1}z_2^{-r_2}\ldots z_n^{-r_n} \tag{1.51}$$

Because $g(r_1, r_2, \ldots, r_n)$, as assumed, is a causal signal; that is, it has zero value if any of its arguments is negative, the expression in Equation 1.51 can be rewritten as

$$U(z_1, z_2, \ldots, z_n) = H^{(n)}(z_1, z_2, \ldots, z_n) \cdot \sum_{r_1=0}^{\infty} \sum_{r_2=0}^{\infty} \ldots \sum_{r_n=0}^{\infty} g(r_1, r_2, \ldots, r_n) z_1^{-r_1} z_2^{-r_2} \ldots z_n^{-r_n} \tag{1.52a}$$

and finally, as

$$U(z_1, z_2, \ldots, z_n) = H^{(n)}(z_1, z_2, \ldots, z_n) \cdot G(z_1, z_2, \ldots, z_n) \tag{1.52b}$$

Note from Equation 1.52b that the representation of the n-dimensional convolution summation in the Z domain is the product of two corresponding Z transforms. Moreover, note that, when we identify $u(k_1, k_2, \ldots, k_n)$, in Equation 1.48 with the partial response $y^{(n)}(k_1, k_2, \ldots, k_n)$, in the Volterra series for time-independent and causal systems, and $g(k_1-i_1, k_2-i_2, \ldots, k_n-i_n)$, in Equation 1.48 with the product $x(k_1-i_1)x(k_2-i_2)\ldots x(k_n-i_n)$, then we will be able to express the nth partial response of the Volterra series in the Z domain by

$$Y^{(n)}(z_1, z_2, \ldots, z_n) = H^{(n)}(z_1, z_2, \ldots, z_n) X(z_1) X(z_2) \cdots X(z_n) \tag{1.53a}$$

because

$$\sum_{r_1=0}^{\infty} \sum_{r_2=0}^{\infty} \ldots \sum_{r_n=0}^{\infty} x(r_1) x(r_2) \cdots x(r_n) z_1^{-r_1} z_2^{-r_2} \cdots z_n^{-r_n} = X(z_1) X(z_2) \cdots X(z_n) \tag{1.53b}$$

holds. According to Equation 1.11, we assumed the following form:

$$y^{(n)}(k_1, k_2, \ldots, k_n) = \sum_{i_1=0}^{\infty} \sum_{i_2=0}^{\infty} \ldots \sum_{i_n=0}^{\infty} h^{(n)}(i_1, i_2, \ldots, i_n) x(k_1 - i_1) x(k_2 - i_2) \cdots x(k_n - i_n),\ n \geq 1 \tag{1.53c}$$

of $y^{(n)}(k_1, k_2, \ldots, k_n)$ in derivation of Equation 1.53a.

Note that, using the same arguments, it can be shown that the relation 1.52b holds also for the n-dimensional convolution with noncausal impulse responses and signals. Then the Z transforms in Equation 1.52b should be understood as the two-sided ones.

Some other useful properties related with the multidimensional Z transform, specific and analogous to those known from the theory of one-dimensional Z transform, can be proved.[6, 7, 8] However, we do not intend to discuss these properties in this book because of one fundamental reason: the multidimensional Z transform is not as widely used as its one-dimensional counterpart in the theory of linear discrete systems.

To calculate the inverse multidimensional Z transform, one of the following ways can be chosen:

1. By the direct use of the inverse formula given by Equation 1.41
2. By applying the properties such as linearity, right-shifting in time, convolution-summation, and others, and using the known transform pairs, both one- and multi-dimensional

1.4 DISCRETE VOLTERRA SERIES IN THE Z DOMAIN

The response $y(k)$ of a nonlinear system, given by Equations 1.1 and 1.3, can be considered as a discrete signal, dependent upon one-dimensional discrete time k. So, we can apply the Z transform to this signal, which gives

$$Y(z)=Z\{y(k)\}=Z\left\{h^{(0)}+\sum_{i=-\infty}^{\infty} h^{(0)}(i)x(k-i)+\sum_{i_1=-\infty}^{\infty}\sum_{i_2=-\infty}^{\infty} h^{(2)}(i_1,i_2)x(k-i_1)\cdot x(k-i_2)+\cdots\right\} \tag{1.54}$$

Because of linearity of the Z transform, Equation 1.54 can be rewritten as

$$Y(z)=Z\{h^{(0)}\}+Z\left\{\sum_{i=-\infty}^{\infty} h^{(1)}(i)x(k-i)\right\}+Z\left\{\sum_{i_1=-\infty}^{\infty}\sum_{i_2=-\infty}^{\infty} h^{(2)}(i_1,i_2)x(k-i_1)\cdot x(k-i_2)\right\}+\cdots \tag{1.55}$$

assuming that $y(k)$ is absolutely convergent.

Calculation of the Z transforms on the right-hand side of Equation 1.55 is not so simple. Note first, that the two-sided Z transform of the constant component $h^{(0)}$ does not exist, as shown below

$$Z\{h^{(0)}\} = \sum_{k=-\infty}^{\infty} h^{(0)}z^{-k} = \underbrace{h^{(0)}\sum_{k=-\infty}^{-1} z^{-k}}_{\substack{\text{converges to}\\ \frac{h^{(0)}z}{1-z} \text{ for } |z|<1}} + \underbrace{h^{(0)}\sum_{k=0}^{\infty} z^{-k}}_{\substack{\text{converges to}\\ \frac{h^{(0)}z}{z-1} \text{ for } |z|>1}} \tag{1.56}$$

However, the one-sided Z transform eventually exists. Hence, we restrict ourselves in further consideration in this section to causal sequences. That is, we assume the impulse response of the zero-order to be equal to $h^{(0)}\varepsilon(k)$ with $h^{(0)}$ being some constant, and so on. Further, note then that the counterpart of Equation 1.55 can be written in the form

$$Y(z) = h^{(0)}\frac{z}{z-1} + H^{(1)}(z)X(z) + Z\left\{\sum_{i_1=0}^{\infty}\sum_{i_2=0}^{\infty} h^{(2)}(i_1, i_2)x(k-i_1)\cdot x(k-i_2)\right\} + \cdots \tag{1.57a}$$

because

$$Z\{h^{(0)}\varepsilon(k)\} = h^{(0)}\frac{z}{z-1} \qquad \text{for } |z| > 1 \tag{1.57b}$$

according to the result in Equation 1.56, and because of the convolution-summation property of the Z transform, i.e.,

$$Z\left\{\sum_{i=0}^{\infty} h^{(1)}(i)x(k-i)\right\} = H^{(1)}(z)X(z) \tag{1.57c}$$

where $H^{(1)}(z)$ and $X(z)$ are the Z transforms of the first-order impulse response and of the input signal, respectively.

In Equation 1.57a, the problem remains of how to calculate the third, fourth, and all following components appearing on the right-hand side of this expression. We shall show now that this problem can be tackled by application of the so-called association of variables. We will explain this notion by means of examples.

Let us start with associating two variables. And for this purpose, let us consider the second-order partial response in the Volterra series with causal impulse responses. This partial response possesses the following form:

$$y^{(2)}(k) = \sum_{i_1=0}^{\infty}\sum_{i_2=0}^{\infty} h^{(2)}(i_1, i_2)x(k-i_1)x(k-i_2) \tag{1.58a}$$

Note that the partial response given by Equation 1.58a can be made two-dimensional in time by naming the time k in the signal $x(k - i_1)$ k_1, and similarly, the time k in the signal $x(k - i_2)$ k_2. This gives the two-dimensional partial response of the second-order as

$$y^{(2)}(k_1, k_2) = \sum_{i_1=0}^{\infty}\sum_{i_2=0}^{\infty} h^{(2)}(i_1, i_2)x(k_1-i_1)x(k_2-i_2) \tag{1.58b}$$

Note that the variables k_1 and k_2 are artificial variables, which enable us to transform $y^{(2)}(k_1, k_2)$ into the Z domain, according to the definition 1.40. This gives

$$Y^{(2)}(z_1, z_2) = \sum_{k_1=0}^{\infty}\sum_{k_2=0}^{\infty}\left(\sum_{i_1=0}^{\infty}\sum_{i_2=0}^{\infty} h^{(2)}(i_1, i_2)x(k_1 - i_1)\cdot x(k_2 - i_2)\right)\cdot z_1^{-k_1} z_2^{-k_2} \tag{1.58c}$$
$$= H^{(2)}(z_1, z_2)X(z_1)X(z_2)$$

when using also the convolution–summation property as given by Equation 1.53a. Furthermore, note that recovering, from the function $y^{(2)}(k_1, k_2)$ with two artificial time variables k_1 and k_2, the function $y^{(2)}(k)$ with the true time variable k follows according to the following relation:

$$k_1 = k_2 = k \tag{1.58d}$$

That is by substitution of $k_1 = k_2 = k$ in Equation 1.58b.

We recall now the formula 1.41 for the inverse multidimensional Z transform. So, in the case of two time variables k_1 and k_2, as in $y^{(2)}(k_1, k_2)$, we have

$$y^{(2)}(k_1, k_2) = \frac{1}{(2\pi j)^2}\oint_{C_1}\oint_{C_2} Y^{(2)}(z_1, z_2) z_1^{k_1-1} z_2^{k_2-1} dz_1 dz_2 \tag{1.59}$$

Association of variables k_1 and k_2 in Equation 1.59 means that we apply Equation 1.58d to it. This gives

$$y^{(2)}(k, k) = \frac{1}{2\pi j}\oint_{C_2}\left\{\frac{1}{2\pi j}\oint_{C_1} Y^{(2)}(z_1, z_2) z_1^{k-1} z_2^{k-1} dz_1\right\} dz_2 \tag{1.60}$$

Let us introduce a new auxiliary variable relating z_1 with z_2 by assuming that

$$z_1 z_2 = u \tag{1.61a}$$

Differentiating in Equation 1.61a with respect to z_1, we get

$$z_2 dz_1 = du \tag{1.61b}$$

Substituting then Equations 1.61a and 1.61b in Equation 1.60 gives

$$y^{(2)}(k, k) = \frac{1}{2\pi j}\oint_{C_2}\left\{\frac{1}{2\pi j}\oint_{C_u} Y^{(2)}(u z_2^{-1}, z_2) u^{k-1} z_2^{-1} du\right\} dz_2 \tag{1.62}$$

where C_u means an appropriate contour in the complex plane u.

Redefining the variables $u = z$ and $z_2 = u_1$ in Equation 1.62, and rearranging the resulting expression, leads finally to

$$y^{(2)}(k) = \frac{1}{2\pi j}\oint_{C_z}\left\{\frac{1}{2\pi j}\oint_{C_{u_1}} Y^{(2)}(zu_1^{-1}, u_1)u_1^{-1}du_1\right\}z^{k-1}dz \tag{1.63}$$

where C_z and C_{u_1} are appropriate contours in the complex planes z and u_1, respectively. Furthermore, note that Equation 1.63 can be rewritten in the following form:

$$y^{(2)}(k) = \frac{1}{2\pi j}\oint_{C_z} Y_z^{(2)}(z)z^{k-1}dz \tag{1.64a}$$

with

$$Y_z^{(2)}(z) = \frac{1}{2\pi j}\oint_{C_{u_1}} Y^{(2)}(zu_1^{-1}, u_1)u_1^{-1}du_1 \tag{1.64b}$$

According to the inverse formula for the one-dimensional Z transform (see Equation 1.38), $Y_z^{(2)}(z)$ in Equation 1.64a is the one-dimensional Z transform of $y^{(2)}(k)$. The relations 1.64a and 1.64b express the association of variables, in this case of two of them, in the Z domain. Its counterpart in the time domain, according to the relation 1.58d, has the form

$$y^{(2)}(k) = y^{(2)}(k_1 = k, k_2 = k) \tag{1.65}$$

Furthermore, note that by the use of Equation 1.58c to express $Y^{(2)}(z_1, z_2)$, we obtain finally from Equation 1.64b

$$Y_z^{(2)}(z) = \frac{1}{2\pi j}\oint_{C_{u_1}} H^{(2)}(zu_1^{-1}, u_1)X(zu_1^{-1})X(u_1)u_1^{-1}du_1 \tag{1.66}$$

Note now that the transform $Y_z^{(2)}(z)$ given by Equation 1.66 is expressed by means of the Z transform $H^{(2)}(\cdot,\cdot)$, called in the literature the transfer function of the second order of a system, and the Z transforms of an input signal, $X(\cdot)$.

Let us repeat the procedure of association of variables once again with three variables. For this purpose, we consider the third-order partial response in the Volterra series with causal impulse responses, having the form

$$y^{(3)}(k) = \sum_{i_1=0}^{\infty}\sum_{i_2=0}^{\infty}\sum_{i_3=0}^{\infty} h^{(3)}(i_1, i_2, i_3)x(k-i_1)x(k-i_2)x(k-x_3) \tag{1.67a}$$

Observe that the partial response given by Equation 1.67a can be made three-dimensional in time by naming the time k in the signal $x(k - i_1)$ k_1, the time k in the signal $x(k - i_2)$ k_2, and finally, the time k in the signal $x(k - i_3)$ k_3. This leads to

$$y^{(3)}(k_1, k_2, k_3) = \sum_{i_1=0}^{\infty} \sum_{i_2=0}^{\infty} \sum_{i_3=0}^{\infty} h^{(3)}(i_1, i_2, i_3)x(k_1 - i_1)x(k_2 - i_2)x(k_3 - x_3) \quad (1.67b)$$

Once again, we stress that the variables k_1, k_2, and k_3 in Equation 1.67b are artificial variables, which allow transformation of $y^{(3)}(k_1, k_2, k_3)$, into the Z domain. Using the property expressed by Equation 1.53a to Equation 1.67b, we get

$$Y^{(3)}(z_1, z_2, z_3) = H^{(3)}(z_1, z_2, z_3)X(z_1)X(z_2)X(z_3) \quad (1.67c)$$

Note also that, for getting the function $y^{(3)}(k)$ from $y^{(3)}(k_1, k_2, k_3)$, the following substitution

$$y^{(3)}(k) = y^{(3)}(k_1 = k, k_2 = k, k_3 = k) \quad (1.67d)$$

is needed.

We recall now the formula 1.41 for the inverse multidimensional Z transform. For three variables, we can write

$$y^{(3)}(k_1, k_2, k_3) = \frac{1}{(2\pi j)^3} \oint_{C_1} \oint_{C_2} \oint_{C_3} Y^{(3)}(z_1, z_2, z_3) z_1^{k_1-1} z_2^{k_2-1} z_3^{k_3-1} dz_1 dz_2 dz_3 \quad (1.68)$$

when considering the partial response $y^{(3)}(k_1, k_2, k_3)$.

Having the expression 1.68, we can start association of variables by performing the substitution $k_1 = k_2 = k$. Note that such substitution means in the time domain, association of first two time variables k_1 and k_2. Thereby, we obtain

$$y^{(3)}(k, k, k_3) = \frac{1}{(2\pi j)^2} \oint_{C_3} \oint_{C_2} \left\{ \frac{1}{2\pi j} \oint_{C_1} Y^{(3)}(z_1, z_2, z_3) z_1^{k-1} z_2^{k-1} z_3^{k_3-1} dz_1 \right\} dz_2 dz_3 \quad (1.69)$$

Let us introduce a new auxiliary variable u_1, relating z_1 with z_2 such that

$$z_1 z_2 = u_1 \quad (1.70a)$$

Differentiating in Equation 1.70a with respect to z_1 gives

$$z_2 dz_1 = du_1 \quad (1.70b)$$

Applying then Equations 1.70a and 1.70b in Equation 1.69 leads to

$$y^{(3)}(k,k,k_3) = \frac{1}{(2\pi j)^2}\oint_{C_3}\oint_{C_2}\left\{\frac{1}{2\pi j}\oint_{C_{u_1}} Y^{(3)}(u_1 z_2^{-1}, z_2, z_3) u_1^{k-1} z_2^{-1} z_3^{k-1} du_1\right\} dz_2 dz_3 \quad (1.71)$$

where C_{u_1} means an appropriate contour in the complex plane u_1.

Now we associate the time variables k and k_3 in Equation 1.71 by performing the substitution $k_3 = k$. This gives

$$y^{(3)}(k,k,k) = \frac{1}{(2\pi j)^2}\oint_{C_3}\oint_{C_2}\left\{\frac{1}{2\pi j}\oint_{C_{u_1}} Y^{(3)}(u_1 z_2^{-1}, z_2, z_3) u_1^{k-1} z_2^{-1} z_3^{k-1} du_1\right\} dz_2 dz_3 \quad (1.72)$$

Then, we introduce the second auxiliary variable u relating u_1 with z_3 in the following way:

$$u_1 z_3 = u \quad (1.73a)$$

Differentiating in Equation 1.73a with respect to u_1 gives

$$z_3 du_1 = du \quad (1.73b)$$

Substituting u_1 and du_1 given by Equation 1.73 into Equation 1.72, we get

$$y^{(3)}(k,k,k) = \frac{1}{(2\pi j)^2}\oint_{C_3}\oint_{C_2}\left\{\frac{1}{2\pi j}\oint_{C_u} Y^{(3)}(u z_3^{-1} z_2^{-1}, z_2, z_3) u_1^{k-1} z_2^{-1} z_3^{-1} du\right\} dz_2 dz_3 \quad (1.74)$$

where C_u means an appropriate contour in the complex plane u. Finally, we redefine the variables $u = z$, $z_2 = u_1$, and $z_3 = u_2$ in Equation 1.74, and rearrange the resulting expression. This results in

$$y^{(3)}(k) = \frac{1}{2\pi j}\oint_{C_z}\left\{\frac{1}{(2\pi j)^2}\oint_{C_{u_1}}\oint_{C_{u_2}} Y^{(3)}(z u_1^{-1} u_2^{-1}, u_1, u_2) u_1^{-1} u_2^{-1} du_1 du_2\right\} z^{k-1} dz \quad (1.75)$$

where C_z, C_{u_1}, and C_{u_2} mean appropriate contours in the complex planes z, u_1, and u_2, respectively. Furthermore, note that Equation 1.75 can be rewritten in the following form:

$$y^{(3)}(k) = \frac{1}{2\pi j}\oint_{C_z} Y_z^{(3)}(z) z^{k-1} dz \quad (1.76a)$$

with

$$Y_z^{(3)}(z) = \frac{1}{(2\pi j)^2}\oint_{C_{u_1}}\oint_{C_{u_2}} Y^{(3)}(zu_1^{-1}u_2^{-1}, u_1, u_2)u_1^{-1}u_2^{-1}du_1du_2 \tag{1.76b}$$

The relations 1.76a and 1.76b for the case of association of three variables are the counterparts of the relations 1.64a and 1.64b derived for two variables. Using the same arguments, one can easily show that the relations:

$$y^{(n)}(k) = \frac{1}{2\pi j}\oint_{C_z} Y_z^{(n)}(z)z^{k-1}dz \tag{1.77a}$$

and

$$Y_z^{(n)}(z) = \frac{1}{(2\pi j)^{n-1}}\oint_{C_{u_1}}\cdots\oint_{C_{u_{n-1}}} Y^{(n)}(zu_1^{-1}\cdots u_{n-1}^{-1}, u_1,\ldots, u_{n-1})\cdot u_1^{-1}\cdots u_{n-1}^{-1}du_1\cdots du_{n-1} \tag{1.77b}$$

hold in general. Moreover, we can express $Y^{(n)}(\cdot, \cdot, \ldots, \cdot)$ in Equation 1.77b for the Volterra series (see Equation 1.53a) through the transfer function of the nth order and Z transforms of the input signal as

$$Y^{(n)}(zu_1^{-1}\ldots u_{n-1}^{-1}, u_1,\ldots,u_{n-1}) = H^{(n)}(zu_1^{-1}\ldots u_{n-1}^{-1}, u_1,\ldots,u_{n-1})\cdot X(zu_1^{-1}\ldots u_{n-1}^{-1})\ldots X(u_{n-1}) \tag{1.77c}$$

Knowing the general formula for the one-dimensional Z transform of the nth partial response in the Volterra series, we are now able to complete Equation 1.57a. So, with the above knowledge, this equation can be rewritten as

$$Y(z) = h^{(0)}\frac{z}{z-1} + H^{(1)}(z)X(z) + \frac{1}{2\pi j}\oint_{C_{u_1}} H^{(2)}(zu_1^{-1}, u_1)X(zu_1^{-1})X(u_1)u_1^{-1}du_1$$
$$+\frac{1}{(2\pi j)^2}\oint_{C_{u_1}}\oint_{C_{u_2}} H^{(3)}(zu_1^{-1}u_2^{-1}, u_1, u_2)X(zu_1^{-1}u_2^{-1})X(u_1)X(u_2)\cdot u_1^{-1}u_2^{-1}du_1du_2 + \cdots \tag{1.78}$$

Note that Equation 1.78 is the equation we have looked for in this section. It expresses the discrete Volterra series in the Z domain.

Using the representation of the discrete Volterra series presented in Figure 1.1b and Equation 1.14b (with the lower summation limit equal to zero for causal $h^{(1)}(i)$), we can separate the linear part (linear component) from the nonlinear part (nonlinear components) in the Volterra series. Formally, we then write

$$y(k) = y_{lin}(k) + y_{non}(k) \tag{1.79a}$$

Note from Equation 1.78 that the counterpart in the Z domain of the relation 1.79a can be expressed as

$$Y(z) = Y_{LIN}(z) + Y_{NON}(z) \tag{1.79b}$$

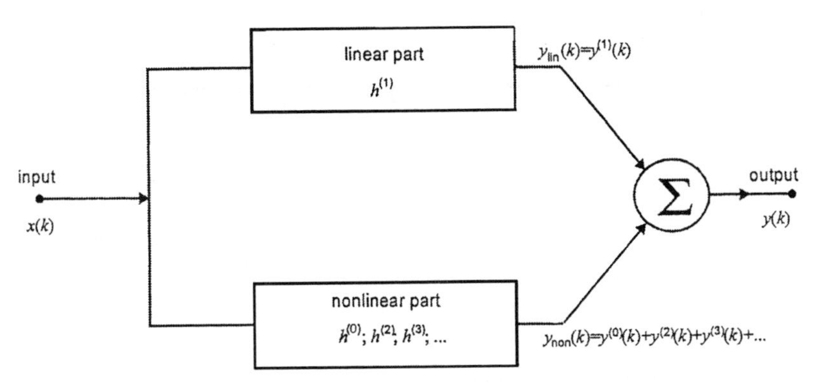

(a) Separation of linear and nonlinear parts in the discrete Volterra series in the time domain

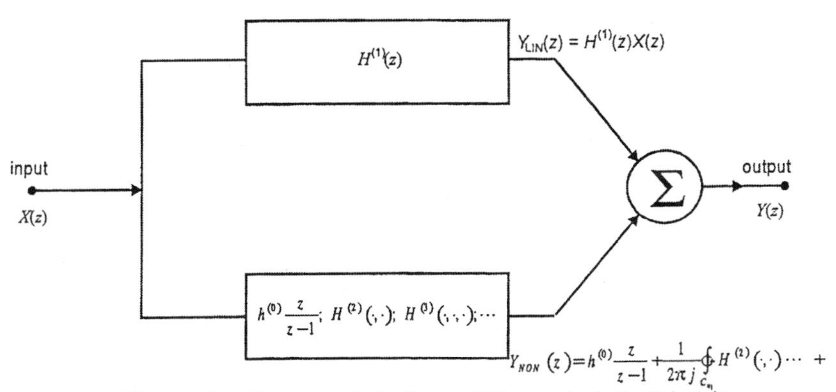

(b) Separation of linear and nonlinear parts in the discrete Volterra series in the Z domain

FIGURE 1.15 Illustration of linear and nonlinear parts in the discrete Volterra series in the time and Z domains.

Relations 1.79a and 1.79b are illustrated in Figure 1.15.

The same rules relating the Z transform with the Fourier and Laplace transforms as in the case of the input signal $x(k)$, apply to the sampled and digitalized output signal $y(k)$. This is illustrated in Figure 1.16.

At the end of this section, we shall point out a role that causal and noncausal sequences play in our considerations. Note that we touched for the first time on the problem of causality and noncausality by consideration of the notion of impulse responses (see Figure 1.3). Then the causal sequences applied to the input of a nonlinear system were implicitly introduced by reduction of the expression 1.11 to the form given by Equation 1.12.

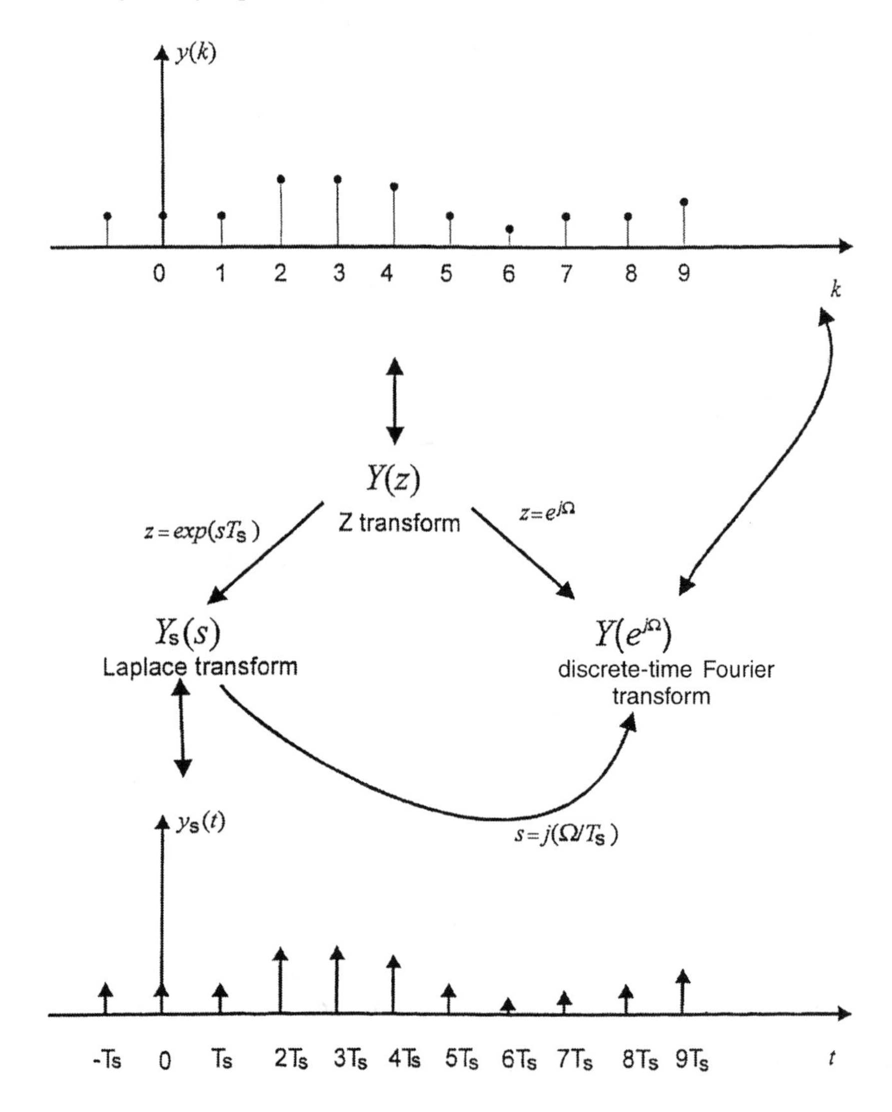

FIGURE 1.16 Illustration of the relationship of the Z transform with the discrete Fourier and Laplace transforms, Figure 1.14 repeated for $y(k)$.

It turns out that the notion of causality is also very useful in considerations of the Z transform. In fact, the one-sided Z transform applies only to causal sequences. When a sequence is noncausal, it must be transformed with the use of the two-sided Z transform. Moreover, note that the Z transform does not distinguish between the causal impulse responses and causal input (output) sequences, as defined schematically in Figure 1.17a for the one-dimensional case. Incidentally, note that the causal sequences are also called one-sided sequences in the literature.[5]

With regard to the Volterra series, we point out that the output sequence $y(k) = y^{(0)}(k) + y^{(1)}(k) + y^{(2)}(k) + \ldots$ of a system described by this series is causal only if both the impulse responses and input signal represent causal sequences. To see this, consider for example the first- and the second-order partial responses with causal impulse responses for the time instant $k = -1$, i.e.,

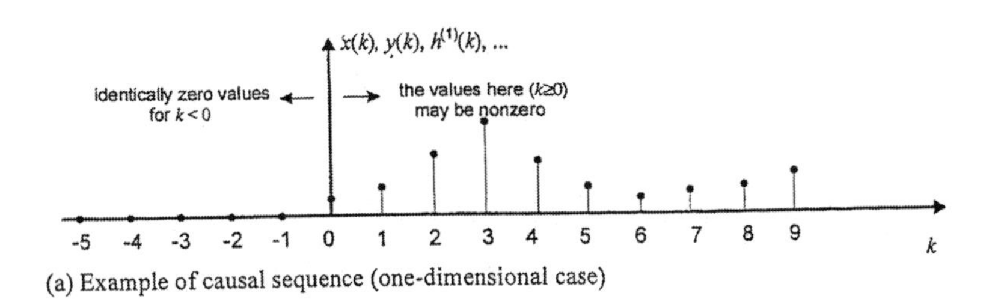

(a) Example of causal sequence (one-dimensional case)

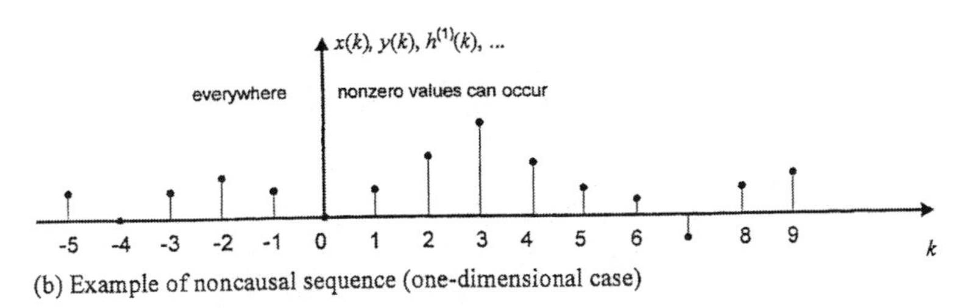

(b) Example of noncausal sequence (one-dimensional case)

FIGURE 1.17 Causal and noncausal sequences (one-dimensional case).

$$y^{(1)}(-1) = \sum_{i=0}^{\infty} h^{(1)}(i)x(-1-i) \tag{1.80a}$$

and

$$y^{(2)}(-1) = \sum_{i_1=0}^{\infty}\sum_{i_2=0}^{\infty} h^{(2)}(i_1, i_2)x(-1-i_1)x(-1-i_2) \tag{1.80b}$$

Note from 1.80a for $i = 0$ that when $h^{(1)}(0) \neq 0$ and $x(k)$ is such that $x(-1) \neq 0$ and $x(k) = 0$ for all $k \neq -1$, then $y^{(1)}(-1) \neq 0$. So, $y^{(1)}(k)$ is not causal. Similarly, we have from 1.80b for $i_1 = 0$ and $i_2 = 0$ the following result: $y^{(2)}(-1) \neq 0$ when $h^{(2)}(0, 0) \neq 0$ and $x(-1) \neq 0$. Hence, $y^{(2)}(k)$ is not causal as well. However, note that putting $x(-1) = 0$ immediately makes $y^{(1)}(-1)$ and $y^{(2)}(-1)$ equal to zero.

Consider now the sequence $\varepsilon(k)$ shown in Figure 1.5. This sequence is, of course, a causal one. If we shift this sequence by, for example, two time units to the right, we get the right-shifted sequence $\varepsilon(k-2)$, as shown in Figure 1.18a. Similarly, if we shift the sequence $\varepsilon(k)$ by, for example, two time units to the left, we shall get the left-shifted sequence $\varepsilon(k + 2)$, as shown in Figure 1.18b.

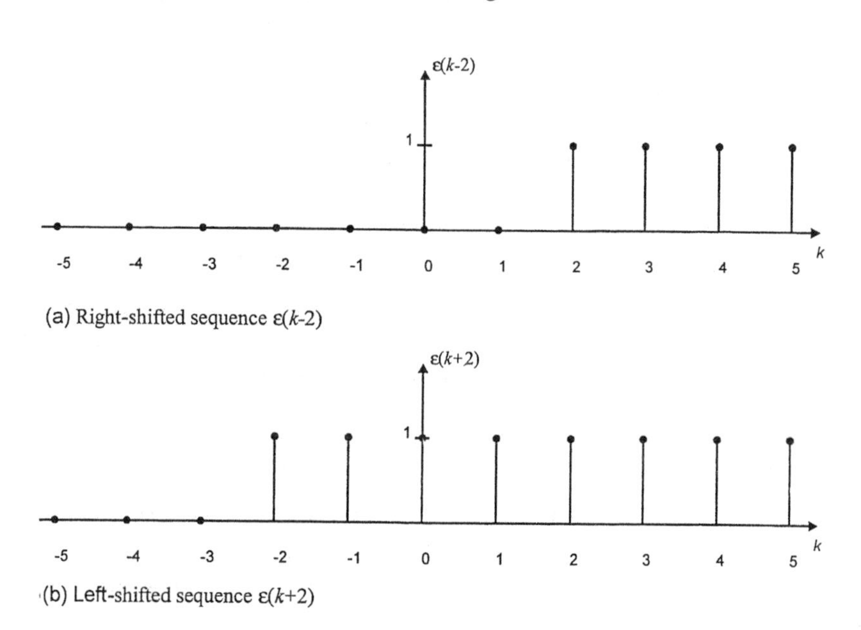

FIGURE 1.18 Example of shifting a causal signal to the right and to the left.

Note from Figure 1.18 that the right-shifting in time of a causal sequence leaves it causal. In contrast to this, the left-shifting in time of a causal sequence makes it noncausal.

1.5 CONDITIONS FOR CONVERGENCE AND STABILITY

The problem of convergence is inherently associated with the Volterra series, as with every other series having an infinite number of elements. In this section, we shall show one of the means of dealing with this problem. However, before going further, one general remark with regard to the relationship between stability and convergence in the context of the Volterra series: by saying that a nonlinear system described by the discrete Volterra series is stable, we mean its stability in the BIBO (bounded input produces bounded output) sense. That is, a bounded input sequence $x(k)$ gives a bounded sequence $y(k)$ at the output. However, in the case of the Volterra series, not every bounded $x(k)$ gives a bounded $y(k)$. So, we conclude that a nonlinear system described by the Volterra series can be BIBO stable only for some set of bounded sequences $x(k)$. And this set is exactly the set of $x(k)$ for which the Volterra series converges. Note that this is in contrast to linear systems which, when BIBO stable, are stable for all bounded input sequences. To see this, consider a linear part in the Volterra series given by Equation 1.14b, for which we can write

$$\left|y^{(1)}(k)\right| = \left|\sum_{i=-\infty}^{\infty} h^{(1)}(i)x(k-i)\right| \leq \sum_{i=-\infty}^{\infty} \left|h^{(1)}(i)\right|\left|x(k-i)\right| \tag{1.81}$$

where $|\cdot|$ means the absolute value of. Then, let us introduce quite formally something that will turn out to be very useful, which is called the norm of a sequence $x(k)$, by saying simply that

$$\|x\| \overset{df}{=} \underset{-\infty \leq k \leq \infty}{(\sup)}|x(k) \tag{1.82a}$$

where sup means the supremum of. Note that having the norm $\|x\|$, we are able to give a bound on all the elements of the sequence $x(k)$ as shown below:

$$|x(k-i)| \leq \|x\|,\ k,\ i \in \mathbb{Z} \tag{1.82b}$$

Hence, applying Equation 1.82b to Equation 1.81 leads to

$$\left|y^{(1)}(k)\right| \leq \|x\| \cdot \sum_{i=-\infty}^{\infty} \left|h^{(1)}(i)\right| \tag{1.83}$$

Note that it follows from the inequality 1.83 that if the impulse response $h^{(1)}(i)$ fulfils the condition

$$\sum_{i=-\infty}^{\infty} \left|h^{(1)}(i)\right| < \infty \tag{1.84a}$$

then, for every bounded input sequence

$$\|x\| < \infty \tag{1.84b}$$

the output sequence $y^{(1)}(k)$ will be bounded, that is

$$\left|y^{(1)}(k)\right| < \infty,\ k \in \mathbb{Z} \tag{1.84c}$$

will hold.

Moreover, note that there are also nonlinear systems that are BIBO stable for all the bounded input sequences $x(k)$. One example is a system described by a truncated Volterra series of the form

$$y(k) = \sum_{i=-\infty}^{\infty} h^{(1)}(i)x(k-i) + \sum_{i_1=-\infty}^{\infty}\sum_{i_2=-\infty}^{\infty} h^{(2)}(i_1, i_2)x(k-i_1)x(k-i_2) \quad (1.85)$$

Taking absolute values of the components in Equation 1.85, we can write

$$|y(k)| \le \sum_{i=-\infty}^{\infty} |h^{(1)}(i)||x(k-i)| + \sum_{i_1=-\infty}^{\infty}\sum_{i_2=-\infty}^{\infty} |h^{(2)}(i_1, i_2)||x(k-i_1)||x(k-i_2)| \quad (1.86)$$

By application of inequality 1.82b to 1.86, we then get

$$|y(k)| \le \|x\| \cdot \sum_{i=-\infty}^{\infty} |h^{(1)}(i)| + \|x\|^2 \cdot \sum_{i_1=-\infty}^{\infty}\sum_{i_2=-\infty}^{\infty} |h^{(2)}(i_1, i_2)| \quad (1.87)$$

Note finally that it follows from inequality 1.87 that, if the impulse response $h^{(1)}(i)$ fulfills inequality 1.84a and the impulse response $h^{(2)}(i_1, i_2)$ fulfills the condition

$$\sum_{i_1=-\infty}^{\infty}\sum_{i_2=-\infty}^{\infty} |h^{(2)}(i_1, i_2)| < \infty \quad (1.88)$$

then for every bounded input sequence $\|x\|<\infty$, the output sequence $y(k)$ will be bounded, $|y(k)| < \infty$, $k \in \mathbb{Z}$.

The inequality 1.87 may suggest that the convergence of the Volterra series is, in some way, related to the convergence of a power series. This is true: the methods used in investigating and calculating the convergence radius for a power series can also be applied in the case of the Volterra series. Every procedure used here relies upon finding a power series with nonnegative coefficients, that converges absolutely, and of which the absolute value is greater or equal to the absolute value of the Volterra series considered. To get such a power series we shall start with the following inequality:

$$|y(k)| = |y^{(0)}(k) + y^{(1)}(k) + y^{(2)}(k) + \cdots| \le |y^{(0)}(k)| + |y^{(1)}(k)| + |y^{(2)}(k)| + \cdots \quad (1.89)$$

Let the partial responses in inequality 1.89 be given by Equations 1.3. Then, we can rewrite 1.89 as

$$|y(k)| \le |h^{(0)}| + \left|\sum_{i=-\infty}^{\infty} h^{(1)}(i)x(k-i)\right| + \left|\sum_{i_1=-\infty}^{\infty}\sum_{i_2=-\infty}^{\infty} h^{(2)}(i_{1,i_2})x(k-i_1)x(k-i_2)\right| + \cdots \quad (1.90)$$

Applying the triangle inequality also to the components of sums in inequality 1.90, in a similar way as in inequality 1.89, we obtain

$$|y(k)| \leq |h^{(0)}| + \sum_{i=-\infty}^{\infty} |h^{(1)}(i)||x(k-i)| + \sum_{i_1=-\infty}^{\infty} \sum_{i_2=-\infty}^{\infty} |h^{(2)}(i_1, i_2)||x(k-i_1)||x(k-i_2)| + \cdots \tag{1.91}$$

Furthermore, using expressions 1.82a and 1.82b in inequality 1.91, we get

$$|y(k)| \leq |h^{(0)}| + \|x\| \cdot \sum_{i=-\infty}^{\infty} |h^{(1)}(i)| + \|x\|^2 \cdot \sum_{i_1=-\infty}^{\infty} \sum_{i_2=-\infty}^{\infty} |h^{(2)}(i_1,i_2)| + \|x\|^3 \sum_{i_1=-\infty}^{\infty} \sum_{i_2=-\infty}^{\infty} \sum_{i_3=-\infty}^{\infty} |h(i_1,i_2,i_3)| + \cdots \tag{1.92a}$$

We now have a power series of the form

$$S(\|x\|) = a_0 + a_1\|x\| + a_2\|x\|^2 + a_3\|x\|^2 + \cdots \tag{1.92b}$$

on the right-hand side of inequality 1.92a, of which coefficients are given by

$$a_0 = |h^{(0)}| \geq 0 \tag{1.92c}$$

$$a_1 = \sum_{i=-\infty}^{\infty} |h^{(1)}(i)| \geq 0 \tag{1.92d}$$

$$a_2 = \sum_{i_1=-\infty}^{\infty} \sum_{i_2=-\infty}^{\infty} |h^{(2)}(i_1, i_2)| \geq 0 \tag{1.92e}$$

$$\vdots$$

and in general,

$$a_n = \sum_{i_1=-\infty}^{\infty} \cdots \sum_{i_n=-\infty}^{\infty} |h^{(n)}(i_1, \ldots, i_n)| \geq 0 \tag{1.92f}$$

All the coefficients a_o, a_1, a_2, … are nonnegative. Moreover, note that when the series $S(\|x\|)$ converges, it converges absolutely because $\|x\| \geq 0$.

Let us now calculate the radius of convergence of the series 1.92b. For this purpose, we use the Cauchy criterion,[9] which says that the power series 1.92b is convergent when

$$\lim_{n \to \infty} \sqrt[n]{a_n\|x\|^n} < 1 \tag{1.93}$$

Substituting a_n given by inequality 1.92f into 1.93 and solving for $\|x\|$ gives

$$\|x\| < \frac{1}{\lim_{n\to\infty} \sqrt[n]{\sum_{i_1=-\infty}^{\infty} \cdots \sum_{i_n=-\infty}^{\infty} \left|h^{(n)}(i_1, \ldots, i_n)\right|}} \tag{1.94a}$$

From inequality 1.94a, we obtain the radius of convergence of the series 1.92a as

$$r = \frac{1}{\lim_{n\to\infty} \sqrt[n]{\sum_{i_1=-\infty}^{\infty} \cdots \sum_{i_n=-\infty}^{\infty} \left|h^{(n)}(i_1, \ldots, i_n)\right|}} \tag{1.94b}$$

Having the expression for the radius of convergence in the form given by Equation 1.94b, we shall now show one interesting relationship between the series convergence and stability of its components. To this end, note that the partial response of the nth order, $y^{(n)}(k)$, is stable when

$$\left|y^{(n)}(k)\right| = \left|\sum_{i_1=-\infty}^{\infty} \cdots \sum_{i_n=-\infty}^{\infty} h^{(n)}(i_1, \ldots, i_n) x(k-i_1) \cdots x(k-i_n)\right| < \infty \tag{1.95}$$

Furthermore, note that we can write the following inequalities:

$$\begin{aligned} &\left|\sum_{i_1=-\infty}^{\infty} \cdots \sum_{i_n=-\infty}^{\infty} h^{(n)}(i_1, \ldots, i_n) x(k-i_1) \cdots x(k-i_n)\right| \\ &\le \sum_{i_1=-\infty}^{\infty} \cdots \sum_{i_n=-\infty}^{\infty} \left|h^{(n)}(i_1, \ldots, i_n)\right| \left|x(k-i_1)\right| \cdots \left|x(k-i_n)\right| \\ &\le \|x\|^n \sum_{i_1=-\infty}^{\infty} \cdots \sum_{i_n=-\infty}^{\infty} \left|h^{(n)}(i_1, \ldots, i_n)\right| \end{aligned} \tag{1.96}$$

with $\|x\|$ given by Equation 1.82a. Assuming then that the value of the expression on the right-hand side of the last inequality in 1.96 is always finite, we have

$$\|x\|^n \cdot \sum_{i_1=-\infty}^{\infty} \cdots \sum_{i_n=-\infty}^{\infty} \left|h^{(n)}(i_1, \ldots, i_n)\right| < \infty \tag{1.97}$$

When the input sequence is bounded, that is $\|x\| < \infty$, then to fulfil inequality 1.97, we must have

$$\sum_{i_1=-\infty}^{\infty} \cdots \sum_{i_n=-\infty}^{\infty} \left| h^{(n)}(i_1, \ldots, i_n) \right| < \infty \tag{1.98}$$

Note now that inequality 1.97 is fulfilled, then inequality 1.95 is also fulfilled. So concluding, we can say that the inequality 1.98 is a sufficient condition of stability of the nth order partial response. Furthermore, it follows from Equation 1.94b that, to have a finite convergence radius different from zero, the limit of the root of the expression 1.98 used in the stability condition must be a finite number, not infinity. Moreover, observe that when the limit of the root in Equation 1.94b equals zero, then the convergence radius $r = \infty$. This means that the corresponding nonlinear system is BIBO stable for all the bounded input sequences (as in linear systems).

At this point, we stress that one very important conclusion follows from the considerations underlying inequalities (1.89 to 1.94): the Volterra series describing the system output response $y(k)$ is absolutely convergent for input sequences for which $|x|<r$ holds. This is clear when looking at the right-hand side of inequality 1.89, keeping in mind the bound on it found afterwards.

Summarizing the main result of this section and returning to the problems related with series convergence, which were tacitly omitted in the previous section, we can say:

1. The series 1.1, which began our considerations, makes sense only when it converges, and best if it converges absolutely. Then, the order of summation of its components is of no importance.
2. The sums occurring in partial responses, as, for example, in Equations 1.2, must converge. As shown, this can be expressed as a problem of stability (in the BIBO sense) of partial responses. And in this context, we have shown that sufficient conditions of the stability of partial responses of the first-order, of the second-order and, generally, of the nth order, are expressed by the inequalities 1.84a, 1.88, and 1.98, respectively. In the case of the first-order partial response, the condition 1.84a is also a necessary one.[10] However, for the higher order partial responses, it has been shown by Sandberg[11] for causal impulse responses that the necessary condition looks like

$$\sup_{\mathbf{J}} \left| \sum_{J_n} \cdots \sum_{J_1} h_{sym}^{(n)}(i_1, \ldots, i_n) \right| < \infty, \qquad n \geq 2 \tag{1.99}$$

where $\mathbf{J}$ is a general n-vector of which elements $J_1, \ldots, J_n$ are nonempty finite subsets of $\mathbb{Z}_+$, with $\mathbb{Z}_+$ meaning the set of nonnegative integers. Observe that

$$\sup_{\mathbf{J}} \left| \sum_{J_n} \cdots \sum_{J_1} h_{sym}^{(n)}(i_1, \ldots, i_n) \right| \leq \sum_{i_1=0}^{\infty} \cdots \sum_{i_n=0}^{\infty} \left| h^{n}(i_1, \ldots, i_n) \right| \tag{1.100}$$

so, really, the bound on the right-hand side of 1.100 is broader than that resulting from the necessity condition.

3. All the series related with the Z transform, discrete Fourier transform, or Laplace transform, which were discussed in Section 1.3, have to converge. Only then do the results presented make sense. One should be fully aware of this fact when calculating some transforms, especially Z transforms, along the lines presented. So, in each case, the convergence of a series must be checked and the radius of convergence must be determined.
4. The series 1.78 expressing the Volterra series in the Z domain makes sense only when all of its components possess Z transforms and the whole series for $Y(z)$ converges. The above must be checked for each concrete case of transforming $y(k)$, and the radius of convergence for $Y(z)$ must be found, taking into account the convergence radii of the Z transforms of all the components of $Y(z)$.

1.6 MATRIX REPRESENTATION FOR MULTIPLE-INPUT AND MULTIPLE-OUTPUT SYSTEMS

The form of the Volterra series described by Equations 1.1 and 1.2 or 1.3 takes into account only one input and one output. Because of this, the series form is sometimes called in the literature a scalar one (for scalar nonlinear systems). And it cannot be used when a nonlinear system possesses more than one input and/or more than one output. The scalar form of the Volterra series must be modified to take into account the above fact. This can be achieved by a technique that is illustrated as follows. For illustration purposes, we consider an example of a nonlinear discrete system possessing three inputs and two outputs, as shown in Figure 1.19. Note that the system in Figure 1.19 can be called the vector nonlinear system, complementing the notion of scalar nonlinear systems defined previously.

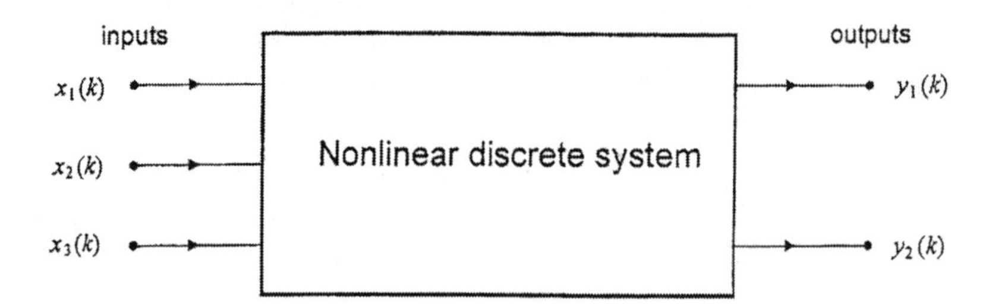

FIGURE 1.19 Example of a nonlinear discrete system with three inputs and two outputs.

For the system of Figure 1.19, we can write

$$\begin{bmatrix} y_1(k) \\ y_2(k) \end{bmatrix} = \begin{bmatrix} y_1^{(0)}(k) \\ y_2^{(0)}(k) \end{bmatrix} + \begin{bmatrix} y_1^{(1)}(k) \\ y_2^{(1)}(k) \end{bmatrix} + \begin{bmatrix} y_1^{(2)}(k) \\ y_2^{(2)}(k) \end{bmatrix} + \begin{bmatrix} y_1^{(3)}(k) \\ y_2^{(3)}(k) \end{bmatrix} + \cdots \tag{1.101}$$

in analogy to Equation 1.1. Note that in more compact form, with the use of the shortened notation for vectors, Equation 1.101 can be rewritten as

$$\mathbf{y}(k) = \mathbf{y}^{(0)}(k) + \mathbf{y}^{(1)}(k) + \mathbf{y}^{(2)}(k) + \mathbf{y}^{(3)}(k) + \cdots = \sum_{n=0}^{\infty} \mathbf{y}^{(n)}(k) \qquad 1.102)$$

where $\mathbf{y}(k)$, $\mathbf{y}^{(0)}(k)$, $\mathbf{y}^{(1)}(k)$, $\mathbf{y}^{(2)}(k)$, … are the corresponding vectors related with the output, the partial response of the zeroth order, the partial response of the first order, the partial response of the second order, and so on. That is, the system output signal and the partial responses of the corresponding orders are considered to be vector-valued.

We now develop Equation 1.101, according to Equations 1.3, for systems independent of time. So we get

$$\begin{bmatrix} y_1(k) \\ y_2(k) \end{bmatrix} = \begin{bmatrix} h_1^{(0)} \\ h_2^{(0)} \end{bmatrix} + \begin{bmatrix} \sum_{i=-\infty}^{\infty} (h_{11}^{(1)}(i)x_1(k-i) + h_{12}^{(1)}(i)x_2(k-i) + h_{13}^{(1)}(i)x_3(k-i)) \\ \sum_{i=-\infty}^{\infty} (h_{21}^{(1)}(i)x_1(k-i) + h_{22}^{(1)}(i)x_2(k-i) + h_{23}^{(1)}(i)x_3(k-i)) \end{bmatrix}$$

$$+ \begin{bmatrix} \sum_{i_1=-\infty}^{\infty} \sum_{i_2=-\infty}^{\infty} (h_{111}^{(2)}(i_1, i_2)x_1(k-i_1)x_1(k-i_2) + h_{121}^{(2)}(i_1, i_2)x_2(k-i_1)x_1(k-i_2) \\ \sum_{i_1=-\infty}^{\infty} \sum_{i_2=-\infty}^{\infty} (h_{211}^{(2)}(i_1, i_2)x_1(k-i_1)x_1(k-i_2) + h_{221}^{(2)}(i_1, i_2)x_2(k-i_1)x_1(k-i_2) \end{bmatrix}$$

$$\begin{matrix} +h_{131}^{(2)}(i_1, i_2)x_3(k-i_1)x_1(k-i_2) + h_{112}^{(2)}(i_1, i_2)x_1(k-i_1)x_2(k-i_2) \\ +h_{231}^{(2)}(i_1, i_2)x_3(k-i_1)x_1(k-i_2) + h_{212}^{(2)}(i_1, i_2)x_1(k-i_1)x_2(k-i_2) \end{matrix}$$

$$\begin{matrix} +h_{122}^{(2)}(i_1, i_2)x_2(k-i_1)x_2(k-i_2) + h_{132}^{(2)}(i_1, i_2)x_3(k-i_1)x_2(k-i_2) \\ +h_{222}^{(2)}(i_1, i_2)x_2(k-i_1)x_2(k-i_2) + h_{232}^{(2)}(i_1, i_2)x_3(k-i_1)x_2(k-i_2) \end{matrix}$$

$$\begin{matrix} +h_{113}^{(2)}(i_1, i_2)x_1(k-i_1)x_3(k-i_2) + h_{123}^{(2)}(i_1, i_2)x_2(k-i_1)x_3(k-i_2) \\ +h_{213}^{(2)}(i_1, i_2)x_1(k-i_1)x_3(k-i_2) + h_{223}^{(2)}(i_1, i_2)x_2(k-i_1)x_3(k-i_2) \end{matrix}$$

$$\left. \begin{matrix} +h_{133}^{(2)}(i_1, i_2)x_3(k-i_1)x_3(k-i_2)) \\ +h_{233}^{(2)}(i_1, i_2)x_3(k-i_1)x_3(k-i_2)) \end{matrix} \right] + \cdots \qquad (1.103)$$

Note that Equation 1.103 can be rewritten in the more compact form:

$$
\begin{bmatrix} y_1(k) \\ y_2(k) \end{bmatrix} = \begin{bmatrix} h_1^{(0)} \\ h_2^{(0)} \end{bmatrix} + \begin{bmatrix} \sum\limits_{i=-\infty}^{\infty} [h_{11}^{(1)}\ h_{12}^{(1)}\ h_{13}^{(1)}]_{(i)} \cdot \begin{bmatrix} x_1 \\ x_2 \\ x_3 \end{bmatrix}_{(k-i)} \\ \sum\limits_{i=-\infty}^{\infty} [h_{21}^{(1)}\ h_{22}^{(1)}\ h_{23}^{(1)}]_{(i)} \cdot \begin{bmatrix} x_1 \\ x_2 \\ x_3 \end{bmatrix}_{(k-i)} \end{bmatrix}
$$

$$
+ \begin{bmatrix} \sum\limits_{i_1=-\infty}^{\infty} \sum\limits_{i_2=-\infty}^{\infty} [h_{111}^{(2)}\ h_{121}^{(2)}\ h_{131}^{(2)}\ h_{112}^{(2)}\ h_{122}^{(2)}\ h_{132}^{(2)}\ h_{113}^{(2)}\ h_{123}^{(2)}\ h_{133}^{(2)}]_{(i_1, i_2)} \cdot \left(\begin{bmatrix} x_1 & 0 & 0 & 0 & 0 & 0 & 0 & 0 & 0 \\ 0 & x_2 & 0 & 0 & 0 & 0 & 0 & 0 & 0 \\ 0 & 0 & x_3 & 0 & 0 & 0 & 0 & 0 & 0 \\ 0 & 0 & 0 & x_1 & 0 & 0 & 0 & 0 & 0 \\ 0 & 0 & 0 & 0 & x_2 & 0 & 0 & 0 & 0 \\ 0 & 0 & 0 & 0 & 0 & x_3 & 0 & 0 & 0 \\ 0 & 0 & 0 & 0 & 0 & 0 & x_1 & 0 & 0 \\ 0 & 0 & 0 & 0 & 0 & 0 & 0 & x_2 & 0 \\ 0 & 0 & 0 & 0 & 0 & 0 & 0 & 0 & x_3 \end{bmatrix}_{(k-i_1)} \begin{bmatrix} x_1 \\ x_1 \\ x_1 \\ x_2 \\ x_2 \\ x_2 \\ x_3 \\ x_3 \\ x_3 \end{bmatrix}_{(k-i_2)} \right) \\ \sum\limits_{i_1=-\infty}^{\infty} \sum\limits_{i_2=-\infty}^{\infty} [h_{211}^{(2)}\ h_{221}^{(2)}\ h_{231}^{(2)}\ h_{212}^{(2)}\ h_{222}^{(2)}\ h_{232}^{(2)}\ h_{213}^{(2)}\ h_{223}^{(2)}\ h_{233}^{(2)}]_{(i_1, i_2)} \cdot \left(\begin{bmatrix} x_1 & 0 & 0 & 0 & 0 & 0 & 0 & 0 & 0 \\ 0 & x_2 & 0 & 0 & 0 & 0 & 0 & 0 & 0 \\ 0 & 0 & x_3 & 0 & 0 & 0 & 0 & 0 & 0 \\ 0 & 0 & 0 & x_1 & 0 & 0 & 0 & 0 & 0 \\ 0 & 0 & 0 & 0 & x_2 & 0 & 0 & 0 & 0 \\ 0 & 0 & 0 & 0 & 0 & x_3 & 0 & 0 & 0 \\ 0 & 0 & 0 & 0 & 0 & 0 & x_1 & 0 & 0 \\ 0 & 0 & 0 & 0 & 0 & 0 & 0 & x_2 & 0 \\ 0 & 0 & 0 & 0 & 0 & 0 & 0 & 0 & x_3 \end{bmatrix}_{(k-i_1)} \begin{bmatrix} x_1 \\ x_1 \\ x_1 \\ x_2 \\ x_2 \\ x_2 \\ x_3 \\ x_3 \\ x_3 \end{bmatrix}_{(k-i_2)} \right) \end{bmatrix} + \ldots \tag{1.104}
$$

where the notation $[]_{(i)}$, $[]_{(i_1, i_2)}$, and $[]_{(k-i_1)}$ means that the elements of the vector or matrix considered (all the elements of that vector or matrix) depend upon arguments i, i_1, and i_2, and $k - i_1$, respectively.

Further simplification of Equation 1.104 leads to

$$\begin{bmatrix} y_1(k) \\ y_2(k) \end{bmatrix} = \begin{bmatrix} h_1^{(0)} \\ h_2^{(0)} \end{bmatrix} + \sum_{i=-\infty}^{\infty} \begin{bmatrix} h_{11}^{(1)} & h_{12}^{(1)} & h_{13}^{(1)} \\ h_{21}^{(1)} & h_{22}^{(1)} & h_{23}^{(1)} \end{bmatrix}_{(i)} \cdot \begin{bmatrix} x_1 \\ x_2 \\ x_3 \end{bmatrix}_{(k-i)}$$

$$+ \sum_{i_1=-\infty}^{\infty} \sum_{i_2=-\infty}^{\infty} \begin{bmatrix} h_{111}^{(2)} & h_{121}^{(2)} & h_{131}^{(2)} & h_{112}^{(2)} & h_{122}^{(2)} & h_{132}^{(2)} & h_{113}^{(2)} & h_{123}^{(2)} & h_{133}^{(2)} \\ h_{211}^{(2)} & h_{221}^{(2)} & h_{231}^{(2)} & h_{212}^{(2)} & h_{222}^{(2)} & h_{232}^{(2)} & h_{213}^{(2)} & h_{223}^{(2)} & h_{233}^{(2)} \end{bmatrix}_{(i_1, i_2)}$$

$$\bullet \begin{bmatrix} x_1(k-i_1)\ x_1(k-i_2) \\ x_2(k-i_1)\ x_1(k-i_2) \\ x_3(k-i_1)\ x_1(k-i_2) \\ x_1(k-i_1)\ x_2(k-i_2) \\ x_2(k-i_1)\ x_2(k-i_2) \\ x_3(k-i_1)\ x_2(k-i_2) \\ x_1(k-i_1)\ x_3(k-i_2) \\ x_2(k-i_1)\ x_3(k-i_2) \\ x_3(k-i_1)\ x_3(k-i_2) \end{bmatrix} + \cdots = \mathbf{h}^{(0)} + \sum_{i=-\infty}^{\infty} \mathbf{h}^{(1)}(i) \cdot \mathbf{x}(k-i \tag{1.105}$$

$$+ \sum_{i_1=-\infty}^{\infty} \sum_{i_2=-\infty}^{\infty} \mathbf{h}^{(2)}(i_1, i_2) \cdot (\mathbf{x}(k-i_1) \otimes \mathbf{x}(k-i_2)) + \cdots$$

where the definitions of the vector $\mathbf{h}^{(0)}$, and the matrices $\mathbf{h}^{(1)}$ and $\mathbf{h}^{(2)}$ follow from the comparison of the corresponding expressions determining the components of the series 1.105. Moreover, the symbol $\otimes$ used in Equation 1.105 means the Kronecker product; it will be explained in detail in what follows.

Let **A** and **B** be matrices of order $M_A \times N_A$ and $M_B \times N_B$, respectively. That is

$$\mathbf{A} = \begin{bmatrix} a_{11} & a_{12} & \cdots\cdots & a_{1N_A} \\ a_{21} & a_{22} & \cdots\cdots & a_{2N_A} \\ \vdots & & \vdots & \vdots \\ a_{M_A1} & a_{M_A2} & \cdots\cdots & a_{M_AN_A} \end{bmatrix} \tag{1.106a}$$

and

$$\mathbf{B} = \begin{bmatrix} b_{11} & b_{12} & \cdots\cdots & b_{1N_B} \\ b_{21} & b_{22} & \cdots\cdots & b_{2N_B} \\ \vdots & & \vdots & \vdots \\ b_{M_B1} & b_{M_B2} & \cdots\cdots & b_{M_BN_B} \end{bmatrix} \tag{1.106b}$$

The Kronecker product of the matrices **A** and **B** is then matrix **C**, given by

$$\mathbf{C} = \mathbf{A} \otimes \mathbf{B} \stackrel{df}{=} \begin{bmatrix} \mathbf{A}b_{11} & \mathbf{A}b_{12} & \cdots\cdots & \mathbf{A}b_{1N_B} \\ \mathbf{A}b_{21} & \mathbf{A}b_{22} & \cdots\cdots & \mathbf{A}b_{2N_B} \\ \vdots & & \vdots & \vdots \\ \mathbf{A}b_{M_B1} & \mathbf{A}b_{M_B2} & \cdots\cdots & \mathbf{A}b_{M_BN_B} \end{bmatrix} \tag{1.107a}$$

The order of the matrix $\mathbf{C}$ is $M_AM_B \times N_AN_B$.

Observe from Equation 1.107a that elements of $\mathbf{C}$ can be expressed as

$$c_{i_cj_c} = a_{i_aj_a}b_{i_bj_b} \tag{1.107b}$$

where $i_c, j_c; i_a, j_a; i_b, j_b$ are the corresponding indices of elements of the matrices $\mathbf{C}$, $\mathbf{A}$, and $\mathbf{B}$, respectively. The indices i_c and j_c fulfill the following two equations:

$$i_c = i_a + M_A(i_b - 1) \tag{1.108a}$$

and

$$j_c = j_a + N_A(j_b - 1) \tag{1.108b}$$

The Kronecker product defined by Equations 1.107 and 1.108 is called in the literature[12] the left Kronecker product. More about the Kronecker product of matrices can be found in Reference 12 and references cited there. Here, we give its most important properties as

$$\mathbf{A} \otimes \mathbf{B} \neq \mathbf{B} \otimes \mathbf{A} \qquad \text{(in general)} \tag{1.109}$$

$$\mathbf{A} \otimes (\mathbf{B} \otimes \mathbf{C}) = (\mathbf{A} \otimes \mathbf{B}) \otimes \mathbf{C} = \mathbf{A} \otimes \mathbf{B} \otimes \mathbf{C} \tag{1.110}$$

$$(\mathbf{A} + \mathbf{B}) \otimes \mathbf{C} = (\mathbf{A} \otimes \mathbf{C}) + (\mathbf{B} \otimes \mathbf{C}) \tag{1.111}$$

$$(\mathbf{A} \cdot \mathbf{B}) \otimes (\mathbf{C} \cdot \mathbf{D}) = (\mathbf{A} \otimes \mathbf{C}) \cdot (\mathbf{B} \otimes \mathbf{D}) \tag{1.112}$$

$$(\mathbf{A} \otimes \mathbf{B})^{-1} = \mathbf{A}^{-1} \otimes \mathbf{B}^{-1} \tag{1.113}$$

$$(\mathbf{A} \otimes \mathbf{B})^T = \mathbf{A}^T \otimes \mathbf{B}^T \tag{1.114}$$

The symbol "-1" in Equation 1.113 means the inverse of, but the symbol "T" in Equation 1.114 refers to the matrix transposition.

Note now applying the definition of the Kronecker product Equation 1.107a to the product of the vectors $\mathbf{x}(k - i_1)$ and $\mathbf{x}(k - i_2)$ in Equation 1.105, we get

$$\mathbf{x}(k-i_1)\otimes\mathbf{x}(k-i_2) = \begin{bmatrix} \begin{bmatrix} x_1(k-i_1) \\ x_2(k-i_1) \\ x_3(k-i_1) \end{bmatrix} x_1(k-i_2) \\ \\ \begin{bmatrix} x_1(k-i_1) \\ x_2(k-i_1) \\ x_3(k-i_1) \end{bmatrix} x_2(k-i_2) \\ \\ \begin{bmatrix} x_1(k-i_1) \\ x_2(k-i_1) \\ x_3(k-i_1) \end{bmatrix} x_3(k-i_2) \end{bmatrix} = \begin{bmatrix} x_1(k-i_1)\ x_1(k-i_2) \\ x_2(k-i_1)\ x_1(k-i_2) \\ x_3(k-i_1)\ x_1(k-i_2) \\ x_1(k-i_2)\ x_2(k-i_2) \\ x_2(k-i_1)\ x_2(k-i_2) \\ x_3(k-i_1)\ x_2(k-i_2) \\ x_1(k-i_1)\ x_3(k-i_2) \\ x_2(k-i_1)\ x_3(k-i_2) \\ x_3(k-i_1)\ x_3(k-i_2) \end{bmatrix} \tag{1.115}$$

And the resulting vector in Equation 1.115 is identical to the corresponding vector written down explicitly in Equation 1.105.

Incidentally, observe that Equation 1.105 could be formulated otherwise, using the right Kronecker product. Then, for example, the vector determining the second-order partial response would look like

$$\begin{bmatrix} y_1^{(2)}(k) \\ y_2^{(2)}(k) \end{bmatrix} = \sum_{i_1=-\infty}^{\infty}\sum_{i_2=-\infty}^{\infty} \begin{bmatrix} h_{111}^{(2)} & h_{112}^{(2)} & h_{113}^{(2)} & h_{121}^{(2)} & h_{122}^{(2)} & h_{123}^{(2)} & h_{131}^{(2)} & h_{132}^{(2)} & h_{133}^{(2)} \\ h_{211}^{(2)} & h_{212}^{(2)} & h_{213}^{(2)} & h_{221}^{(2)} & h_{222}^{(2)} & h_{223}^{(2)} & h_{231}^{(2)} & h_{232}^{(2)} & h_{233}^{(2)} \end{bmatrix}_{(i_1,i_2)} \cdot \begin{bmatrix} x_1(k-i_1)\ x_1(k-i_2) \\ x_1(k-i_1)\ x_2(k-i_2) \\ x_1(k-i_1)\ x_3(k-i_2) \\ x_2(k-i_1)\ x_1(k-i_2) \\ x_2(k-i_1)\ x_2(k-i_2) \\ x_2(k-i_1)\ x_3(k-i_2) \\ x_3(k-i_1)\ x_1(k-i_2) \\ x_3(k-i_1)\ x_2(k-i_2) \\ x_3(k-i_1)\ x_3(k-i_2) \end{bmatrix} \tag{1.116}$$

In this book we use the same notational convention as proposed by Saleh,[12] that is, the formulation with the use of the left Kronecker product.

Now return to Equation 1.105 and rewrite its final form as

$$\mathbf{y}(k) = \mathbf{h}^{(0)} + \sum_{i=-\infty}^{\infty} \mathbf{h}^{(1)}(i) \cdot \mathbf{x}(k-i) + \sum_{i_1=-\infty}^{\infty} \sum_{i_2=-\infty}^{\infty} \mathbf{h}^{(2)}(i_1, i_2) \cdot (\mathbf{x}(k-i_1) \otimes \mathbf{x}(k-i_2))$$
$$+ \sum_{i_1=-\infty}^{\infty} \sum_{i_2=-\infty}^{\infty} \sum_{i_3=-\infty}^{\infty} \mathbf{h}^{(3)}(i_1, i_2, i_3) \cdot (\mathbf{x}(k-i_1) \otimes \mathbf{x}(k-i_2) \otimes \mathbf{x}(k-i_3)) + \cdots \tag{1.117}$$

where all the Kronecker products indicated are the left ones.

Keeping in mind the considerations just presented, we will now explain the general formula 1.117 for a nonlinear discrete system with N inputs and M outputs. Note that such a system is presented schematically in Figure 1.20. For the purpose of explanation, we refer in what follows to the

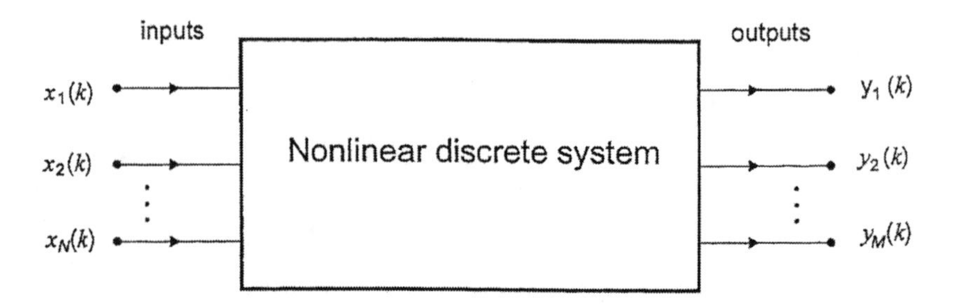

FIGURE 1.20 Nonlinear discrete system with N inputs and M outputs.

nth order component in 1.117, having the following form:

$$\sum_{i_1=-\infty}^{\infty} \cdots \sum_{i_n=-\infty}^{\infty} \mathbf{h}^{(n)} \cdot \Big(\underbrace{\mathbf{x} \otimes \cdots \otimes \mathbf{x}}_{\text{n times}}\Big), \qquad n \geq 2 \tag{1.118}$$

where the arguments are omitted in the matrix $\mathbf{h}^{(n)}$ and vectors $\mathbf{x}$ to simplify the notation. Moreover, we call the reader's attention to the fact that the small n here, being the order of the nonlinear impulse response, has nothing to do with the capital N, being the number of inputs.

It is clear from the definition of the Kronecker product given by Equation 1.107a that the n-fold product $\mathbf{x} \otimes \ldots \otimes x$ in Equation 1.118 results in the vector of order $N^n \times 1$, which looks like

$$\begin{bmatrix} x_1 \\ x_2 \\ \vdots \\ x_N \end{bmatrix} \overset{\otimes \text{applied } n \text{ times}}{\Rightarrow} \begin{bmatrix} \underbrace{x_1 x_1 \dots x_1}_{N \text{ times}} \\ \vdots \\ x_{j_1} x_{j_2} \dots x_{j_n} \\ \vdots \\ x_N x_N \dots x_N \end{bmatrix} \begin{matrix} \leftarrow \text{the first element} \\ \\ \leftarrow \text{the } j\text{th element} \\ j_1, j_2, \dots, j_n \in \{1, 2, \dots, N\} \\ \leftarrow \text{the } N^n\text{th element} \end{matrix} \tag{1.119a}$$

Observe from Equation 1.119a that the jth element of the vector is equal to

$$[\mathbf{x} \otimes \cdots \otimes \mathbf{x}]_j = x_{j_1} \cdots x_{j_n} \tag{1.119b}$$

where the index j is related to the indices $j_1, \dots, j_n$ in the following way:

$$j = j_1 + N(j_2 - 1) + \cdots + N^{n-1}(j_n - 1) \tag{1.119c}$$

The matrices $\mathbf{h}^{(n)}$ have the following form:

$$\mathbf{h}^{(n)} = \overset{\downarrow N^n \text{ columns}}{\begin{bmatrix} h_{11 \cdot\cdot 1} & \cdots\cdots & h_{1N \cdot\cdot N} \\ \vdots & \vdots & \vdots \\ h_{M1 \cdot\cdot 1} & \cdots\cdots & h_{MN \cdot\cdot N} \end{bmatrix}} \leftarrow M \text{ rows} \tag{1.120a}$$

These matrices are of order $M \times N^n$.

Observe from Equation 1.120a that the ij-th element in the matrix $\mathbf{h}^{(n)}$ can be expressed as

$$[\mathbf{h}^{(n)}]_{ij} = h^{(n)}_{j_1 \cdots j_n} \tag{1.120b}$$

with j given by the relation 1.119c.

At the end of this section, we shall illustrate application of the representation 1.117, developed for nonlinear multiple-input and multiple-output systems, to the stability and convergence problems encountered in such systems. We start with some useful definitions. First, let us rewrite Equation 1.119b in the following symbolic form:

$$\Big[\underbrace{\mathbf{x} \otimes \cdots \otimes \mathbf{x}}_{n \text{ times}}\Big]_j = (x^n)_j \tag{1.121}$$

The symbolic notation 1.121 means the jth element of the vector $\mathbf{x} \otimes \ldots \otimes x$ (n times) consists of n elements of the vector x.

Second, we extend the definition of the norm for scalar sequences, given by Equation 1.82a, to vector-valued sequences. We do this for the input vector $\mathbf{x}(k)$ and the output vector $\mathbf{y}(k)$ (more precisely, these, of course are the vector-valued input and output sequences) as:

$$\|\mathbf{x}\|_I \overset{df}{=} \max_{1 \le i \le N} (\sup_{-\infty \le k \le \infty} |x_i(k)|) \tag{1.122a}$$

and

$$\|\mathbf{y}\|_I \overset{df}{=} \max_{1 \le i \le M} (\sup_{-\infty \le k \le \infty} |y_i(k)|) \tag{1.122b}$$

Note from definitions 1.122a and 1.122b that to calculate the norm of the vector $\mathbf{x}$ or $\mathbf{y}$, one calculates first the norms of all the components of these vectors according to the definition 1.82a. Then one picks up the maximal value from them.

Let us now illustrate the usage of the vector representation of the Volterra series as in Equation 1.117 to find the stability condition of the linear part in it, having the form

$$\mathbf{y}^{(1)}(k) = \sum_{i=-\infty}^{\infty} \mathbf{h}^{(1)}(i) \cdot \mathbf{x}(k-i) \tag{1.123}$$

We proceed here similarly as before (see inequality 1.81 and the following ones). Hence, we get from Equation 1.123

$$|y_i^{(1)}(k)| = \left| \sum_{i_1=-\infty}^{\infty} [\mathbf{h}^{(1)}(i_1) \cdot \mathbf{x}(k-i_1)]_i \right| \le \sum_{i_1=-\infty}^{\infty} |[\mathbf{h}^{(1)}(i_1) \cdot \mathbf{x}(k-i_1)]_i|,\ 1 \le i \le M \tag{1.124a}$$

where the argument i_1 instead of i is used. The letter i is used in the inequality 1.124a and following inequalities to denote the row index of the elements of the matrix $\mathbf{h}$ and of the vector $\mathbf{y}$.

For better understanding of inequality 1.124a, let us write it explicitly for all row indices i; that is,

$$\begin{aligned}
\left|y_1^{(1)}(k)\right| &\le \sum_{i_1=-\infty}^{\infty}\left|\sum_{j=1}^{N} h_{1j}^{(1)}(i_1)x_j(k-i_1)\right| \le \sum_{i_1=-\infty}^{\infty}\sum_{j=1}^{N}\left|h_{1j}^{(1)}(i_1)\right|\left|x_j(k-i_1)\right| \\
\left|y_2^{(1)}(k)\right| &\le \sum_{i_1=-\infty}^{\infty}\left|\sum_{j=1}^{N} h_{2j}^{(1)}(i_1)x_j(k-i_1)\right| \le \sum_{i_1=-\infty}^{\infty}\sum_{j=1}^{N}\left|h_{2j}^{(1)}(i_1)\right|\left|x_j(k-i_1)\right| \\
&\vdots \\
\left|y_M^{(1)}(k)\right| &\le \sum_{i_1=-\infty}^{\infty}\left|\sum_{j=1}^{N} h_{Mj}^{(1)}(i_1)x_j(k-i_1)\right| \le \sum_{i_1=-\infty}^{\infty}\sum_{j=1}^{N}\left|h_{Mj}^{(1)}(i_1)\right|\left|x_j(k-i_1)\right|
\end{aligned} \tag{1.124b}$$

Note that inequalities 1.124b can be written in more compact form as

$$\left|y_i^{(1)}(k)\right| \le \sum_{i_1=-\infty}^{\infty}\sum_{j=1}^{N}\left|h_{ij}^{(1)}(i_1)\right|\left|x_j(k-i_1)\right|, \ 1 \le i \le M \tag{1.124c}$$

It is clear that the following inequality

$$\left|x_j(k-i_1)\right| \le \|\mathbf{x}\|_I \tag{1.125}$$

is always true. So, taking into account the above fact in inequality 1.124c leads to

$$\left|y_i^{(1)}(k)\right| \le \|\mathbf{x}\|_I \sum_{j=1}^{N}\sum_{i_1=-\infty}^{\infty}\left|h_{ij}^{(1)}(i_1)\right|, \ 1 \le i \le M \tag{1.126a}$$

Note that inequality 1.126a holds for every k. Hence, the following:

$$\sup_{-\infty \le k \le \infty}\left|y_i^{(1)}(k)\right| \le \|\mathbf{x}\|_I \sum_{j=1}^{N}\sum_{i_1=-\infty}^{\infty}\left|h_{ij}^{(1)}(i_1)\right|, \ 1 \le i \le M \tag{1.126b}$$

is also fulfilled.

Consider now the index $i = i'$, for which

$$\max_{1 \le i \le M}\left(\sup_{-\infty \le k \le \infty}\left|y_i^{(1)}(k)\right|\right) \tag{1.127a}$$

occurs. We rewrite inequality 1.126b for this index

$$\sup_{-\infty \le k \le \infty}\left|y_{i'}^{(1)}(k)\right| \le \|\mathbf{x}\|_I \sum_{j=1}^{N}\sum_{i_1=-\infty}^{\infty}\left|h_{i'j}^{(1)}(i_1)\right| \tag{1.127b}$$

Then taking into account equality 1.122b in inequality 1.127b, the latter can be rewritten as

$$\left\|\mathbf{y}^{(1)}\right\|_I \le \|\mathbf{x}\|_I \sum_{j=1}^{N} \sum_{i_1=-\infty}^{\infty} \left|h_{ij}^{(1)}(i_1)\right| \tag{1.127c}$$

To make inequality 1.127c more general, let us denote by

$$K_1 = \max_{1 \le i \le M} \left(\sum_{j=1}^{N} \sum_{i_1=-\infty}^{\infty} \left|h_{ij}^{(1)}(i_1)\right| \right) \tag{1.128a}$$

Applying then K_I in inequality 1.127c allows us to write

$$\left\|\mathbf{y}^{(1)}\right\|_I \le K_1 \|\mathbf{x}\|_I \tag{1.128b}$$

So with

$$K_1 < \infty \tag{1.129a}$$

and bounded $\mathbf{x}$, that is $\|\mathbf{x}\|_I < \infty$, it is evident from inequality 1.128b that $y^{(1)}$ is bounded:

$$\left\|\mathbf{y}^{(1)}\right\|_I < \infty \tag{1.129b}$$

Hence, inequality 1.129a with K_1 given by Equation 1.128a represents the stability condition we looked for.

Equivalently, the stability condition for $\mathbf{y}^1(k)$ given by Equation 1.123 can be found using another form of the norm for the input and output vector-valued sequences. Maybe this is a more elegant way. It assumes defining first the absolute value for vectors, as written below for the input and output vectors

$$|\mathbf{x}| \overset{df}{=} \max_i |x_i| = \max_{1 \le i \le N} |x_i| \tag{1.130a}$$

$$|\mathbf{y}| \overset{df}{=} \max_i |y_i| = \max_{1 \le i \le M} |y_i| \tag{1.130b}$$

Note from the definitions 1.130a and 1.130b that the absolute value of a vector is defined as the maximal value of all its elements.

When a vector depends upon the discrete time k, then it is a vector-valued sequence. For such a sequence, using the absolute value of a vector just defined, we define a norm. Using the examples of the input and output vectors (more precisely, the input and output vector-valued sequences) **x** and **y**, this is formulated as follows

$$\|\mathbf{x}\| \stackrel{df}{=} \sup_{-\infty \le k \le \infty} |\mathbf{x}(k)| \tag{1.131a}$$

$$\|\mathbf{y}\| \stackrel{df}{=} \sup_{-\infty \le k \le \infty} |\mathbf{y}(k)| \tag{1.131b}$$

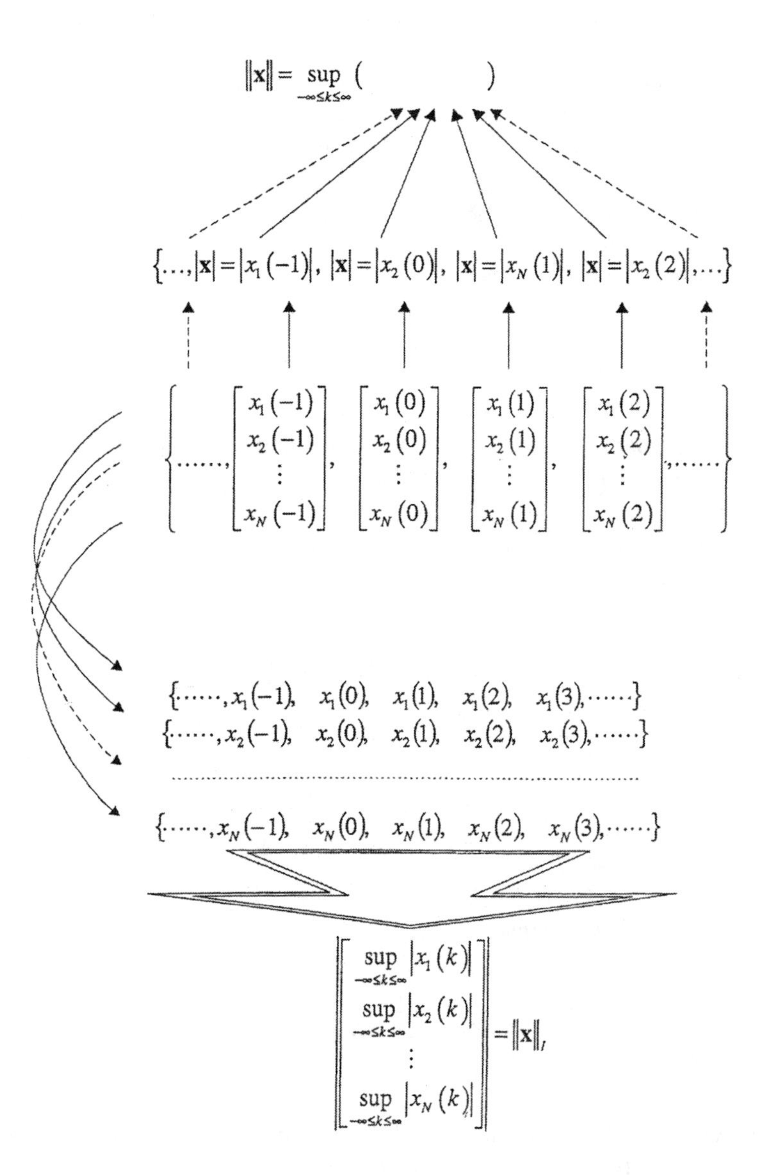

FIGURE 1.21 Illustration of calculation of the norms $\|\cdot\|$ and $\|\cdot\|_l$ for vector-valued sequences.

Note that $\mathbf{x}(k)$ and $\mathbf{y}(k)$ in Equation 1.131a and 1.131b, respectively, with k given explicitly, mean the vector-valued sequences. We use here the notational convention similarly used before with regard to scalars $x(k)$ and $y(k)$. Recall that $x(k)$ (or similarly, $y(k)$) was used with two meanings: as the element $x(k)$ of the sequence x, or as the whole sequence; that is, $x(k) = \{\cdots, x(-1), x(0), x(1), x(2), x(3), \cdots\}$ of which meaning was valid at the moment, it followed from the context. And similarly here, $\mathbf{x}(k)$ (or $\mathbf{y}(k)$) is used: as the element $\mathbf{x}(k)$ of the vector-valued sequence x, or as the whole vector-valued sequence: that is, $\mathbf{x}(k) = \{\cdots, \mathbf{x}(-1), \mathbf{x}(0), \mathbf{x}(1), \mathbf{x}(2), \mathbf{x}(3), \cdots\}$. Furthermore, the context determines which of the above meanings is valid at the moment.

To point out the differences in calculating the norms $\|\cdot\|$ and $\|\cdot\|_l$ for vector-valued sequences, we illustrate the calculation of both in Figure 1.21.

Let us now come back to calculating the stability condition for $\mathbf{y}^{(1)}(k)$ given by Equation 1.123 with the use of the norm $\|\cdot\|$ defined for vector-valued sequences. We start by applying the definition 1.130b to the vector-valued output sequence of Equation 1.123. This gives

$$\left|\mathbf{y}^{(1)}(k)\right| = \left|\sum_{i_1=-\infty}^{\infty} \mathbf{h}^{(1)}(i_1)\cdot \mathbf{x}(k-i_1)\right| \le \sum_{i_1=-\infty}^{\infty} \left|\mathbf{h}^{(1)}(i_1)\cdot \mathbf{x}(k-i_1)\right| \qquad (1.132)$$

where the argument i_1 instead of i is used. The letter i is now reserved for row indexing in the vector $\mathbf{y}^{(1)}(k)$ (precisely, for indexing rows in the given vector-valued sequence).

Note that inequality 1.132 can be rewritten as

$$\left|\mathbf{y}^{(1)}(k)\right| \le \sum_{i_1=-\infty}^{\infty} \max_i \left|\sum_{j=1}^{N} h_{ij}^{(1)}(i_1)\cdot x_j(k-i_1)\right| \le \sum_{i_1=-\infty}^{\infty} \max_i \left(\sum_{j=1}^{N} \left|h_{ij}^{(1)}(i_1)\right|\cdot\left|x_j(k-i_1)\right|\right),\ 1 \le i \le M \qquad (1.133)$$

Because

$$\left|x_j(k-i_1)\right| \le \left|\mathbf{x}(k-i_1)\right| \le \|\mathbf{x}\| \qquad (1.134)$$

holds, we can continue inequality 1.133 as

$$\left|\mathbf{y}^{(1)}(k)\right| \le \|\mathbf{x}\| \cdot \sum_{i_1=-\infty}^{\infty} \max_i \left(\sum_{j=1}^{N} \left|h_{ij}^{(1)}(i_1)\right|\right) \qquad (1.135)$$

Furthermore, note that inequality 1.135 will also hold, if we use the supremum function on the left-hand side of inequality 1.135; that is,

$$\sup_{-\infty \le k \le \infty} \left|\mathbf{y}^{(1)}(k)\right| \le \|\mathbf{x}\| \cdot \sum_{i_1=-\infty}^{\infty} \max_i \left(\sum_{j=1}^{N} \left|h_{ij}^{(1)}(i_1)\right| \right) \tag{1.136a}$$

So, finally, with the definition 1.131b in mind, we get from inequality 1.136a

$$\left\|\mathbf{y}^{(1)}\right\| \le \|\mathbf{x}\| \cdot \sum_{i_1=-\infty}^{\infty} \max_i \left(\sum_{j=1}^{N} \left|h_{ij}^{(1)}(i_1)\right| \right) \tag{1.136b}$$

If we introduce

$$K = \sum_{i_1=-\infty}^{\infty} \max_i \left(\sum_{j=1}^{N} \left|h_{ij}^{(1)}(i_1)\right| \right) \tag{1.136c}$$

in inequality 1.136b, we shall get the inequality 1.136b in the following equivalent form:

$$\left\|\mathbf{y}^{(1)}\right\| \le K\|\mathbf{x}\| \tag{1.136d}$$

Note that inequality 1.136d is the counterpart of inequality 1.128b derived previously. And, as before, when

$$K < \infty \tag{1.137a}$$

and the vector-valued sequence $\mathbf{x}$ is bounded, then it follows from Equation 1.136d that inequality

$$\left\|\mathbf{y}^{(1)}\right\| < \infty \tag{1.137b}$$

So inequality 1.137a represents the stability condition for $\mathbf{y}^{(1)}(k)$ Equation 1.123, derived with the use of the norms $\|\cdot\|$ defined by Equations 1.131a and 1.131b.

Let us now consider the convergence of the series 1.117. For this purpose, we write

$$\begin{aligned} |y_i| \le \left|h_i^{(0)}\right| &+ \left| \sum_{i_1=-\infty}^{\infty} \sum_{j=1}^{N} h_{ij}^{(1)} x_j \right| + \left| \sum_{i_1=-\infty}^{\infty} \sum_{i_2=-\infty}^{\infty} \sum_{j=1}^{N^2} h_{ij}^{(2)} (x^2)_j \right| \\ &+ \left| \sum_{i_1=-\infty}^{\infty} \sum_{i_2=-\infty}^{\infty} \sum_{i_3=-\infty}^{\infty} \sum_{j=1}^{N^3} h_{ij}^{(3)} (x^3)_j \right| + \cdots \end{aligned} \qquad 1 \le i \le M \tag{1.138}$$

for the ith element of the vector $\mathbf{y}$. The arguments in y_i, $h_{ij}^{(\cdot)}$, x_j, related to the time in inequaltiy 1.138, i.e., k, i_1, $k - i_1$, i_2, $\cdots$, are omitted for simplicity of notation. Note the use of the symbolic notation 1.121, and the fact that the letter i is exclusively retained for the row indexing of matrices and vectors encountered. Inequality 1.138 follows from Equation 1.117 by applying the triangle inequality to its components.

Further application of the triangle inequality in inequality 1.138 leads to

$$\begin{aligned} |y_i| &\le |h_i^{(0)}| + \sum_{i_1=-\infty}^{\infty} \left| \sum_{j=1}^{N} h_{ij}^{(1)} x_j \right| + \sum_{i_1=-\infty}^{\infty} \sum_{i_2=-\infty}^{\infty} \left| \sum_{j=1}^{N^2} h_{ij}^{(2)} (x^2)_j \right| \\ &\quad + \sum_{i_1=-\infty}^{\infty} \sum_{i_2=-\infty}^{\infty} \sum_{i_3=-\infty}^{\infty} \left| \sum_{j=1}^{N^3} h_{ij}^{(3)} (x^3)_j \right| + \cdots \\ &\le |h_i^{(0)}| + \sum_{i_1=-\infty}^{\infty} \sum_{j=1}^{N} |h_{ij}^{(1)}| |x_j| \sum_{i_1=-\infty}^{\infty} \sum_{i_2=-\infty}^{\infty} \sum_{j=1}^{N^2} |h_{ij}^{(2)}| |(x^2)_j| \\ &\quad + \sum_{i_1=-\infty}^{\infty} \sum_{i_2=-\infty}^{\infty} \sum_{i_3=-\infty}^{\infty} \sum_{j=1}^{N^3} |h_{ij}^{(3)}| |(x^3)_j| + \cdots \end{aligned} \qquad 1 \le i \le M \quad (1.139)$$

Moreover, note that the inequalities

$$|x_j| \le \|\mathbf{x}\|_l, \; |(x^2)_j| \le \|\mathbf{x}\|_l^2, \; |(x^3)_j| \le \|\mathbf{x}\|_l^3, \; \ldots \qquad (1.140a)$$

hold. So using these inequalities in inequality 1.139, we get

$$\begin{aligned} |y_i| &\le |h_i^{(0)}| + \|\mathbf{x}\|_l \sum_{i_1=-\infty}^{\infty} \sum_{j=1}^{N} |h_{ij}^{(1)}| + \|\mathbf{x}\|_l^2 \sum_{i_1=-\infty}^{\infty} \sum_{i_2=-\infty}^{\infty} \sum_{j=1}^{N^2} |h_{ij}^{(2)}| \\ &\quad + \|\mathbf{x}\|_l^3 \sum_{i_1=-\infty}^{\infty} \sum_{i_2=-\infty}^{\infty} \sum_{i_3=-\infty}^{\infty} \sum_{j=1}^{N^3} |h_{ij}^{(3)}| + \cdots \end{aligned} \qquad 1 \le i \le M \quad (1.140b)$$

Note that inequality 1.140b holds for every k. Hence, the following inequality:

$$\begin{aligned} \sup_{-\infty \le k \le \infty} |y_i| &\le |h_i^{(0)}| + \|\mathbf{x}\|_l \sum_{i_1=-\infty}^{\infty} \sum_{j=1}^{N} |h_{ij}^{(1)}| + \|\mathbf{x}\|_l^2 \sum_{i_1=-\infty}^{\infty} \sum_{i_2=-\infty}^{\infty} \sum_{j=1}^{N^2} |h_{ij}^{(2)}| \\ &\quad + \|\mathbf{x}\|_l^3 \sum_{i_1=-\infty}^{\infty} \sum_{i_2=-\infty}^{\infty} \sum_{i_3=-\infty}^{\infty} \sum_{j=1}^{N^3} |h_{ij}^{(3)}| + \cdots \end{aligned} \qquad 1 \le i \le M \quad (1.140c)$$

is also fulfilled.

Let us now take into account the index $i = i'$ for which

$$\max_{1 \le i \le M} \left(\sup_{-\infty \le k \le \infty} |y_i| \right) \tag{1.141a}$$

occurs. For this index, we rewrite inequality 1.140c as

$$\sup_{-\infty \le k \le \infty} |y_{i'}| \le \left|h_{i'}^{(0)}\right| + \|\mathbf{x}\|_I \sum_{i_1=-\infty}^{\infty} \sum_{j=1}^{N} \left|h_{i'j}^{(1)}\right| + \|\mathbf{x}\|_I^2 \sum_{i_1=-\infty}^{\infty} \sum_{i_2=-\infty}^{\infty} \sum_{j=1}^{N^2} \left|h_{i'j}^{(2)}\right| + \|\mathbf{x}\|_I^3 \sum_{i_1=-\infty}^{\infty} \sum_{i_2=-\infty}^{\infty} \sum_{i_3=-\infty}^{\infty} \sum_{j=1}^{N^3} \left|h_{i'j}^{(3)}\right| + \cdots \tag{1.141b}$$

Furthermore, taking into account definition 1.122b and the fact that

$$\left|h_{i'}^{(0)}\right| \le \max_{1 \le i \le N} \left|h_i^{(0)}\right| = \left|\mathbf{h}^{(0)}\right| \tag{1.141c}$$

(see definition 1.130b) holds, inequality 1.141b can be rewritten as

$$\|y\|_I \le \left|\mathbf{h}^{(0)}\right| + \|\mathbf{x}\|_I \sum_{i_1=-\infty}^{\infty} \sum_{j=1}^{N} \left|h_{i'j}^{(1)}\right| + \|\mathbf{x}\|_I^2 \sum_{i_1=-\infty}^{\infty} \sum_{i_2=-\infty}^{\infty} \sum_{j=1}^{N^2} \left|h_{i'j}^{(2)}\right| + \|\mathbf{x}\|_I^3 \sum_{i_1=-\infty}^{\infty} \sum_{i_2=-\infty}^{\infty} \sum_{i_3=-\infty}^{\infty} \sum_{j=1}^{N^3} \left|h_{i'j}^{(3)}\right| + \cdots \tag{1.141d}$$

Note now the following inequalities:

$$\sum_{i_1=-\infty}^{\infty} \sum_{j=1}^{N} \left|h_{i'j}^{(1)}\right| = \sum_{j=1}^{N} \sum_{i_1=-\infty}^{\infty} \left|h_{i'j}^{(1)}\right| \le \max_{1 \le i \le M} \left(\sum_{j=1}^{N} \sum_{i_1=-\infty}^{\infty} \left|h_{ij}^{(1)}\right| \right) = a_1 \tag{1.142a}$$

$$\sum_{i_1=-\infty}^{\infty} \sum_{i_2=-\infty}^{\infty} \sum_{j=1}^{N^2} \left|h_{i'j}^{(2)}\right| = \sum_{j=1}^{N^2} \sum_{i_1=-\infty}^{\infty} \sum_{i_2=-\infty}^{\infty} \left|h_{i'j}^{(2)}\right| \le \max_{1 \le i \le M} \left(\sum_{j=1}^{N^2} \sum_{i_1=-\infty}^{\infty} \sum_{i_2=-\infty}^{\infty} \left|h_{ij}^{(2)}\right| \right) = a_2 \tag{1.142b}$$

$$\sum_{i_1=-\infty}^{\infty} \sum_{i_2=-\infty}^{\infty} \sum_{i_3=-\infty}^{\infty} \sum_{j=1}^{N^3} \left|h_{i'j}^{(3)}\right| = \sum_{j=1}^{N^3} \sum_{i_1=-\infty}^{\infty} \sum_{i_2=-\infty}^{\infty} \sum_{i_3=-\infty}^{\infty} \left|h_{i'j}^{(3)}\right| \le \max_{1 \le i \le M} \left(\sum_{j=1}^{N^3} \sum_{i_1=-\infty}^{\infty} \sum_{i_2=-\infty}^{\infty} \sum_{i_3=-\infty}^{\infty} \left|h_{ij}^{(3)}\right| \right) = a_3 \tag{1.142c}$$

and in general,

$$\sum_{i_1=-\infty}^{\infty}\cdots\sum_{i_n=-\infty}^{\infty}\sum_{j=1}^{N^n}\left|h_{i'j}^{(n)}\right|=\sum_{j=1}^{N^n}\sum_{i_1=-\infty}^{\infty}\cdots\sum_{i_n=-\infty}^{\infty}\left|h_{i'j}^{(n)}\right|\leq\max_{1\leq i\leq M}\left(\sum_{j=1}^{N^n}\sum_{i_1=-\infty}^{\infty}\cdots\sum_{i_n=-\infty}^{\infty}\left|h_{ij}^{(n)}\right|\right)=a_n \tag{1.142d}$$

hold. So the use of these inequalities in 1.141d gives

$$\|\mathbf{y}\|_I\leq a_0+a_1\|\mathbf{x}\|_I+a_2\|\mathbf{x}\|_I^2+a_3\|\mathbf{x}\|_I^3+\cdots+a_n\|\mathbf{x}\|_1^n+\cdots \tag{1.143}$$

with the coefficient a_0 equal to $|\mathbf{h}^{(0)}|$ (compare inequality 1.143 with 1.141d). Incidentally, note that all the coefficients a_n, $n = 0, 1, 2, \cdots$, in inequality 1.143 are nonnegative.

Now calculate the radius of convergence of the power series on the right-hand side of inequality 1.143 in the same way as before; that is, by the use of the Cauchy criterion 1.93, with $||\mathbf{x}||_1$ instead of $||\mathbf{x}||$. So, substituting a_n given by Equation 1.142d into 1.93 and solving for $||\mathbf{x}||_1$ gives

$$\|\mathbf{x}\|_I<\frac{1}{\lim\limits_{n\to\infty}\sqrt[n]{\max\limits_{1\leq i\leq M}\left(\sum_{j=1}^{N^n}\sum_{i_1=-\infty}^{\infty}\cdots\sum_{i_n=-\infty}^{\infty}\left|h_{ij}^{(n)}\right|\right)}} \tag{1.144a}$$

Finally, it follows from inequality 1.144a that the radius of convergence of the power series in inequality 1.143 is given by the expression

$$r=\frac{1}{\lim\limits_{n\to\infty}\sqrt[n]{\max\limits_{1\leq i\leq M}\left(\sum_{j=1}^{N^n}\sum_{i_1=-\infty}^{\infty}\cdots\sum_{i_n=-\infty}^{\infty}\left|h_{ij}^{(n)}\right|\right)}} \tag{1.144b}$$

In conclusion, we can say that the series given by Equation 1.117 converges absolutely for the input vector-valued sequences fulfilling

$$\|\mathbf{x}\|_I<r \tag{1.144c}$$

We also point out that similar results to those obtained with the use of the norm $||\cdot||_I$ for vector-valued sequences can be achieved using the norm $||\cdot||$ defined by Equations 1.131. The methodology of arriving at such results is the same as that used in expressions from 1.138 to 1.144.

1.7 NONLINEAR SYSTEMS WITH FADING MEMORY

It has been shown in Section 1.5 that the notion of the norm of a sequence is very useful in handling expressions used in convergence and stability considerations. The norm of a sequence turned out to express, in some defined way, the magnitude of a sequence — in terms of the functional analysis.[13,14] And because of the usefulness of the functional analysis, we will use often its notions and tools in this and in the following sections. Hence, for the purposes of further considerations, we recall now some of the fundamentals of this analysis. We start with the definitions of mapping and functional.

So, let X and Y be some sets. A mapping is then defined as a rule by which the elements of the set X are assigned to the elements of the set Y. Moreover, the assignment is unique; that is, each of the elements of X is mapped to only one of the elements of Y.

To express the above definition symbolically, the three kinds of notation given below are used in the literature. They are

$$f : X \rightarrow Y \tag{1.145a}$$

$$y = f(x), x \in X, y \in Y \tag{1.145b}$$

$$y = Fx, x \in X, y \in Y \tag{1.145c}$$

Moreover, note that the element y as defined in Equations 1.145 is the image of the element x under the mapping f. Furthermore, the set X is the domain of the mapping, and the set of all images y is the range of that mapping. Of course, the range of the mapping is contained in the set Y, or is equal with this set. In the latter case, we say that the mapping f maps X onto Y, and we write

$$f : X \overset{onto}{\rightarrow} Y \tag{1.146}$$

When this is not the case, we simply say that the mapping f maps X into Y.

The different kinds of notation in Equations 1.145 point out some particular aspects of the mapping. So notation 1.145a expresses the mapping of one set into (onto) another set. Notation 1.145b is used, when one wants to point out that the mapping is identical with the notion of the function. And finally, 1.145c with the capital F instead of the small f, shows that the mapping is also nothing else than the operation (operator). Hence, we see that the notions of the mapping, function, operation (operator), transformation mean the same in the functional analysis. There is, however, one terminological exception: we do not say that a functional is identical to an operator. Quite the contrary. These notions are disjunctive, but both are types of mappings.

The general illustration of the notion of mapping is presented in Figure 1.22. Moreover,

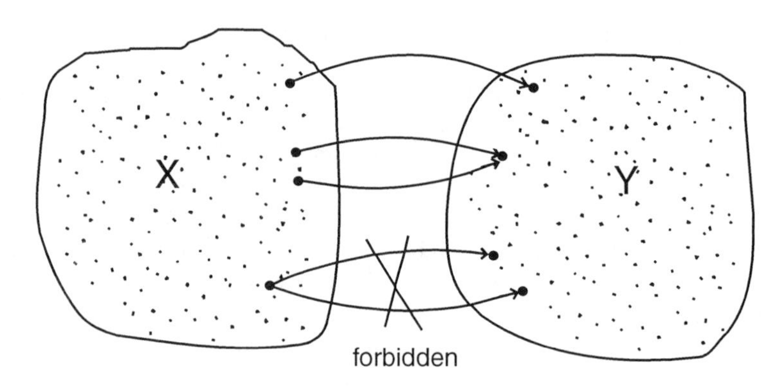

FIGURE 1.22 Illustration of the definition of mapping.

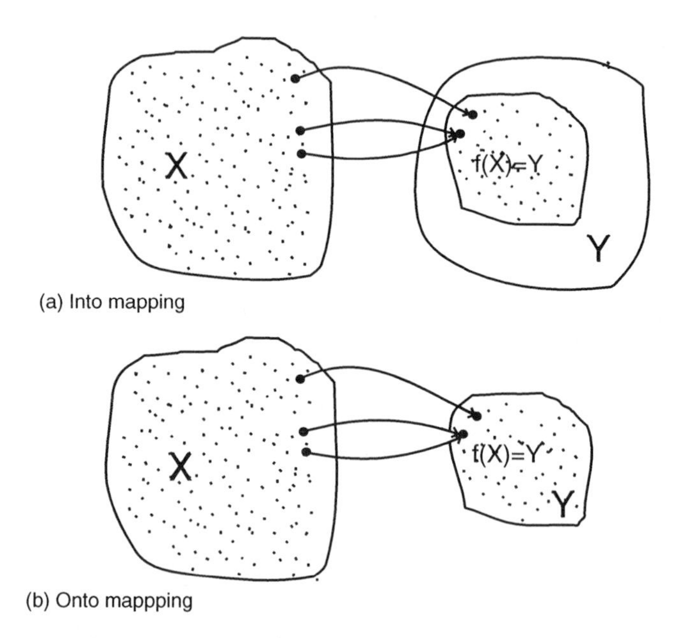

FIGURE 1.23 Mapping of the set X into the set Y, or onto the set Y.

Figure 1.23 illustrates the notion of into and onto mappings. Note that the set of images y in Figure 1.23 is denoted by $f(X)$.

If the images y of all the elements of the set X are distinct in Y, as shown in Fig. 1.24, then the mapping f is one-to-one. Moreover, when the mapping f is also onto, then there exists the inverse mapping f^{-1} with the following properties:

$$f^{-1}(y) = f^{-1}(f(x)) = x \tag{1.147a}$$

$$f(x) = f(f^{-1}(y)) = y \tag{1.147b}$$

for every $x \in X$ and $y \in Y$.

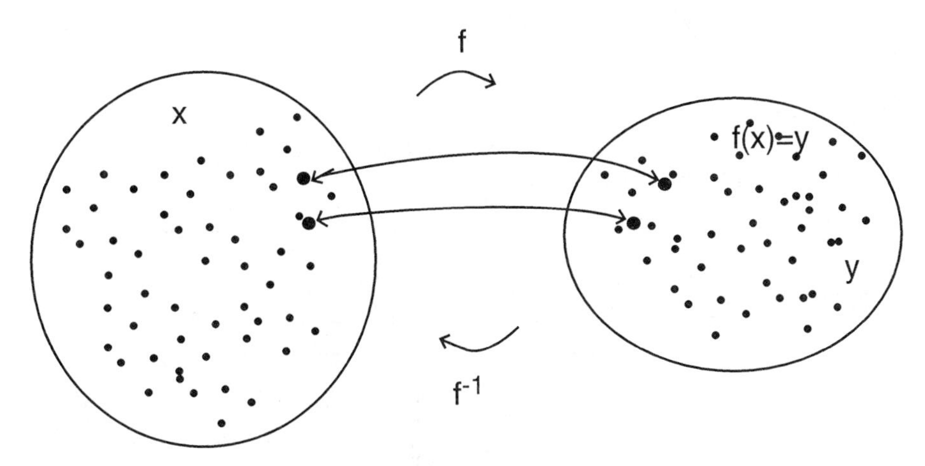

FIGURE 1.24 One-to-one and onto mapping f with its inverse f^{-1}.

When the set resulting from one mapping is mapped into the next, we have a composition of mappings, and, of course, this process can be continued. The resulting mapping is called a composite mapping, and is illustrated in Figure 1.25 for the case of using two functions. We write then

$$f = f_2 f_1 : X \rightarrow Y_2 \tag{1.148a}$$

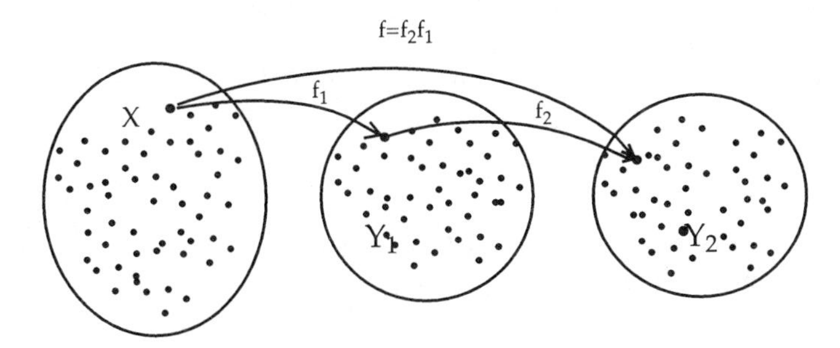

FIGURE 1.25 Illustration of construction of the composite mapping $f = f_2 f_1$.

with

$$f_1 : X \rightarrow Y_1 \text{ and } f_2 : Y_1 \rightarrow Y_2 \tag{1.148b}$$

which means that

$$y_2 = f_2(y_1) = f_2(f_1(x)) = f(x), x \in X \tag{1.148c}$$

The operation most often used in this book is that which operates on signals that are sequences. In this case, the elements of the sets X and Y are the sequences. This fact is illustrated in Figure 1.26.

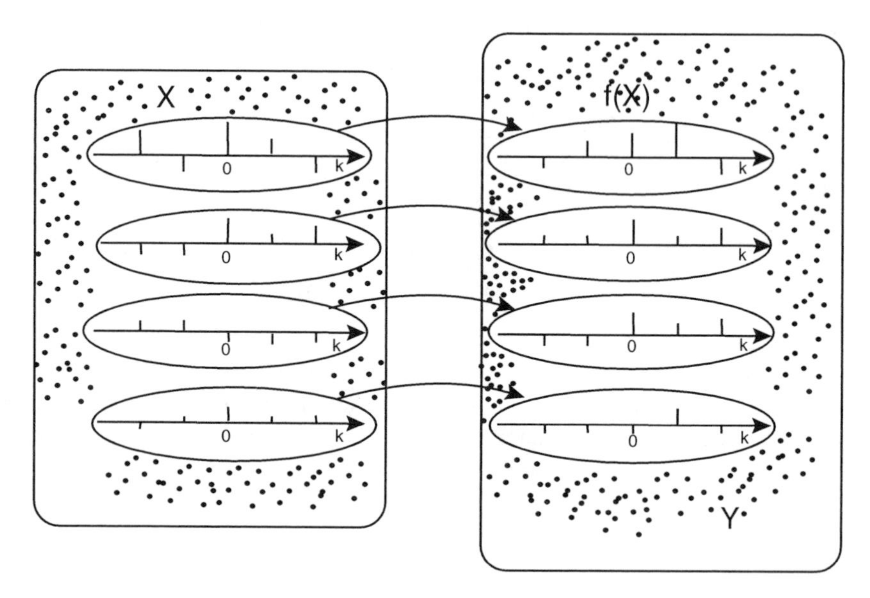

FIGURE 1.26 Illustration of the mapping of sequences belonging to the set X into sequences belonging to the set Y.

We will call a functional such a mapping, which maps the signals (sequences) into the set of real numbers, $\mathbb{R}$. For illustration, see Figure 1.27.

Explanation 1.5

Note that the Volterra series given by Equations 1.1 and 1.2 or 1.3 is a mapping, which maps the element $x(k)$ (input sequence) into the element $y(k)$ (output sequence).

On the other hand, the norm $\|x\|$ given by Equation 1.82a is a functional converting the sequence $x(k)$ into a real number.

Let us now introduce some algebraic structure into the sets considered. This structure is added by defining a linear space, determining operations of summation of elements of the set considered and of their multiplication by scalars. So we say

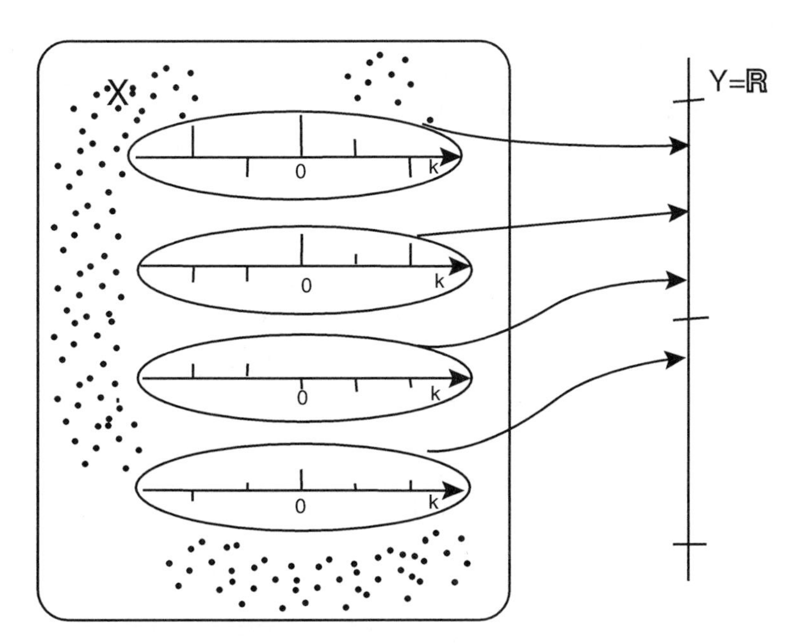

FIGURE 1.27 Illustration of the notion of a functional.

that a given set X is linear, or, in other words, is a linear space if for any pair of its elements x_1 and x_2 there exists the sum $x_1+x_2 \in X$ having the following four properties:

1a) $x_1 + x_2 = x_2 + x_1$ (summation is commutative) (1.149a)

2a) $(x_1 + x_2) + x_3 = x_1 + (x_2 + x_3)$ where $x_3 \in X$, too (summation is associative) (1.149b)

3a) there exists in X a unique zero element θ (sometimes called the origin) such that, for any $x \in X$, $x + \theta = x$ holds (1.149c)

4a) there exists for each $x \in X$ such a unique element $-x$ that $x + (-x) = \theta$ holds (1.149d)

Moreover, the multiplication by scalar must fulfill the following:

0b) $\alpha x \in X$ where α is an element of the set of scalars (the resulting element belongs to the space X) (1.150a)

1b) $\alpha_1(\alpha_2 x) = (\alpha_1\alpha_2)x$ where α_1 and α_2 are scalars (multiplication by scalar is associative) (1.150b)

2b) $1x = x$ for each $x \in X$ (1.150c)

3b) $0x = \theta$ for each $x \in X$; (1.150d)

4b) $(\alpha_1 + \alpha_2)x = \alpha_1 x + \alpha_2 x$ (distributive law) (1.150e)

5b) $\alpha(x_1+ x_2) = \alpha x_1 + \alpha x_2$ (distributive law) (1.150f)

Note that when the scalars α, α_1, α_2, ... are real numbers, then the linear space X is called real; but when they are complex, the space X is also called complex.

In the functional analysis, the elements (points) of a linear space are also called vectors. Note that this book deals with spaces whose elements are scalar sequences or sequences of vectors (vector-valued sequences). See, for example, Figure 1.21. When we consider the latter sequences, we usually indicate them by boldface type symbols.

Explanation 1.6

Consider now whether our sets of sequences can be considered as linear spaces. Let us, for example, check the property 1a) from the list given. So, taking into account two sequences

$$x_1(k) = \{\ldots, x_1(-1), x_1(0), x_1(1), \ldots\}$$

and

$$x_2(k) = \{\ldots, x_2(-1), x_2(0), x_2(1), \ldots\}$$

we write their sum as

$$x_1(k) + x_2(k) = \{\ldots, x_1(-1) + x_2(-1), x_1(0) + x_2(0), x_1(1) + x_2(1), \ldots\}$$

Changing then the order of occurrence of components in each of the elements of the above sequence, we get

$$x_1(k) + x_2(k) = \{\ldots, x_2(-1) + x_1(-1), x_2(0) + x_1(0), x_2(1) + x_1(1), \ldots\}$$

which finally allows us to write

$$x_1(k) + x_2(k) = x_2(k) + x_1(k)$$

So property 1a) is proved.

Note also that we can similarly prove the property 2a).

Now, if we take into account the following two sequences,

$$x(k) = \{\ldots, x(-1), x(0), x(1), x(2), \ldots\}$$

and

$$\theta(k) = \{\ldots, 0, 0, 0, 0, \ldots\}$$

we can write

$$\begin{aligned} x(k) + \theta(k) &= \{\ldots, x(-1) + 0, x(0) + 0, x(1) + 0, x(2) + 0, \ldots\} \\ &= \{\ldots, x(-1), x(0), x(1), x(2), \ldots\} = x(k) \end{aligned}$$

The above equality proves the property 3a) and shows what the zero element (zero sequence) looks like. Furthermore, see that for the sequences

$$x(k) = \{\ldots, x(-1), x(0), x(1), x(2), \ldots\}$$

and

$$-x(k) = \{\dots, -x(-1), -x(0), -x(1), -x(2), \dots\}$$

their sum can be expressed as

$$\begin{aligned} x(k) + (-x(k)) &= \{\dots, x(-1) - x(-1), x(0) - x(0), x(1) - x(1), x(2) - x(2), \dots\} \\ &= \{\dots, 0, 0, 0, 0, \dots\} = \theta(k) \end{aligned}$$

And this proves property 4a).

Let us now observe that

$$\alpha x(k) = \{\dots, \alpha x(-1), \alpha x(0), \alpha x(1), \alpha x(2), \dots\}$$

and the result of multiplication of $x(k)$ by a scalar α is a sequence, too. So property 0b) is thereby proved. Furthermore, note that, using the above result, it is easy to show that the properties 1b), 4b), and 5b) hold. Also see then that properties 2b) and 3b), as

$$1x(k) = \{\dots, 1x(-1), 1x(0), 1x(1), 1x(2), \dots\} = x(k)$$

and

$$0x(k) = \{\dots, 0x(-1), 0x(0), 0x(1), 0x(2), \dots\} = \{\dots, 0, 0, 0, 0, \dots\} = \theta(k)$$

hold, too. Altogether, this means that the set of scalar sequences forms a linear space.

It is clear that by using the same arguments, we are able to prove all the properties of linear spaces for the set of vector-valued sequences. In what follows, we point out briefly some interesting points related to these proofs. So, with regard to the property 3a), see that

$$\mathbf{x}(k) + \boldsymbol{\theta}(k) = \left\{\dots, \begin{bmatrix} x_1(-1) \\ x_2(-1) \\ \vdots \\ x_n(-1) \end{bmatrix}, \begin{bmatrix} x_1(0) \\ x_2(0) \\ \vdots \\ x_n(0) \end{bmatrix}, \begin{bmatrix} x_1(1) \\ x_2(1) \\ \vdots \\ x_n(1) \end{bmatrix}, \dots\right\} + \left\{\dots, \begin{bmatrix} 0 \\ 0 \\ \vdots \\ 0 \end{bmatrix}, \begin{bmatrix} 0 \\ 0 \\ \vdots \\ 0 \end{bmatrix}, \begin{bmatrix} 0 \\ 0 \\ \vdots \\ 0 \end{bmatrix}, \dots\right\}$$

$$= \left\{\dots, \begin{bmatrix} x_1(-1) + 0 \\ x_2(-1) + 0 \\ \vdots \\ x_n(-1) + 0 \end{bmatrix}, \begin{bmatrix} x_1(0) + 0 \\ x_2(0) + 0 \\ \vdots \\ x_n(0) + 0 \end{bmatrix}, \begin{bmatrix} x_1(1) + 0 \\ x_2(1) + 0 \\ \vdots \\ x_n(1) + 0 \end{bmatrix}, \dots\right\} = \mathbf{x}(k)$$

And it follows from the above how the zero element (zero vector–valued sequence) looks in the case of considering vectors with n rows. Note that the sequence $\boldsymbol{\theta}(k)$ is often denoted by $\boldsymbol{\theta}_n(k)$ (to indicate the number of vector rows). Furthermore, with regard to the next property, 4a), we see that

$$\mathbf{x}(k) + (-\mathbf{x}(k))$$

$$= \left\{\ldots, \begin{bmatrix} x_1(-1) \\ x_2(-1) \\ \vdots \\ x_n(-1) \end{bmatrix}, \begin{bmatrix} x_1(0) \\ x_2(0) \\ \vdots \\ x_n(0) \end{bmatrix}, \begin{bmatrix} x_1(1) \\ x_2(1) \\ \vdots \\ x_n(1) \end{bmatrix}, \ldots \right\} + \left\{\ldots, \begin{bmatrix} -x_1(-1) \\ -x_2(-1) \\ \vdots \\ -x_n(-1) \end{bmatrix}, \begin{bmatrix} -x_1(0) \\ -x_2(0) \\ \vdots \\ -x_n(0) \end{bmatrix}, \begin{bmatrix} -x_1(1) \\ -x_2(1) \\ \vdots \\ -x_n(1) \end{bmatrix}, \ldots \right\}$$

$$= \left\{\ldots, \begin{bmatrix} x_1(-1)-x_1(-1) \\ x_2(-1)-x_2(-1) \\ \vdots \\ x_n(-1)-x_n(-1) \end{bmatrix}, \begin{bmatrix} x_1(0)-x_1(0) \\ x_2(0)-x_2(0) \\ \vdots \\ x_n(0)-x_n(0) \end{bmatrix}, \begin{bmatrix} x_1(1)-x_1(1) \\ x_2(1)-x_2(1) \\ \vdots \\ x_n(1)-x_n(1) \end{bmatrix}, \ldots \right\} = \left\{\ldots, \begin{bmatrix} 0 \\ 0 \\ \vdots \\ 0 \end{bmatrix}, \begin{bmatrix} 0 \\ 0 \\ \vdots \\ 0 \end{bmatrix}, \begin{bmatrix} 0 \\ 0 \\ \vdots \\ 0 \end{bmatrix}, \ldots \right\} = \boldsymbol{\theta}_n(k)$$

As far as the next series of properties is concerned, we see that $\alpha\mathbf{x}(k)$ is given by

$$\alpha\mathbf{x}(k)$$

$$= \alpha\left\{\ldots, \begin{bmatrix} x_1(-1) \\ x_2(-1) \\ \vdots \\ x_n(-1) \end{bmatrix}, \begin{bmatrix} x_1(0) \\ x_2(0) \\ \vdots \\ x_n(0) \end{bmatrix}, \begin{bmatrix} x_1(1) \\ x_2(1) \\ \vdots \\ x_n(1) \end{bmatrix}, \ldots \right\} = \left\{\ldots, \begin{bmatrix} \alpha x_1(-1) \\ \alpha x_2(-1) \\ \vdots \\ \alpha x_n(-1) \end{bmatrix}, \begin{bmatrix} \alpha x_1(0) \\ \alpha x_2(0) \\ \vdots \\ \alpha x_n(0) \end{bmatrix}, \begin{bmatrix} \alpha x_1(1) \\ \alpha x_2(1) \\ \vdots \\ \alpha x_n(1) \end{bmatrix}, \ldots \right\}$$

So the resulting sequence belongs also to the set of the vector–valued sequences, as property 0b) requires. Moreover, note that

$$1\mathbf{x}(k)$$

$$= 1\left\{\ldots, \begin{bmatrix} x_1(-1) \\ x_2(-1) \\ \vdots \\ x_n(-1) \end{bmatrix}, \begin{bmatrix} x_1(0) \\ x_2(0) \\ \vdots \\ x_n(0) \end{bmatrix}, \begin{bmatrix} x_1(1) \\ x_2(1) \\ \vdots \\ x_n(1) \end{bmatrix}, \ldots \right\} = \left\{\ldots, \begin{bmatrix} 1x_1(-1) \\ 1x_2(-1) \\ \vdots \\ 1x_n(-1) \end{bmatrix}, \begin{bmatrix} 1x_1(0) \\ 1x_2(0) \\ \vdots \\ 1x_n(0) \end{bmatrix}, \begin{bmatrix} 1x_1(1) \\ 1x_2(1) \\ \vdots \\ 1x_n(1) \end{bmatrix}, \ldots \right\} = \mathbf{x}(k)$$

and

$$0\mathbf{x}(k) = \left\{\ldots, \begin{bmatrix} 0x_1(-1) \\ 0x_2(-1) \\ \vdots \\ 0x_n(-1) \end{bmatrix}, \begin{bmatrix} 0x_1(0) \\ 0x_2(0) \\ \vdots \\ 0x_n(0) \end{bmatrix}, \begin{bmatrix} 0x_1(1) \\ 0x_2(1) \\ \vdots \\ 0x_n(1) \end{bmatrix}, \ldots \right\} = \left\{\ldots, \begin{bmatrix} 0 \\ 0 \\ \vdots \\ 0 \end{bmatrix}, \begin{bmatrix} 0 \\ 0 \\ \vdots \\ 0 \end{bmatrix}, \begin{bmatrix} 0 \\ 0 \\ \vdots \\ 0 \end{bmatrix}, \ldots \right\} = \theta_n(k)$$

Finally, observe that the linear space of scalar sequences is in fact a subspace of the linear space of vector–valued sequences, when putting $n = 1$.

Having defined the notion of linear spaces, we are now able to define something which determines, in some sense, the size of each element of a given linear space.

This is done by assigning a real nonnegative number to each of the space elements and calling this number a norm of that element. Furthermore, it is assumed that the norm $\|x\|$ of each element x of a linear space X fulfills the following properties:

1. $\|x\| = 0$ if and only if $x = \theta$ (1.151a)
2. $\|\alpha x\| = |\alpha|\|x\|$ where α is a real or complex number (1.151b)
3. $\|x_1 + x_2\| \leq \|x_1\| + \|x_2\|$ where $x_1, x_2 \in X$ (1.151c)

The inequality 1.151c is called the triangle inequality. Moreover, note that a linear space in which a norm is defined is called the normed space. Furthermore, observe also that, without having defined the algebraic structure consisting of the definitions of the summation operation and of the operation of multiplication by a scalar, the norm definition as given by Equations 1.151 would be impossible.

Explanation 1.7

Let us check whether the norm defined by Equation 1.82a fulfils the required properties 1.151.

Observe first that when $x(k) = \theta(k) = \{\ldots,0,0,0,\ldots\}$, then $\sup_{-\infty \leq k \leq \infty} |\theta(k)| = 0$. Hence, $\|\theta(k)\| = 0$ really holds.

Assume now that $x(k) \neq \theta(k)$. So some of the elements of $x(k)$ are different from zero. Then, of course, $\sup_{-\infty \leq k \leq \infty} (|\theta(k)| \neq 0)$ for such a sequence. In conclusion, we see that property 1.151a holds for the norm given by Equation 1.82a.

Note that

$$\|\alpha x(k)\| = \sup_{-\infty \leq k \leq \infty} |\alpha x(k)| = |\alpha| \sup_{-\infty \leq k \leq \infty} |x(k)| = |\alpha| \|x(k)\|$$

holds. So, property 1.151b is also fulfilled.

With regard to property 1.151c, note that we have

$$|x_1(k) + x_2(k)| \leq |x_1(k)| + |x_2(k)|$$

for each k. So

$$|x_1(k) + x_2(k)| \leq \sup_{-\infty \leq k \leq \infty} |x_1(k)| + \sup_{-\infty \leq k \leq \infty} |x_2(k)|$$

will also hold for each k. That is, we can write the supremum symbol on the left-hand side of the above inequality as well. Hence, we get

$$\sup_{-\infty \leq k \leq \infty} |x_1(k) + x_2(k)| \leq \sup_{-\infty \leq k \leq \infty} |x_1(k)| + \sup_{-\infty \leq k \leq \infty} |x_2(k)|$$

which, after using the norm definition of Equation 1.82a, is nothing other than the property 1.151c. So, we can conclude that all the properties 1.151 are really fulfilled for the norm definition of Equation 1.82a.

Closely related to the notion of the norm is the notion of a metric. This is because the metric determines in some way what we can call, using the terminology taken from geometry, the distance between elements of a space. And this distance is determined by assigning to each pair x_1, x_2 of elements of a space X a real nonnegative number. We define thereby such a functional $d:\{x_1, x_2\} \to \mathbb{R}$ that has the properties

1. $d(x_1, x_2) \geq 0$ with $d(x_1, x_2) = 0$ if and only if $x_1 = x_2$ (1.152a)
2. $d(x_1, x_2) = d(x_2, x_1)$ (symmetry of the metric) (1.152b)
3. $d(x_1, x_2) + d(x_2, x_3) \geq d(x_1, x_3)$ (triangle inequality) (1.152c)

where x_3 is another element of the considered space X.

Note that the properties 1.152 reflect those associated with the notion of distance. That is 1.152a says that the distance is a nonnegative number. And, of course, the distance between the same elements is equal to zero. Furthermore, the distance from the element x_1 to the element x_2 is the same as the distance from the element x_2 to the element x_1. And finally, property 1.152c says that the sum of the lengths of two triangle sides is always greater than (or eventually equal to) the length of its third side.

A set of elements of a space X, together with the metric defined in this space, is called a metric space. Note also that the same space X with two different metrics forms, of course, two different metric spaces.

Incidentally, note that the notion of the space metric does not need the space considered to be linear. When a space is linear, it makes sense to relate the space norm with its metric. This can be done, for example, by defining that

$$d(x_1, x_2) \overset{df}{=} \|x_1 - x_2\| \tag{1.153}$$

Explanation 1.8

Let us check whether the definition of the metric in a linear space as given by Equation 1.153 is correct. That is, we ask whether the metric so defined fulfills the properties of a metric expressed in 1.152. For this purpose, we take into account an element $x = x_l - x_2 \in X$.

Note that it follows from property 1.151a that $\|x_1 - x_2\| = 0$ if and only if $x_1 - x_2 = \theta$. Therefore, $x_1 = x_2 + \theta$, and furthermore, $x_1 = x_2$ by applying property 1.149c of a linear space. In conclusion, we see that $d(x_1, x_2) = \|x_1 - x_2\| = 0$ holds if and only if $x_1 = x_2$. Otherwise, $d(x_1, x_2) > 0$. Property 1.152a is fulfilled.

Consider

$$d(x_1, x_2) = \|x_1 - x_2\| = \|-(x_2 - x_1)\| = |-1|\|x_2 - x_1\| = \|x_2 - x_1\| = d(x_2, x_1)$$

where some of the properties of a linear space have been used. See properties 1.149a and 1.150f. Moreover, note that property 1.151b has been applied as well. We conclude that the above is a proof of symmetry property 1.152b.

To prove property 1.152c, take into account an element $x_1 - x_2 = x_1 - x_2 + x_3 - x_3 \in X$. According to property 1.151c, we can write for the above element

$$\|x_1 - x_2 + x_3 - x_3\| = \|x_1 - x_3 + x_3 - x_2\| \\ \leq \|x_1 - x_3\| + \|x_3 - x_2\| = \|x_1 - x_3\| \\ + \|-(x_2 - x_3)\| = \|x_1 - x_3\| + \|x_2 - x_3\|$$

where, as before, some of the properties of a linear space, and the property 1.151b of the norm, have been used. Keeping in mind definition 1.153, we observe that the above inequality is, in fact, a proof of the triangle inequality 1.152c.

Note that the notion of the metric of a space applies in a natural way to properties such as convergence and continuity. We deal with convergence when we are concerned with infinite sequences of elements of some space. We are concerned with continuity when we ask whether mapping from one metric space to another metric space is continuous.

With regard to an infinite sequence $\{x_1, x_2, x_3, \ldots, x_n, \ldots\}$ of elements of a space X ($x_n \in X, n = 1, 2, \ldots$), we say that this sequence is convergent to some element $x \in X$ if, for any $\varepsilon > 0$, there exists such a $n = n_0$ that

$$\text{for each } n \geq n_0 \;\; d(x_n, x) < \varepsilon \tag{1.154a}$$

holds. This relation is also expressed as

$$\lim_{n \to \infty} x_n = x \tag{1.154b}$$

With regard to a mapping f which maps elements x of a metric space X with the metric d_x into elements y of a metric space Y with the metric d_y, we say that this mapping is continuous at a point x_0 if for any $\varepsilon > 0$ there exists a $\delta > 0$, such that

$$d_x(x, x_0) < \delta \text{ implies } d_y(y, y_0) < \varepsilon \tag{1.155}$$

where $y = f(x)$ and $y_0 = f(x_0)$. Furthermore, if f is continuous at each $x \in X$, then we say that this mapping is a continuous mapping.

Having defined the fundamental notions of the functional analysis, we can now continue our main considerations regarding nonlinear systems. Our goal in this section is to formulate a property, which is called the fading memory.[1,15,16] Note that very closely related with the definition of this memory are the definitions of the decaying memory[17] and approximately finite memory,[18,19,20] also formulated in the literature on the Volterra series. The relationship among all the above definitions will be discussed in the next section.

Let us now start with the definition of the space l^∞. First, for scalar sequences, with k the discrete time, taking the values from the set $K = \mathbb{Z}$, or $K = \mathbb{Z}_+$, or $K = \mathbb{Z}_-$, where $\mathbb{Z}$, $\mathbb{Z}_+$, or $\mathbb{Z}_-$, mean the sets of integers, nonnegative, and nonpositive integers, respectively, we define the space $l^\infty(K)$ of bounded sequences as that having the norm defined by

$$\|x\| \stackrel{df}{=} \sup_{k \in K} |x(k)| < \infty \tag{1.156}$$

Note that the definition 1.156 is the same as the definition introduced for the first time in Section 1.5 by Equation 1.82a.

To define the space $l^\infty(K)$ for vector-valued sequences, we recall first from Section 1.6 the norm for vectors (the absolute value of a vector), denoted $|\cdot|$, and given by

$$|\mathbf{x}| \stackrel{df}{=} \max_i |x_i| = \max_{1 \le i \le N} |x_i(k)| \tag{1.157a}$$

where $\mathbf{x}(k)$ for each k is a vector having N elements. In the case of vector-valued sequences, the operation 1.157a "reduces" a given vector-valued sequence to a scalar sequence

$$\{\ldots, |\mathbf{x}(-2)|, |\mathbf{x}(-1)|, |\mathbf{x}(0)|, |\mathbf{x}(1)|, \ldots\}$$

to which, of course, the definition 1.156 can now be applied. Consequently, we get

$$\|\mathbf{x}\| \stackrel{df}{=} \sup_{k \in K} |\mathbf{x}(k)| \tag{1.157b}$$

And summarising, the l^∞ space of vector-valued sequences is a space with the norm given by Equations 1.157. Moreover, note that the defining expressions 1.157 are the same as those presented for the first time in Section 1.6 for vector-valued sequences (Equations 1.130 and 1.131).

Let us now define a delay operator U_τ. So, we say that the delay operator U_τ is an operator, that shifts a scalar or vector signal (a scalar or vector-valued sequence) $x(k)$ or $\mathbf{x}(k)$ by $\tau \in \mathbb{Z}$ on the discrete time axis. Its defining equation is

$$(U_\tau x)(k) \stackrel{df}{=} x(k-\tau) \tag{1.158a}$$

or

$$(U_\tau \mathbf{x})(k) \stackrel{df}{=} \mathbf{x}(k-\tau) \tag{1.158b}$$

Observe on the left-hand side of Equations 1.158 a usage typical in the literature of one pair of parentheses to separate a operator (here $U_\tau x$) from the time argument k put into the second pair of parentheses. Generally, in this book we allow dropping parentheses or separating composite operators by the use of parentheses. Hence, for example, $U_\tau x(k)$, $U_\tau(x)(k)$, and $(U_\tau)(x)(k)$ are exactly the same.

Moreover, we say that an operator N is time-invariant if the equality

$$U_{\tau}N = NU_{\tau} \tag{1.159}$$

holds for all $\tau \in \mathbb{Z}$.

According to Explanation 1.5, the Volterra series given by Equations 1.1 and 1.2 or 1.3 represents an operator. We shall name this operator the Volterra series operator and define it as

$$y(k) = (Vx)(k) \stackrel{df}{=} \sum_{n=0}^{\infty} y^{(n)}(k) \tag{1.160}$$

where the partial responses $y^{(n)}(k)$ are given by Equations 1.2 or 1.3. Note that these responses represent operators themselves. That is, for example, Equation 1.2c can be rewritten as

$$y^{(2)}(k) = (V_2x)(k) = \sum_{i_1=-\infty}^{\infty}\sum_{i_2=-\infty}^{\infty} h^{(2)}(k, i_1, i_2)x(i_1)x(i_2) \tag{1.161}$$

where V_2 represents an operator related to the impulse response of the second-order and twofold summation, as shown in Equation 1.161.

The notions of stationarity and of time-invariance (time-independence) with regard to the Volterra series (Volterra series operator) are equivalent to each other. We shall show this before going ahead with the fading memory definition.

Our goal first is to show that applying the condition 1.159 to the Volterra series operator for time-dependent systems (given by Equations 1.160 and 1.2) leads to its simplification. Then, it takes on the form for time-independent systems (given by Equations 1.160 and 1.3). Moreover, according to what was already shown in Section 1.1, this operator describes then the stationary system, in the sense of stationarity definitions 1.4.

To proceed, assume that the operator N in Equation 1.159 has the form of the Volterra series given by Equations 1.1 and 1.2. That is,

$$\begin{aligned}(Nx)(k) &= h^{(0)}(k) + \sum_{i=-\infty}^{\infty} h^{(1)}(k, i)x(i) + \sum_{i_1=-\infty}^{\infty}\sum_{i_2=-\infty}^{\infty} h^{(2)}(k, i_1, i_2)x(i_1)x(i_2) \\ &\quad + \sum_{i_1=-\infty}^{\infty}\sum_{i_2=-\infty}^{\infty}\sum_{i_3=-\infty}^{\infty} h^{(3)}(k, i_1, i_2, i_3)x(i_1)x(i_2)x(i_3) + \cdots\end{aligned} \tag{1.162}$$

Keeping this in mind, we see that the composite operator $U_{\tau}N$ takes on the value of

$$(U_\tau Nx)(k) = h^{(0)}(k) + \sum_{i=-\infty}^{\infty} h^{(1)}(k, i)x(i) + \sum_{i_1=-\infty}^{\infty}\sum_{i_2=-\infty}^{\infty} h^{(2)}(k, i_1, i_2)x(i_1)x(i_2)$$
$$+ \sum_{i_1=-\infty}^{\infty}\sum_{i_2=-\infty}^{\infty}\sum_{i_3=-\infty}^{\infty} h^{(3)}(k, i_1, i_2, i_3)x(i_1)x(i_2)x(i_3) + \cdots \tag{1.163}$$

at the point $k + \tau$ on the discrete time axis.

Similarly, the composite operator NU_τ possesses the value of

$$(NU_\tau x)(k) = h^{(0)}(k+\tau) + \sum_{i=-\infty}^{\infty} h^{(1)}(k+\tau, i)x(i-\tau)$$
$$+ \sum_{i_1=-\infty}^{\infty}\sum_{i_2=-\infty}^{\infty} h^{(2)}(k+\tau, i_1, i_2)x(i_1-\tau)x(i_2-\tau) \tag{1.164}$$
$$+ \sum_{i_1=-\infty}^{\infty}\sum_{i_2=-\infty}^{\infty}\sum_{i_3=-\infty}^{\infty} h^{(3)}(k+\tau, i_1, i_2, i_3)x(i_1-\tau)x(i_2-\tau)x(i_3-\tau) + \ldots$$

at the point $k + \tau$ on the discrete time axis.

The calculations performed in Equations 1.163 and 1.164 have been illustrated in Figure 1.28.

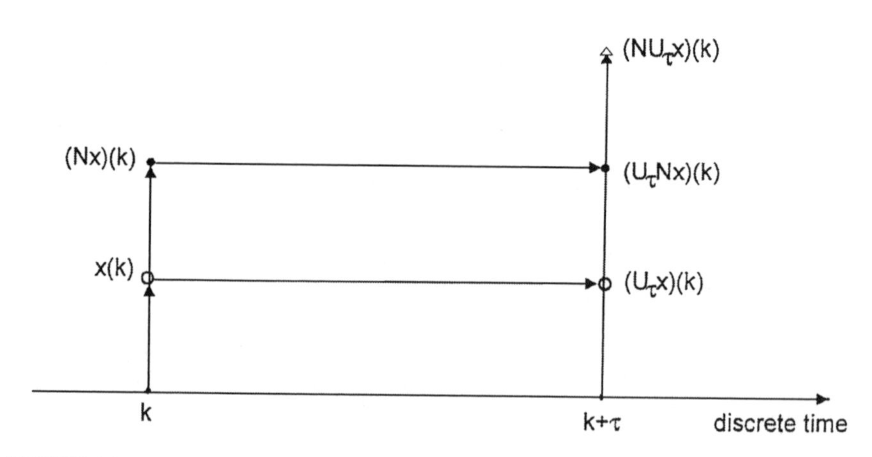

FIGURE 1.28 Illustration of calculations performed in Equations 1.163 and 1.164.

Let us now introduce new variables in Equation 1.163 by making the following substitutions: $i = i'' - \tau$, $i_p = i'_p \tau$, $p = 1, 2, 3\ldots$, and afterwards, let us drop the symbol "'". As a result, we get

$$(U_\tau Nx)(k) = h^{(0)}(k) + \sum_{i=-\infty}^{\infty} h^{(1)}(k, i-\tau)x(i-\tau)$$
$$+ \sum_{i_1=-\infty}^{\infty}\sum_{i_2=-\infty}^{\infty} h^{(2)}(k, i_1-\tau, i_2-\tau)x(i_1-\tau)x(i_2-\tau) \quad (1.165)$$
$$+ \sum_{i_1=-\infty}^{\infty}\sum_{i_2=-\infty}^{\infty}\sum_{i_3=-\infty}^{\infty} h^{(3)}(k, i_1-\tau, i_2-\tau, i_3-\tau)x(i_1-\tau)x(i_2-\tau)x(i_3-\tau) + \ldots$$

To satisfy the condition for time-invariance 1.159, the expressions on the right-hand sides of Equations 1.164 and 1.165 must be equal to each other (referring to Figure 1.28, the points "Δ" and "•" on the right-hand side of it must reduce to one point) for all possible input sequences, all time instants k, and all values of delay τ. This however, is only possible when the corresponding impulse responses in Equations 1.164 and 1.165 are equal to each other. That is, the following equalities

$$h^{(0)}(k+\tau) = h^{(0)}(k) = h^{(0)} \quad (1.166a)$$

$$h^{(1)}(k+\tau, i) = h^{(1)}(k, i-\tau) \quad (1.166b)$$

$$h^{(2)}(k+\tau, i_1, i_2) = h^{(2)}(k, i_1-\tau, i_2-\tau) \quad (1.166c)$$

and, in general,

$$h^{(n)}(k+\tau, i_1, i_2, \ldots, i_n) = h^{(n)}(k, i_1-\tau, i_2-\tau, \ldots, i_n-\tau) \quad (1.166d)$$

hold for all $\tau, k \in \mathbb{Z}$.

Observe now that by substituting $k = 0$ in Equations 1.166, and then renaming the variable τ as k, we arrive finally at the equations identical with Equations 1.4. In conclusion, we can say that application of the condition 1.159 led us to the description for stationary systems, which, as we know from Section 1.1, have the Volterra series description in the form given by Equations 1.1 and 1.3.

Note that the converse — the stationary Volterra series fulfills the condition 1.159 — is also true. To prove this, observe that, when Equations 1.166 are fulfilled, then the expression in 1.165, and thereby, also the expression in 1.163, is equal to the corresponding expression in 1.164. This is nothing other than the condition 1.159. So, in summarizing we say that the notions of stationarity and time-invariance are equivalent.

Now we present two definitions of the fading memory in the form that was presented by Boyd and Chua.[1]

FMD1 (Fading Memory Definition 1): A time-invariant operator N: $l^\infty(\mathbb{Z}) \to l^\infty(\mathbb{Z})$ has fading memory on the subspace B of $l^\infty(\mathbb{Z})$ if there is a decreasing sequence w: $\mathbb{Z}_+ \rightarrow (0,1\rangle$, $\lim_{k\rightarrow\infty} w(k)=0$, such that for each $x \in$ B and $\in > 0$ there is a $\delta > 0$ such that for all $v \in$B the following relation

$$\sup_{k\leq 0}|x(k)-v(k)|w(-k)<\delta \rightarrow |(Nx)(0)-(Nv)(0)|<\varepsilon \tag{1.167}$$

holds.

FMD2 (Fading Memory Definition 2): A time-invariant operator N: $l^{\infty}(\mathbb{Z}_+)\rightarrow l^{\infty}(\mathbb{Z}_+)$ possesses fading memory on the subspace B_+ of $l^{\infty}(\mathbb{Z}_+)$ if there is a decreasing sequence $w: \mathbb{Z}_+ \rightarrow (0,1\rangle$, $\lim_{k\rightarrow\infty} w(k)=0$, such that for each $x \in$ B$_+$ and $\varepsilon > 0$ there is a $\delta > 0$ such that for all $v \in$ B$_+$ the following implication

$$\sup_{0\leq\tau\leq k}|x(\tau)-v(\tau)|w(k-\tau)<\delta \rightarrow |(Nx)(k)-(Nv)(k)|<\varepsilon \tag{1.168}$$

holds.

1.8 FURTHER CONSIDERATIONS ON FADING MEMORY

In this section, a continuation of the previous one, first we show how to associate a functional with any time-invariant operator. Using the following example, the operator N, defined in definition 1.167 is associated with a functional F defined on $l^{\infty}(\mathbb{Z}_-)$ by

$$Fx \overset{df}{=} (Nx_e)(0) \tag{1.169a}$$

where the sequence x_e is given by

$$x_e(k) \overset{df}{=} \begin{cases} x(k) \text{ for } k\leq 0 \\ x(0) \text{ for } k>0 \end{cases} \tag{1.169b}$$

Note that Equation 1.169b defines an extension of the sequence $x \in l^{\infty}(\mathbb{Z}_-)$ to the sequence $x_e \in l^{\infty}(\mathbb{Z})$. The mapping F can be interpreted as one that maps the past input to the operator N, which is an element of $l^{\infty}(\mathbb{Z}_-)$, into the present output of N at the discrete time $k = 0$, which is an element of the set of real numbers $\mathbb{R}$.

It is possible to recover the operator N knowing its associated functional F. We show how to do this. For this purpose, we define a truncation operator P: $l^{\infty}(\mathbb{Z})\rightarrow l^{\infty}(\mathbb{Z}_-)$, by

$$(Px)(k) \overset{df}{=} x(k) \text{ for } (k\leq 0) \tag{1.170}$$

It follows clearly from definition 1.170 that P is such an operator that truncates an element of $l^{\infty}(\mathbb{Z})$ into an element of $l^{\infty}(\mathbb{Z}_-)$.

Note now that, by applying the delay operator U_{-k} to the input sequence $x(l)$, we get

$$(U_{-k}x)(l) = x(l+k) \in l^{\infty}(\mathbb{Z}) \tag{1.171a}$$

Using the truncation operator P in the sequence given by Equation 1.171a results in

$$(PU_{-k}x)(l) = (Px)(l+k) = x(l+k) \text{ for } l \leq 0 \tag{1.171b}$$

which is an element of $l^{\infty}(\mathbb{Z}_{-})$.

Finally, applying the functional F to the element of $l^{\infty}(\mathbb{Z}_{-})$ given by Equation 1.171b, we obtain

$$(FPU_{-k}x)(l) = (Nx)(0+k) = (Nx)(k) \tag{1.171c}$$

From Equation 1.171c, we conclude that the operator N can be recovered from its associated functional F, and the corresponding relationship is

$$(Nx)(k) = (FPU_{-k}x)(l) \tag{1.172}$$

The operations described in Equations 1.169 to 1.172 are illustrated in Figure 1.29. Note that the starting sequence $x(k)$ in Figure 1.29a is assumed to belong to $l^{\infty}(\mathbb{Z})$ (not to $l^{\infty}(\mathbb{Z}_{-})$). The truncated sequence $z = PU_{-k}x$ belongs, however, to $l^{\infty}(\mathbb{Z}_{-})$. When this sequence is extended according to the formula 1.169b, it assumes the form z_e given in Figure 1.29e. Observe that the form of the starting sequence $x(k)$ is different from the form of the sequence $z_e(l)$.

Now we show that the second definition of the fading memory (FMD2) presented in the previous section, which was proposed for systems of which behavior is considered only for nonnegative times ($k \geq 0$), follows from the first definition of the fading memory (FMD1). To do this, we refer to the definition FMD2 with the time-invariant operator N: $l^{\infty}(\mathbb{Z}_{+}) \rightarrow l^{\infty}(\mathbb{Z}_{+})$, and the sequences x and v belonging to $B_{+} \in l^{\infty}(\mathbb{Z}_{+})$. Note that to be able to use the definition FMD1, we must redefine in some way the above operator N and the sequences on which it operates. We start by defining the sequences x_f and v_f, with the discrete time arguments taking on the values from the whole set $\mathbb{Z}$, as

$$x_f(k) \stackrel{df}{=} \begin{cases} x(k) & \text{for } k \geq 0 \\ 0 & \text{for } k < 0 \end{cases} \tag{1.173a}$$

and similarly,

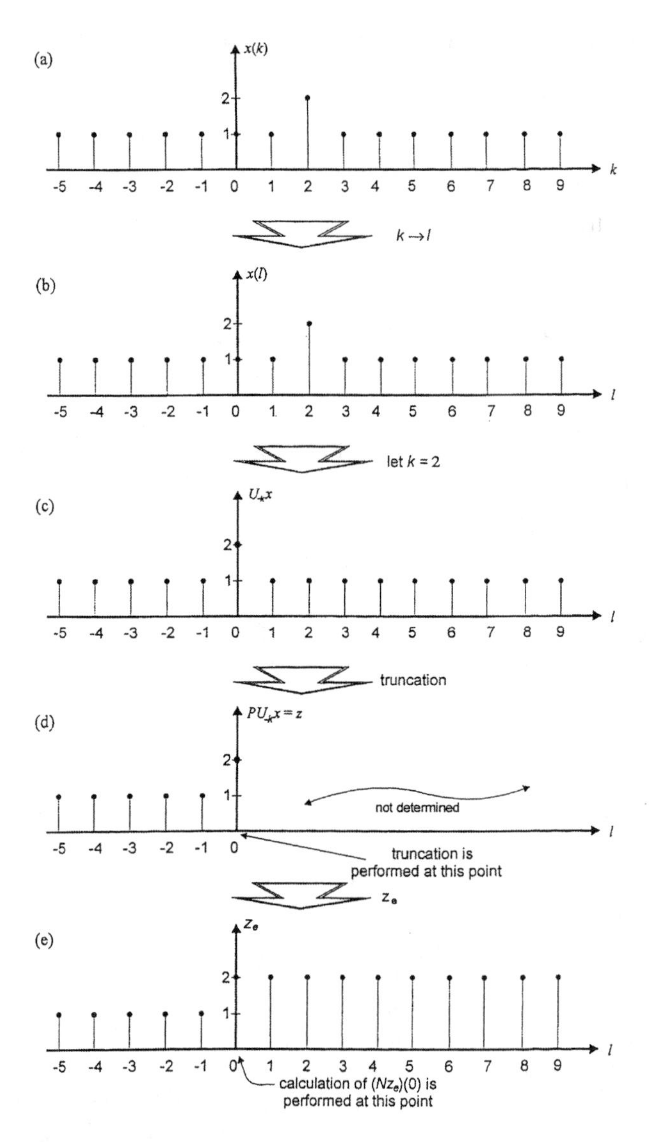

FIGURE 1.29 Illustration of operations described in Equations 1.169 to 1.172: (a) starting sequence $x(k)$, (b) sequence $x(k)$ in which $k \to l$, (c) delayed sequence $U_{-k}x$, (d) truncated sequence $z = PU_{-k}x$, (e) extended sequence z_e.

$$v_f(k) \stackrel{df}{=} \begin{cases} v(k) & \text{for } k \geq 0 \\ 0 & \text{for } k < 0 \end{cases} \tag{1.173b}$$

Using the above definitions of the extended sequences, we define a time-invariant operator N_f working on such sequences by

$$(N_f x_f)(k) \stackrel{df}{=} \begin{cases} (Nx)(k) & \text{for } k \geq 0 \\ 0 & \text{for } k < 0 \end{cases} \tag{1.174}$$

where x_f stands for all the sequences of type 1.173, which form a subspace B_f of $l^\infty(\mathbb{Z})$.

Assume now that the operator N_f defined by Equation 1.174, possesses a fading memory. Moreover, let us take into account time-shifted sequences $U_{-k}x_f(l)$ and $U_{-k}v_f(l)$, as illustrated in Figure 1.30 for the first of these sequences.

Now, having in mind the definition 1.167, we can write the following relation:

$$\begin{aligned} &\sup_{l \leq 0} |U_{-k}x_f(l) - U_{-k}v_f(l)| w(-l) < \delta \rightarrow \\ &\rightarrow |(N_f U_{-k} x_f)(0) - (N_f U_{-k} v_f)(0)| < \varepsilon \end{aligned} \tag{1.175}$$

for the operator N_f. Furthermore, note that, by the use of the definition of the delay operator U_{-k}, the first part of the relation 1.175 can be rewritten as

$$\sup_{l \leq 0} |x_f(l+k) - v_f(l+k)| w(-l) < \delta \tag{1.176a}$$

Then, using a new variable $l + k = \tau$ in inequality 1.176a, we get

$$\sup_{\tau - k \leq 0} |x_f(\tau) - v_f(\tau)| w(k-\tau) < \delta \tag{1.176b}$$

Because $x_f(\tau) = v_f(\tau) = 0$ for $\tau < 0$ (see Figure 1.30), inequality 1.176b takes on the form

$$\sup_{0 \leq \tau \leq k} |x_f(\tau) - v_f(\tau)| w(k-\tau) < \delta \tag{1.176c}$$

Moreover, taking into account the definitions 1.173 in inequality 1.176c, we arrive finally at

$$\sup_{0 \leq \tau \leq k} |x(k) - v(k)| w(k-\tau) < \delta \tag{1.176d}$$

After getting the result 1.176d, consider the right-hand side of the relation 1.175. Note that it can be rewritten as

$$|(N_f x_f)(k+0) - (N_f v_f)(k+0)| < \varepsilon \tag{1.177a}$$

or

$$|(N_f x_f)(k) - (N_f v_f)(k)| < \varepsilon \tag{1.177b}$$

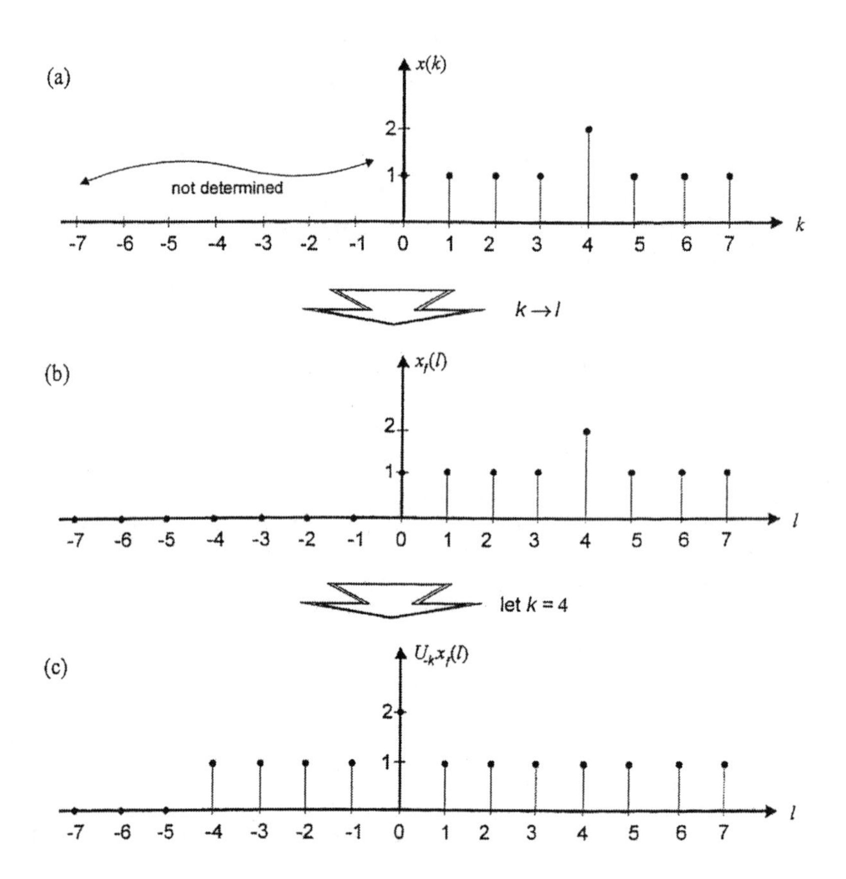

FIGURE 1.30 (a) Original sequence belonging to the space $l^\infty(\mathbb{Z}_+)$, (b) the sequence $x_f \in l^\infty(\mathbb{Z})$ created from $x(k)$, (c) the time-shifted sequence $U_{-k}x_f(l)$.

Note that in our considerations, we have chosen the time-shifting parameter k to be greater than or equal to zero. With that in mind and definition 1.174, we rewrite inequality 1.177b as

$$|(Nx)(k) - (Nv)(k)| < \varepsilon, \quad k \geq 0 \tag{1.177c}$$

Taking into account the achieved results 1.176d and 1.177c, and comparing them to definition 1.168, we conclude that really the definition FMD2 can be derived from the FMD1. So the definition FMD2 is, in fact, nothing more than a specific variant of the definition FMD1, derived for sequences defined only for nonnegative times ($k \geq 0$).

Now let us define what we mean by the notion of causality of an operator. We call the operator N to be causal if the equality of sequences $x(\tau) = v(\tau)$ for times $\tau \leq k$ implies equality $(Nx)(k) = (Nv)(k)$.

Using the above definition of operator causality, we will show now that every time-invariant operator N: $l^{\infty}(\mathbb{Z}) \rightarrow l^{\infty}(\mathbb{Z})$, possessing the fading memory on the subspace B of $l^{\infty}(\mathbb{Z})$, is causal. To do this, we take into account sequences $(PU_{-k}x)(m)$ and $(PU_{-k}v)(m)$, where $k,m \in \mathbb{Z}$. Note that these sequences are elements of the space $l^{\infty}(\mathbb{Z}_{-})$. Furthermore, applying the extension formula 1.169b to them, we arrive at the sequences belonging to the space $l^{\infty}(\mathbb{Z})$. So we get

$$x_{ePU_{-k}}(m) = \begin{cases} x(k+m) & \text{for} \quad m \leq 0 \\ x(k) & \text{for} \quad m > 0 \end{cases} \tag{1.178a}$$

and

$$v_{ePU_{-k}}(m) = \begin{cases} v(k+m) & \text{for} \quad m \leq 0 \\ v(k) & \text{for} \quad m > 0 \end{cases} \tag{1.178b}$$

Moreover, we assume that the sequences given by Equations 1.178 belong to the subspace B of $l^{\infty}(\mathbb{Z})$. Hence, we can use the definition 1.167 of the fading memory, to get

$$\begin{aligned} \sup_{m \leq 0} |x_{ePU_{-k}}(m) - v_{ePU_{-k}}(m)| w(-m) &= \sup_{m+k \leq k} |x(m+k) - v(m+k)| w(-m) < \delta \\ \rightarrow |(Nx_{ePU_{-k}})(0) - (Nv_{ePU_{-k}})(0)| &= |(Nx)(k) - (Nv)(k)| < \varepsilon \end{aligned} \tag{1.179}$$

Let us now introduce a new variable $\tau = m + k$ in (1.179). This leads to

$$\sup_{\tau \leq k} |x(\tau) - v(\tau)| w(k - \tau) < \delta \rightarrow |(Nx)(k) - (Nv)(k)| < \varepsilon \tag{1.180}$$

Furthermore, note that we use sequences, that are equal to each other, i.e., $x(\tau) = v(\tau)$ for $\tau \leq k$, in the definition of causality of an operator N. Taking this into account, in Equation 1.180, we get, finally

$$\sup_{\tau \leq k} |x(\tau) - v(\tau)| w(k - \tau) = 0 < \delta \rightarrow |(Nx)(k) - (Nv)(k)| < \varepsilon \tag{1.181}$$

Observe that the inequality on the right-hand side of the relation 1.181 holds independently of the choice of δ. Furthermore, because this inequality must hold for any small $\varepsilon > 0$, it implies $(Nx)(k) = (Nv)(k)$. Any time-invariant operator N possessing the fading memory is causal. The inverse is not true. That is, not every time-invariant and causal operator has the property of fading memory. This is because nothing is said in the definition of causality about the behavior of the operator N for signals $x(\tau) \neq v(\tau)$, for $\tau \leq k$ and $\tau,k \in \mathbb{Z}$.

In considering causality of an operator N in this section, we recall we have already tackled a similar problem in Section 1.1. We explained in Section 1.1 what the nonlinear causal responses $h^{(n)}(i_1,i_2,\ldots,i_n)$ are. For illustration, see again Figure 1.3. Now, if we define a causal time-invariant Volterra series operator given by definition 1.160 as one that has nonlinear responses of the kind just mentioned, we can ask about its relationship with the definition presented in this section. We shall show that the two definitions of causality of the Volterra series operator are equivalent to each other.

First, we show that the definition of Section 1.1 follows from the definition of this section. And to this end, we consider the following difference:

$$(Vx)(k)-(Vv)(k)=\sum_{i=-\infty}^{\infty} h^{(1)}(i)[x(k-i)-v(k-i)]$$
$$+\sum_{i_1=-\infty}^{\infty}\sum_{i_2=-\infty}^{\infty} h^{(2)}(i_1,i_2)[x(k-i_1)x(k-i_2)-v(k-i_1)v(k-i_2)]$$
$$+\sum_{i_1=-\infty}^{\infty}\sum_{i_2=-\infty}^{\infty}\sum_{i_3=-\infty}^{\infty} h^{(3)}(i_1,i_2,i_3)[x(k-i_1)x(k-i_2)x(k-i_3)-v(k-i_1)v(k-i_2)v(k-i_3)]+\ldots$$

(1.182)

Note that if we assume that the operator V in Equation 1.182 is causal, it follows from the causality definition of this section that the difference 1.182 equals zero when $x(\tau) = v(\tau)$, $\tau \le k$. So, for $k - i$, $k - i_1$, $k - i_2$, etc., less than or equal to k, the corresponding components in Equation 1.182 are equal to zero because $x(k - i) = v(k - i)$, $x(k - i_1) = v(k - i_1)$, $x(k - i_2) = v(k - i_2)$, etc., according to the assumption $x(\tau) = v(\tau)$, $\tau \le k$. On the other hand, when $k - i > k$, $k - i_1 > k$, $k - i_2 > k$, etc., the differences $x(k - i) - v(k - i)$, $x(k - i_1)x(k - i_2) - v(k - i_1)v(k - i_2)$, etc., can take on any values different from zero. So, to make the difference on the left-hand side of Equation 1.182 equal to zero in each case, we must postulate fulfillment of the following equalities:

$$h^{(1)}(i) = 0 \quad \text{for} \quad i<0 \tag{1.183a}$$

$$h^{(2)}(i_1,i_2) = 0 \quad \text{for} \quad i_1<0 \quad \text{and/or} \quad i_2<0 \tag{1.183b}$$

$$h^{(3)}(i_1,i_2,i_3) = 0 \quad \text{for} \quad i_1<0 \quad \text{and/or} \quad i_2<0 \quad \text{and/or} \quad i_3<0 \tag{1.183c}$$

and so on. Further, observe that Equations 1.183, in fact, express nothing other than the definition of the nonlinear response causality presented in Section 1.1.

To prove the converse, let us take again into account Equation 1.182. Under the assumption that the equalities 1.183 are fulfilled, this equation can be rewritten as

$$(Vx)(k) - (Vv)(k) = \sum_{i=0}^{\infty} h^{(1)}(i)[x(k-i) - v(k-i)]$$
$$+\sum_{i_1=0}^{\infty}\sum_{i_2=0}^{\infty} h^{(2)}(i_1, i_2)[x(k-i_1)x(k-i_2) - v(k-i_1)v(k-i_2)]$$
$$+\sum_{i_1=0}^{\infty}\sum_{i_2=0}^{\infty}\sum_{i_3=0}^{\infty} h^{(3)}(i_1, i_2, i_3)[x(k-i_1)x(k-i_2)x(k-i_3) - v(k-i_1)v(k-i_2)v(k-i_3)] + \ldots$$
$$(1.184)$$

Note now that the left-hand side of Equation 1.184 is identically equal to zero, if we assume additionally $x(\tau) = v(\tau)$ for $\tau \leq k$, where $\tau, k \in \mathbb{Z}$, $\tau = k - i, k - i_1, k - i_2, \cdots$, with $i, i_1, i_2, \cdots \in \mathbb{Z}$. This means that the causality definition of this section follows from that of Section 1.1.

It is worth noting that the property of fading memory can be expressed in another way using the notion of operator continuity. That is this can be expressed in the following form: a time-invariant operator N: $l^{\infty}(\mathbb{Z}) \rightarrow l^{\infty}(\mathbb{Z})$ possesses fading memory on the subspace B of $l^{\infty}(\mathbb{Z})$ if, and only if, its associated functional F, defined by Equation 1.169a, is continuous on the subspace

$$PB \stackrel{df}{=} \{Px | x \in B\} \qquad (1.185a)$$

with respect to the weighted norm

$$\|x\|_{\omega} \stackrel{df}{=} \|x(k)w(-k)\| = \sup_{k \leq 0} |x(k)| w(-k) \qquad (1.185b)$$

where $x \in l^{\infty}(\mathbb{Z})$ and the nonnegative weighting sequence w is a mapping w : $\mathbb{Z}_{+} \rightarrow (0,1>)$, $\lim_{m \to \infty} w(m) = 0$.

In what follows, we show that the above definition of fading memory is true. To this end, note first that the definition FMD1 of fading memory is so constructed that it follows immediately from it, and from the definition 1.169 that the following relation:

$$\|Px - Pv\|_{\omega} < \delta \rightarrow |(FP)(x) - (FP)(v)| < \varepsilon \qquad (1.186)$$

holds. Furthermore, observe that relation 1.186 is nothing other than the continuity definition expressed in definition 1.155. In other words, the fact that the distance between the elements Px and Pv of the subspace PB, i.e.,

$$d_{PB}(Px, Pv) = \|Px - Pv\|_{\omega}$$

is less than δ, implies the distance between the elements FPx and FPv, i.e.,

$$d_{FPB}((FP)(x), (FP)(v)) = |(FP)(x) - (FP)(v)|$$

is less than ε. We conclude that if a time-invariant operator, N, has fading memory property, its associated functional F is continuous on the subspace PB.

To prove the converse, we consider the relation 1.186, which expresses the continuity property of the associated functional F. This relation can be expressed in the following form:

$$\|(PU_{-0})(x) - (PU_{-0})(v)\|_{\omega} < \delta \rightarrow |(FPU_{-0})(x) - (FPU_{-0})(v)| < \varepsilon \quad (1.187)$$

because $(P)(x) = (PU_{-0})(x)$ and $(P)(v) = (PU_{-0})(v)$.

Then, using the definitions 1.185b and 1.172, the relation 1.187 can be rewritten as

$$\sup_{k \leq 0} |x(k) - v(k)| w(-k) < \delta \rightarrow |(Nx)(0) - (Nv)(0)| < \varepsilon \quad (1.188)$$

which is nothing other than the fading memory definition, (see 1.167).

Incidentally, note in the relations 1.186 to 1.188 once again the typical usage of parentheses around the arguments of functionals and operators. Here parentheses are used to stress an argument, as in $(P)(x)$ standing for Px. Sometimes, parentheses are dropped, as in $(FPU_{-0})(x)$, standing for $(F)(P)(U_{-0})(x)$. In $(FPU_{-0})(x)$, the composite operator FPU_{-0} is separated from its argument x.

As mentioned in the previous section, the notions of decaying memory and approximately finite memory, introduced to the literature by Sandberg, are closely related to the notion of fading memory. Now we consider them, starting from the notions of decaying memory, which are, as originally formulated by Sandberg, the definitions of the so-called $M_1(m)$ and $P_1(m)$ sets,[17] specialized here for discrete-time systems.

To be able to define the set $M_1(m)$, we have to introduce first Sandberg's truncation operator Q_τ as

$$(Q_\tau x)(k) = \begin{cases} x(k) & \text{for } k \geq \tau \\ 0 & \text{for } k \geq \tau \end{cases} \quad (1.189)$$

where $x(k) \in l^\infty(\mathbb{Z})$, and $k, \tau \in \mathbb{Z}$. This operator is illustrated in Figure 1.31. It is worth noting at this point two fundamental differences existing between the previously defined truncation operator P and the operator Q : P's "critical" point is zero, but Q's one is any chosen $\tau \in \mathbb{Z}$. Moreover, P is a mapping $l^\infty(\mathbb{Z}) \rightarrow l^\infty(\mathbb{Z}_-)$, in contrast to Q, which is the mapping of $l^\infty(\mathbb{Z})$ into itself, that is $l^\infty(\mathbb{Z}) \rightarrow l^\infty(\mathbb{Z})$.

Having defined the operator Q_τ, we define now the set $M_1(m)$ in the following way: for each positive integer m, after Sandberg, an operator $N : l^\infty(\mathbb{Z}) \rightarrow l^\infty(\mathbb{Z})$ is an element of the set $M_1(m)$ if given any $k_1 \in (-\infty, \infty)$ and any real $\varepsilon > 0$, there exists such a $k_2 \in (-\infty, k_1)$ that

$$|(Nx)(k) - (NQ_{k_2}x)(k)| \leq \varepsilon \|x\|^m \text{ for } k \geq k_1 \quad (1.190)$$

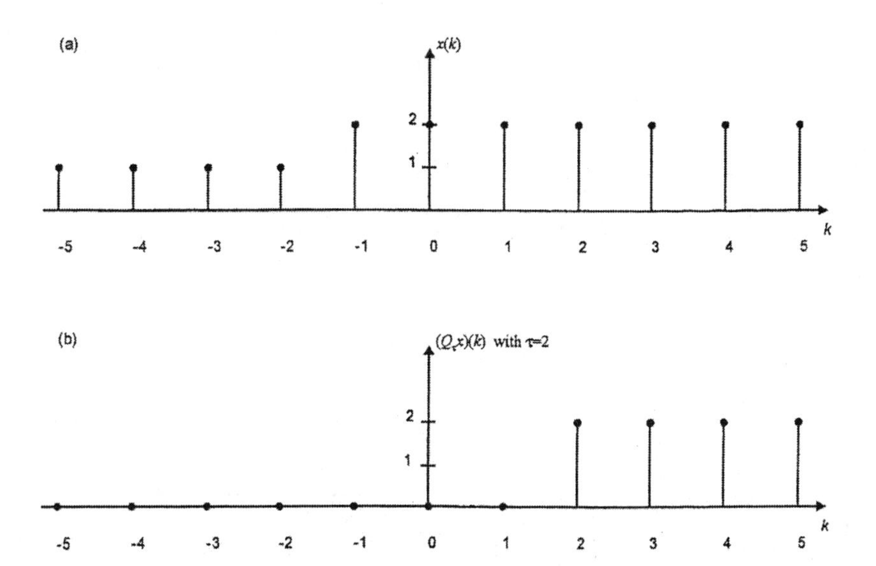

FIGURE 1.31 (a) Original sequence $x(k) \in l^\infty(\mathbb{Z})$, (b) the truncated sequence $(Q_\tau x)(k)$ for $\tau=2$.

holds for $x(k)$ belonging to some subspace B of $l^\infty(\mathbb{Z})$. We call this definition the first Sandberg's Decaying Memory definition (SDM1).

In other words, an operator N possesses the property of decaying memory in the sense of the definition SDM1 if it belongs to the set $M_1(m)$ for which elements the inequality 1.190 holds.

To proceed further, we now need to define two other operators introduced by Sandberg in his paper.[17] For this purpose, we introduce the so-called $l^\infty(\mathbb{Z}_{k_0})$ space, which consists of sequences defined for the discrete times belonging to the set $\langle k_0, \infty)$, where $k_0 \in \mathbb{Z}$, with the usual norm $\|x\|_{k_0} = \sup_{k \geq k_0} |x(k)|$. In other words, the sequences of the space $l^\infty(\mathbb{Z}_{k_0})$ are such sequences that begin at $k = k_0$ and for the time instants k less than k_0 are not determined.

Now having defined the space $l^\infty(\mathbb{Z}_{k_0})$, we are able to define the Sandberg's $T_{k_1k_2}$ operator as such a mapping, that maps elements of the space $l^\infty(\mathbb{Z}_{k_1})$ into elements of the space $l^\infty(\mathbb{Z}_{k_2})$ according to the following rule:

$$(T_{k_1k_2}x)(k) = x(k + k_1 - k_2) \quad k \geq k_2 \tag{1.191}$$

The operator $T_{k_1k_2}$ is illustrated in Figure 1.32.

Similarly, we define Sandberg's R_{k_0} operator, $k_0 \in \mathbb{Z}$, as a mapping from the space $l^\infty(\mathbb{Z})$ into the space $l^\infty(\mathbb{Z}_{k_2})$, which is given by

$$(R_{k_0}x)(k) = x(k) \;\text{ for }\; k \geq k_0 \tag{1.192}$$

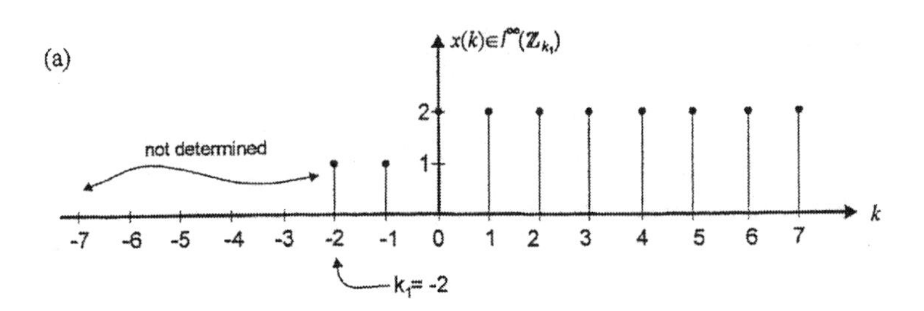

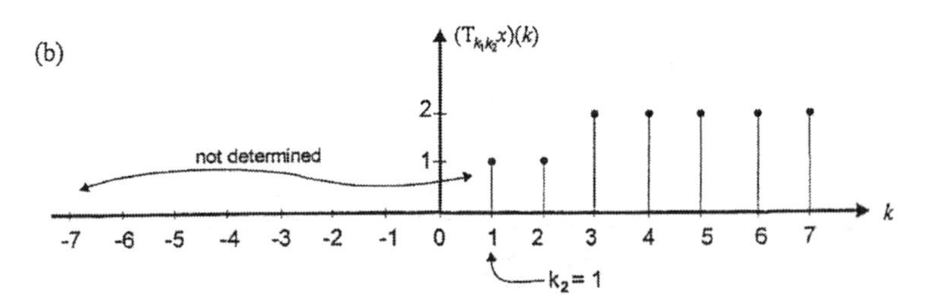

FIGURE 1.32 (a) Original sequence $x(k)\in l^{\infty}\left(\mathbb{Z}_{k_1=-2}\right)$, (b) shifted sequence $\left(T_{k_1k_2}x\right)(k)\in l^{\infty}\left(\mathbb{Z}_{k_2=1}\right)$.

The operator $(R_{k_0}x)(k)$ is illustrated in Figure 1.33.

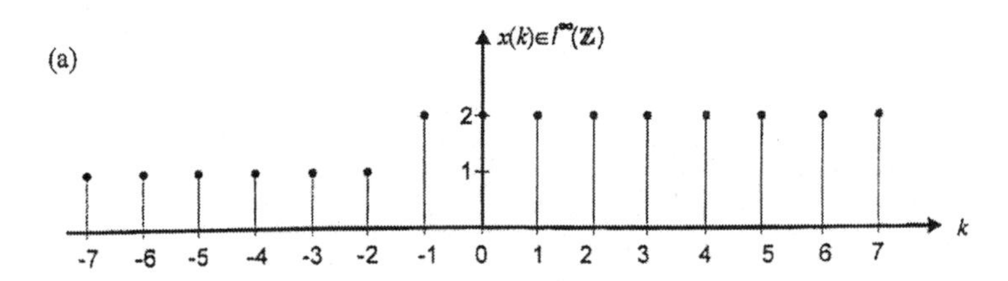

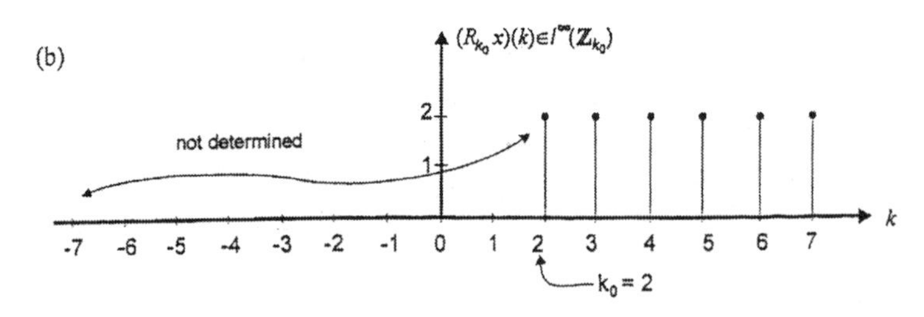

FIGURE 1.33 (a) Original sequence $x(k)\in l^{\infty}(\mathbb{Z})$, (b) truncated sequence $(R_{k_0}x)(k)\in l^{\infty}(\mathbb{Z}_{k_0=2})$.

Note the fundamental difference between the truncation operator Q_τ defined in 1.189 and the just-defined R_{k_0} in 1.192. The first is a mapping from $l^{\infty}(\mathbb{Z})$ into itself, but the latter is a mapping from $l^\infty(\mathbb{Z})$ into $l^\infty(\mathbb{Z}_{k_0})$. This follows clearly from the comparison of Figure 1.31 with Figure 1.33.

Consider now a special kind of mappings from $l^\infty(\mathbb{Z}_+)$ into $l^\infty(\mathbb{Z}_+)$ called K_0 mappings by Sandberg in his paper.[17] These mappings have the following property:

$$\|K_0 x\|_0 \leq \rho(\|x\|_0) \tag{1.193}$$

where ρ is a nondecreasing function ρ : subset $\langle 0,\infty)$ of real numbers $\mathbb{R}$ into itself, and the norm $\|x\|_0$ is $\|x\|_{k_0} = \sup_{k \geq k_0}|x(k)|$, specialized for $k_0 = 0$.

We can associate with the mapping K_0 its k_0-associate K_{k_0}, which is a mapping defined as $T_{0k_0} K_0 T_{k_0 0}$ for any $k_0 \in \mathbb{Z}$. This is, of course, a composite mapping, a mapping from the space $l^\infty(\mathbb{Z}_{k_0})$ into itself.

Now we are able to define Sandberg's[17] $P_1(m)$ set in the following way: a mapping K_0 is an element of the set $P_1(m)$ if, given any $k_1 \in \mathbb{Z}$ and any real $\varepsilon > 0$, there exists such a $k_2 < k_1$ that

$$\left|(K_{k_3} R_{k_3} x)(k) - (K_{k_4} R_{k_4} x)(k)\right| \leq \varepsilon \|x\|^m, \quad k \geq k_1 \tag{1.194}$$

holds for all x belonging to some subspace B of $l^\infty(\mathbb{Z})$, when max $(k_3, k_4) \leq k_2$. We call this definition the second Sandberg's Decaying Memory definition (SDM2). In other words, an operator K_0 with the property 1.193 possesses the decaying memory in the sense of the definition SDM2 if it belongs to the set $P_1(m)$ for which elements the inequality 1.194 holds. Moreover, comparing the definitions SDM1 and SDM2, we see that the latter can be considered as a variant of the first, in which the values of the input signal (sequence) for $k <$ min(k_3, k_4) are not taken into account.

Let us now consider two definitions of the notion of approximately-finite memory, as formulated by Sandberg in another paper.[20] The first definition was originally formulated in the following way:

AFM1 (Approximately-Finite Memory definition 1): let N be a mapping N : $l^\infty(\mathbb{Z}_+) \rightarrow C_R(\mathbb{Z}_+)$, where C_R means the collection of all $\mathbb{R}$-valued mappings defined on $\mathbb{Z}_+$. Moreover, let $\mathbb{N}$ mean the set of integers $\{1, 2, 3, \cdots\}$. Then we say that the mapping N has approximately-finite memory on some subspace B of $l^\infty(\mathbb{Z}_+)$ if, given $\varepsilon > 0$, there exists such an $a \in \mathbb{N}$ that

$$\left|(Nx)(k) - (N W_{k,a} x)(k)\right| < \varepsilon, \quad k \in \mathbb{Z}_+ \tag{1.195a}$$

holds for all $x \in B$, where $W_{k,a}x$ means the sequence $x(k)$ after performing a windowing operation given by

$$(W_{k,a}x)(\tau) = \begin{cases} x(\tau) & \text{for} \quad k-a \leq \tau \leq k \\ 0 & \text{otherwise} \end{cases} \tag{1.195b}$$

The operation expressed by 1.195b is illustrated in Figure 1.34.

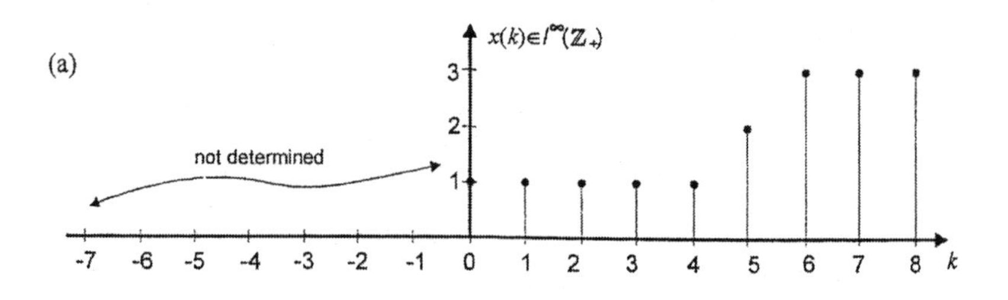

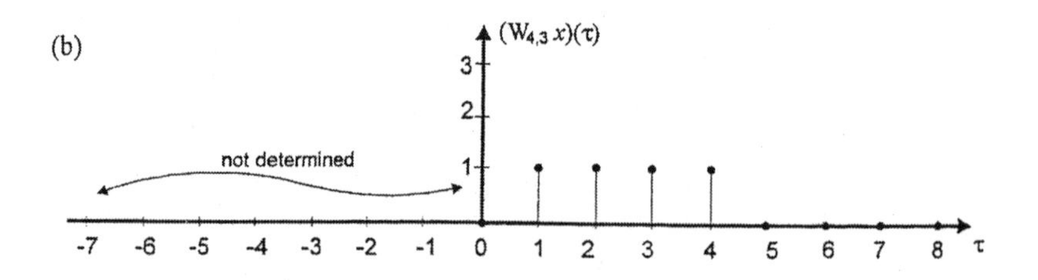

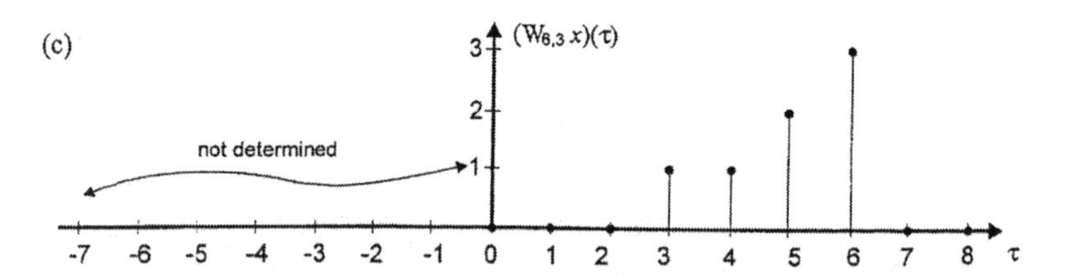

FIGURE 1.34 (a) Original sequence $x(k) \in l^{\infty}(\mathbb{Z}_+)$, (b) example of windowed sequence for $k = 4$ and window width $a + 1 = 4$, (c) another example of windowed sequence for $k = 6$ and the same window width.

The second definition of the notion of approximately-finite memory is closely related to that given by 1.195, and can be formulated, after Sandberg, as:

AFM2 (Approximately-Finite Memory definition 2): let N be a mapping $N : l^{\infty}(\mathbb{Z}_+) \rightarrow C_R(\mathbb{Z}_+)$. We say that this mapping possesses approximately-finite memory on some subspace B of $l^{\infty}(\mathbb{Z}_+)$ if, given $\varepsilon > 0$, there exists such a $\Delta > 0$ that

$$\left|(Nx)(k) - (NW_{k,a}x)(k)\right| < \varepsilon, \qquad k \in \mathbb{Z}_+ \tag{1.196}$$

holds for $a \geq \Delta$.

Comparison of Sandberg's definitions SDM1, SDM2, AFM1, and AFM2 shows that, in contrast to the definitions FMD1 and FMD2 of Boyd and Chua, they concentrate primarily on a proper choice of truncated, or windowed, sequences as inputs for a mapping considered. In other words, they focus on the discrete-time axis and consider two versions of the same input sequence, truncated or windowed, as two different input sequences. The Boyd and Chua definitions 1.167 and 1.168, rather take into account two different input sequences to determine what the weighted size of the difference between them should look like, to ensure the correspondingly small size of the absolute value of the difference between the values of the mapping calculated for these input sequences. This is simply a certain form of the continuity condition. In contrast, all of Sandberg's definitions do not take into account the size of the difference between the sequences for which a mapping is calculated.

Figure 1.35 summarizes the main characteristics of the types of memory described in this and previous section.

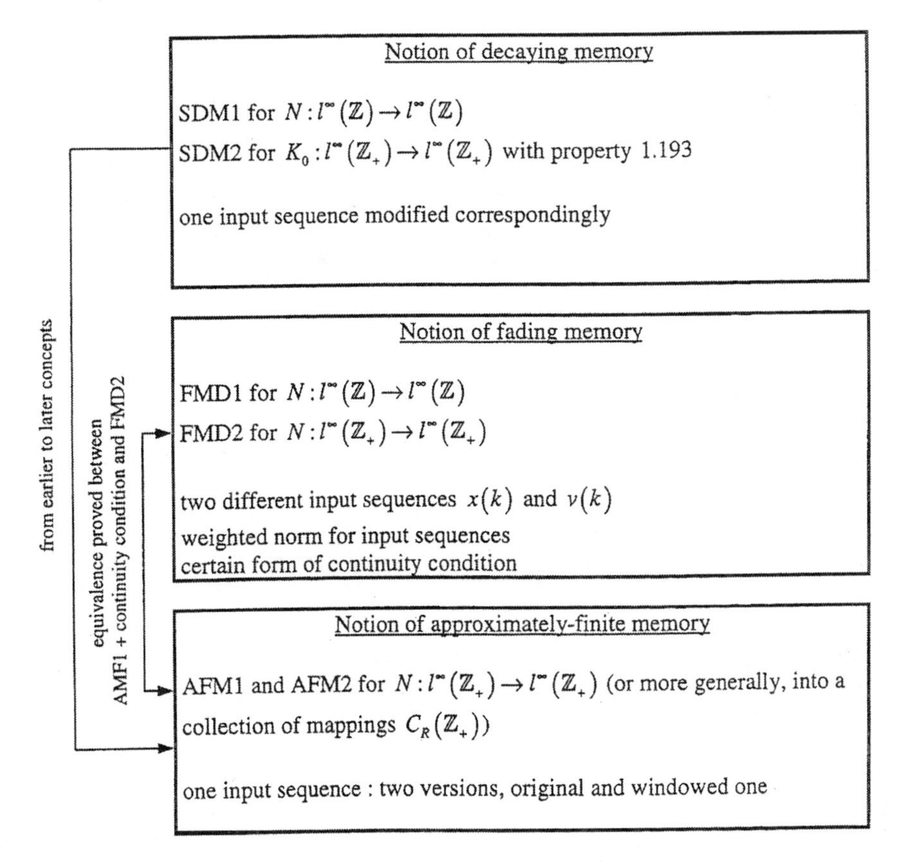

FIGURE 1.35 Summary of the definitions regarding the notions of decaying, fading, and approximately-finite memory.

Park and Sandberg[4] have proven that there is an exact relation between the definitions AMF1 and FMD2. That is, the definition AFM1 with the continuity condition imposed on the mapping N is equivalent to the definition FMD2. This is shown in Figure 1.35. Moreover, we have shown in this section that the definition FMD2 can be derived under some assumptions from the definition FMD1.

Finally, the definitions of the fading memory, decaying memory, and approximately-finite memory presented in this and the previous section, extend easily for the case of vector-valued input and output sequences. Then, we simply exchange the scalar-valued notation $x(k)$ and $v(k)$ in inequalities 1.167, 1.168, 1.190, 1.194, 1.195, and 1.196 for the vector-valued notation $\mathbf{x}(k)$ and $\mathbf{v}(k)$. And we interpret then the symbol $|\cdot|$ of the absolute value of, as given for vectors, according to Equation 1.157a. Moreover, when the output of a system is a vector-valued signal, then the output sequences $(N\mathbf{x})(k)$ and $(N\mathbf{v})(k)$ are the vector-valued sequences, too. This description is used in the next sections in those considerations where the multiple-input and multiple-output systems are discussed.

1.9 APPROXIMATION OF THE SYSTEM RESPONSE BY VOLTERRA SERIES

In this section, we consider nonlinear discrete-time systems, which have the property of fading or approximately-finite memory. We show that the responses of these systems can be approximated by the discrete Volterra series. We start by presenting two theorems published by Boyd and Chua in their paper.[1]

Theorem 1.1

Let $\varepsilon > 0$ and

$$B_{M_1} \stackrel{df}{=} \{x \in l^{\infty}(\mathbb{Z}) | \|x\| \leq M_1\} \tag{1.197a}$$

be a ball in $l^{\infty}(\mathbb{Z})$ of radius M_1. If N: $l^{\infty}(\mathbb{Z}) \rightarrow l^{\infty}(\mathbb{Z})$ is a time-invariant operator possessing the fading memory in the sense of definition FMD1 on the ball B_{M_1}, then there exists such a Volterra series operator V given by Equations 1.1 and 1.11 with a finite number of components, and with nonlinear responses fulfilling the stability condition 1.98 (with lower summation limits equal to zero), that the following inequality

$$\|Nx - Vx\| \leq \varepsilon \tag{1.197b}$$

holds for all $x \in B_{M_1}$.

Theorem 1.2

Let B_{m_1} be a ball in $l^{\infty}(\mathbb{Z})$ of radius M_1 as given by Equation 1.197a. If N: $l^{\infty}(\mathbb{Z}) \rightarrow l^{\infty}(\mathbb{Z})$ is a time-invariant operator possessing the fading memory in the sense of definition FMD1 on the ball B_{M_1} then, for any $\varepsilon > 0$, there exists such a polynomial p of M variables: $\mathbb{R}^M \rightarrow \mathbb{R}$ that the following

$$\|(Nx)(k) - p(x(k), x(k-1), \ldots, x(k-M+1))\| \leq \varepsilon \tag{1.198}$$

holds for all $x \in B_{M_1}$ and with $y(k) = p(x(k), x(k-1), \cdots, x(k-M+1))$ considered as a mapping from $l^{\infty}(\mathbb{Z})$ into itself.

Because of the importance of the above two theorems, we shall present their proofs in this book. We begin with the proof of Theorem 1.2. For this purpose, we show that the two lemmas presented in what follows are true.

Lemma 1.1. A ball

$$B_{-} \overset{df}{=} \{x \in l^{\infty}(\mathbb{Z}_{-}) \,|\, \|x\| \leq M_1\} \tag{1.199}$$

is a compact set with respect to the weighted norm given by Equation 1.185b, with x belonging now to $l^{\infty}(\mathbb{Z}_{-})$.

Note that we have to work with a new notion in the above lemma, namely, the notion of the compactness of a set. So, we must first explain it. However, to do this, we need to consider two other new notions; the notion of Cauchy sequence and the notion of complete space.

We say that a sequence $\{x_n\} = \{x_0, x_1, x_2,\ldots\}$ of a metric space X is a Cauchy sequence if

$$d(x_n, x_m) \rightarrow 0, \quad \text{when} \quad n, m \rightarrow 0 \tag{1.200}$$

Now, using the above definition, we define the complete space. That is, the complete metric space X is a space in which every Cauchy sequence possesses its limit, which belongs to this space.

In this book, we deal with metric spaces, which are complete. Furthermore, knowing what the notion of completeness of a metric space means, we are now able to define the set compactness in the following way. Let X be a complete metric space. Then we say that a set $B_X \subset X$ is compact if every infinite sequence $\{x_n\}$ of elements $x_n \in B_X$ possesses a subsequence $\{x_{n_i}\}$ convergent to some element $x^* \in B_X$.

We begin with the proof of Lemma 1.1. For this purpose, let us denote by x_n a sequence in B_{-}. Afterwards, observe that, $|x_n(0)| \leq M_1$ according to assumption 1.199. To proceed further, recall the well-known Bolzano-Weierstrass[21] theorem, that every finite segment $\langle a,b\rangle$ on the $\mathbb{R}$-line is a compact set. According to this theorem, we are able to find a subsequence of x_n, which converges at point $k = 0$ to some element we call $x_0(0)$. That is,

$$|x_{n_1}(0) - x_0(0)| \rightarrow 0, \quad \text{when} \quad n_1 \rightarrow \infty \tag{1.201}$$

holds. In relation 1.201, n_1 is an index that belongs to a certain set $\mathbb{N}_1 \subset \mathbb{N}$, where $\mathbb{N}$ stands for the set of positive integers.

Note now that $|x_{n_1}(-1)| \leq M_1$ because, of course, x_{n_1} belongs to the ball B_{-}. Using the Bolzano-Weierstrass theorem, as before, we can take a subsequence of $x_{n_1}(k)$ such that it converges at point $k = -1$ to some element we call $x_0(-1)$. Denoting this subsequence $x_{n_2}(k)$, we can write

$$|x_{n_2}(-1) - x_0(-1)| \rightarrow 0, \quad \text{when} \quad n_2 \rightarrow \infty \tag{1.202}$$

with $n_2 \subset \mathbb{N}_2 \subset \mathbb{N}_1 \subset \mathbb{N}$.

Observe that the above process can be continued, and this results in obtaining an element $x_0 = \{\cdots, x_0(-2), x_0(-1), x_0(0)\}$ of the ball B_-. Moreover, observe that the subsequence x_{n2} converges pointwise for $k = -1$ and $k = 0$ as well. Furthermore, see that the latter observation can be generalized, resulting in what we call a diagonal subsequence. So the diagonal subsequence x_{n_m}, $n_m \subset \mathbb{N}_m \subset \mathbb{N}_{m-1} \subset \cdots \subset \mathbb{N}$, is a subsequence of x_n, which is convergent pointwise for discrete-time instants $-(m-1)$, $-(m-2)$, $\cdots$, -1, 0. And this fact can be written as

$$\sup_{-(m-1)\le -k\le 0} \left|x_{n_m}(-k) - x_0(-k)\right| \to 0, \quad \text{when} \quad n_m \to \infty, n_m \in \mathbb{N}_m, k \in \mathbb{Z}_+$$

(1.203)

Let ε be any $\varepsilon > 0$. From the property of the decreasing weighting sequence $w(k)$: $w(k) \to 0$, when $k \to \infty$, it follows that we can find such a k_0 for which the inequality $w(k_0) < \varepsilon/(2M_1)$ holds. Furthermore, because $x_{n_m}, x_0, \in B_-$, we can write

$$\left|x_{n_m}(-k) - x_0(-k)\right| \le \left|x_{n_m}(-k)\right| + \left|x_0(-k)\right| \le 2M_1 \tag{1.204}$$

Having the above inequality in mind, we see that the inequality

$$\sup_{-k\le -k_0} \left|x_{n_m}(-k) - x_0(-k)\right| w(k) \le 2M_1 w(k) \tag{1.205}$$

holds, too. Applying then the relation $w(k) \le w(k_0) < \varepsilon/(2M_1)$ for $k \ge k_0$ in inequality 1.205, we arrive at

$$\sup_{-k\le -k_0} \left|x_{n_m}(-k) - x_0(-k)\right| w(k) \le \varepsilon \tag{1.206}$$

On the other hand, from the relation 1.203 and the fact that the inequality $w(k) \le 1$ holds for the weighting sequence, it follows that

$$\sup_{-(m-1)\le -k\le 0} \left|x_{n_m}(-k) - x_0(-k)\right| w(k) \le \varepsilon \tag{1.207}$$

is valid for all indices $n_m \ge n_0$.

Before going further, let us choose $m - 1 = k_0$ in relation 1.207. Taking into account both inequalities 1.205 and 1.207, we see that

$$\sup_{-k\le 0} \left|x_{n_m}(-k) - x_0(-k)\right| w(k) = \left\|x_{n_m}(-k) - x_0(-k)\right\|_\omega \le \varepsilon \tag{1.208}$$

holds for all $n_m \ge n_0$. This of course means that the subsequence x_{n_m} converges to x_0, so the ball B_- is a compact set with respect to the weighted norm 1.185b.

Lemma 1.2. Take into account the set of functionals

$$G \overset{df}{=} \{G_0, G_1, G_2, \cdots\} \tag{1.209a}$$

of which elements are given by

$$G_\tau x \overset{df}{=} x(-\tau), \quad \tau \in \mathbb{Z}_+ \tag{1.209b}$$

That is, the functional G_τ is associated with the delay operator U_τ given by Equation 1.158a. The functionals G_τ are continuous with respect to the weighted norm $\|\cdot\|_\omega$.

To prove the above lemma, observe first that it follows from $\lim_{\tau\to\infty} w(\tau) = 0$ that the inequality

$$\sup_{\tau\geq 0}|x(-\tau) - v(-\tau)|w(\tau) > \lim_{\tau\to\infty}|x(-\tau) - v(-\tau)|w(\tau) \tag{1.210}$$

holds for bounded sequences considered, belonging to the ball B_-, when the sequence $x(k)$ is not identical with $v(k)$, and is not the zero sequence. The result in inequality 1.210 means that the supremum is not achieved at the point $\tau = \infty$. This allows us to write the inequality

$$\sup_{\tau\geq 0}|x(-\tau) - v(-\tau)|w(\tau) \geq |x(-\tau) - v(-\tau)|w(\tau_1) \tag{1.211}$$

where τ_1 stands for the τ for which the supremum is achieved. Furthermore, inequality 1.211 can be rewritten as

$$|G_\tau x - G_\tau v| \leq \|x - v\|_\omega / w(\tau_1) \tag{1.212}$$

Note that $w(\tau_1) \neq 0$ always holds because τ_1, according to inequality 1.210, cannot lie in infinity, so, for any $\varepsilon > 0$ for which $|G_\tau x - G_\tau v| \leq \|x - v\|_\omega / w(\tau_1) < \varepsilon$ holds, we can choose $\delta = \varepsilon w(\tau_1)$ such that $\|x - v\|_\omega < \delta$ implies, according to inequality 1.212, $|G_\tau x - G_\tau v| < \dfrac{\delta}{w(\tau_1)} = \dfrac{\varepsilon w(\tau_1)}{w(\tau_1)} = \varepsilon$. This proves the continuity of the functionals G_τ.

Lemma 1.3. The functionals of the set G given by Equations 1.209 separate the elements of the ball B_-. The separability property means here that having any two different sequences x and v, we can always find such a functional G_τ, $\tau = 0, 1, 2, 3, \cdots$, that $G_\tau x - G_\tau v \neq 0$.

To prove Lemma 1.3, we assume that the sequences x and v belong to the ball B_-. Moreover, assume that these sequences are not identical; that is, $x(k) \neq v(k)$ holds at least at one point $k \in \mathbb{Z}_-$. We can then find such a functional G_τ, that

$$G_\tau x - G_\tau v = x(-\tau) - v(-\tau) \neq 0 \tag{1.213}$$

holds in the point in which, $x(-\tau) \neq v(-\tau)$. Finally, observe that the inequality 1.213 proves the separability property of the functionals of the set G.

We also need to prove Theorem 1.2 is the Stone-Weierstrass theorem.[21] This theorem can be expressed in the following way:

Theorem 1.3 (Stone-Weierstrass)

Let E be a compact metric space and G a set of continuous functionals on E. Moreover, the functionals of the set G are such that they separate points of E, that is, for any two distinct $x, v \in E$ there exists a functional $G_i \in G$, $i = 0, 1, 2, \cdots$, such that $G_i x \neq G_i v$. Furthermore, let F be any continuous functional on E and $\varepsilon > 0$. Then there is a polynomial $p : \mathbb{R}^M \to \mathbb{R}$ and $G_0, G_1, \cdots, G_{M-1} \in G$ such that for all $x \in E$

$$\left|Fx - p(G_0 x, G_1 x, \cdots, G_{M-1} x)\right| < \varepsilon \tag{1.214}$$

holds.

Using the lemmas just presented and the Stone-Weierstrass theorem, we now prove Theorem 1.2. For this purpose, we assume that N is a time-invariant operator possessing the property of fading memory on the ball B_{M_1} in the sense of the definition FMD1, and that F is a functional associated with the operator N. The associated functional F is given by defining Equations 1.169a and 1.169b. (Caution: note that x means an element of $l^\infty(\mathbb{Z}_-)$ in the definition 1.169 and in the defining Equation 1.199 of the ball B_-. But in Equation 1.197a, it is a different element: an element of $l^\infty(\mathbb{Z})$). Furthermore, the functional F is continuous with respect to the norm $\|\cdot\|_\omega$, which follows from the property proved previously (see the relation 1.186).

Identifying now the ball B_- with the space E in the Stone-Weierstrass theorem, using the lemmas 1, 2, and 3, and the fact that the associated functional F is continuous, we conclude from the Stone-Weierstrass theorem that for any $\varepsilon > 0$, there can be found such a polynomial p of M variables: $\mathbb{R}^M \to \mathbb{R}$ and a subset of functionals $G_0, G_1, \cdots, G_{M-1} \in G$, that for all $x \in B_-$

$$\left|Fx - p(G_0 x, G_1 x, \cdots, G_{M-1} x)\right| < \varepsilon \tag{1.215}$$

holds.

Assume now that $x \in B_{M_1}$ and $k \in \mathbb{Z}$. The element $PU_{-k}x$ is an element of the ball B_- and the inequality 1.215 can be applied. Consequently, we arrive at

$$\left|FPU_{-k}x - p(G_0 PU_{-k}x, G_1 PU_{-k}x, \cdots, G_{M-1} PU_{-k}x)\right| < \varepsilon \tag{1.216}$$

under the assumptions stated, with regard to the operator N.

Finally, we use the general relation 1.172 for recovering the operator N from its associated functional F. We apply this relation to all the functionals $F, G_0, G_1, \cdots, G_{M-1}$ occurring in inequality 1.216. Then we get

$$|(Nx)(k) - p((U_0x)(k), (U_1x)(k), \cdots, (U_{M-1}x)(k))| < \varepsilon \tag{1.217}$$

This is so because $(Nx)(k) = (FPU_{-k}x)(1)$ follows directly from relation 1.172 and, for example, $G_2PU_{-k}x$ results in U_2x because we get successively: $U_{-k}x = x(l + k)$; $PU_{-k}x = x(l + k), l \leq 0$; $G_2PU_{-k}x = x(-2 + k) = x(k - 2)$; $(U_2x)(k) = x(k - 2)$. Finally, note that the inequality 1.217 holds for all k and is identical with inequality 1.198. Hence, this ends the proof of Theorem 1.2.

Observe now that Theorem 1.2 is stronger than Theorem 1.1. That is, if Theorem 1.2 is true, than Theorem 1.1 is also true. This follows from the fact that every polynomial p (of M variables): $\mathbb{R}^M \to \mathbb{R}$ is at the same time a Volterra series operator $V : l^\infty \to l^\infty$ given by Equations 1.1 and 1.11 with a finite number of components, and with nonlinear responses fulfilling the stability condition 1.98 (with lower summation limits equal to zero). To show the above, note that the polynomial $p((U_0x)(k), (U_1x)(k), ..., (U_{M-1}x)(k))$ can be expressed in the following form:

$$\begin{aligned}
&p((U_0x)(k), (U_1x)(k), \cdots, (U_{M-1}x)(k)) \\
&= \alpha_0 + \sum_{n=1}^{L} \sum_{0 \leq i_1, \cdots, i_n \leq M-1} \alpha_{i_1 \ldots i_n}(U_{i_1}x)(k) \cdots (U_{i_n}x)(k) \\
&= \alpha_0 + \sum_{i_1=1}^{M-1} \alpha_{i_1}(U_{i_1}x)(k) + \sum_{i_1=1}^{M-1}\sum_{i_2=1}^{M-1} \alpha_{i_1 i_2}(U_{i_1}x)(k)(U_{i_2}x)(k) \\
&\quad + \cdots + \sum_{i_1=1}^{M-1} \cdots \sum_{i_L=1}^{M-1} \alpha_{i_1 \cdots i_L}(U_{i_1}x)(k) \cdots (U_{i_L}x)(k) \\
&= h^{(0)} + \sum_{i_1=0}^{M-1} h^{(1)}(i_1)x(k - i_1) + \sum_{i_1=0}^{M-1}\sum_{i_2=0}^{M-1} h^{(2)}(i_1, i_2)x(k - i_1)x(k - i_2) \\
&\quad + \cdots + \sum_{i_1=0}^{M-1} \cdots \sum_{i_L=0}^{M-1} h^{(L)}(i_1, \cdots, i_L)x(k - i_1) \cdots x(k - i_L) \\
&= p(x(k), x(k-1), x(k-2), \ldots, x(k - M + 1))
\end{aligned} \tag{1.218}$$

where α_0 is a constant coefficient of the polynomial, and the coefficients $\alpha_{i_1 \ldots i_n}$, $0 \leq i_1, \ldots, i_n \leq \mathrm{M} - 1$, $1 \leq n \leq L$, are the coefficients by the corresponding powers of the signal samples. L in Equation 1.218 means the order of the polynomial.

Comparison of the corresponding coefficients of the polynomial p expressed in two different forms in Equation 1.218 gives

$$\begin{aligned}h^{(0)} &= \alpha_0,\\ h^{(1)}(i_1) &= \alpha_{i_1}, \quad 0 \le i_1 \le M-1\\ &\cdots\cdots\cdots\\ h^{(n)}(i_1, \cdots, i_n) &= \alpha_{i_1 \cdots i_n}, \quad 0 \le i_1, \cdots, i_n \le M-1, \quad 1 \le n \le L\end{aligned} \tag{1.219}$$

The relations 1.219 mean that the constant coefficient of the polynomial p equals the zero-order impulse response and the remaining coefficients $\alpha_{i_1}, \cdots, \alpha_{i_1 \ldots i_n}$ equal the corresponding samples of the nonlinear responses of the first (linear one) and higher orders. Furthermore, observe that the form of the Volterra series operator presented in Equation 1.218 differs from the form in Equations 1.1 and 1.11 only in that there occur finite upper summation limits ($M - 1$) in Equation 1.218 instead of infinite ones as in Equation 1.11. At the same time, this means that the stability conditions such as in inequality 1.98 are fulfilled because of the occurrence of finite lower and upper summation limits. Finally, we conclude that the polynomial $p(x(k), x(k-1), \cdots, x(k-M+1))$ from Theorem 1.2 is simply the Volterra series operator $V: l^\infty \to l^\infty$ with a finite number of components and finite lower and upper summation limits in the defining Equations 1.160 and 1.3.

For further considerations, denote the approximating Volterra series operator V from Theorem 1.1 $\hat{V}$ and rewrite it according to Equation 1.218, as

$$\begin{aligned}(\hat{V}x)(k) &= h^{(0)} + \sum_{i_1=0}^{M-1} h^{(1)}(i_1)x(k-i_1) + \sum_{i_1=0}^{M-1}\sum_{i_2=0}^{M-1} h^{(2)}(i_1, i_2)x(k-i_1)x(k-i_2)\\ &+ \cdots + \sum_{i_1=0}^{M-1}\sum_{i_2=0}^{M-1} \cdots \sum_{i_L=0}^{M-1} h^{(L)}(i_1, i_2, \cdots, i_L)x(k-i_1)x(k-i_2)\cdots x(k-i_L)\end{aligned} \tag{1.220a}$$

and the accompanying stability condition for its nonlinear impulse responses as

$$\sum_{i_1=0}^{M-1}\sum_{i_2=0}^{M-1} \cdots \sum_{i_n=0}^{M-1} \left|h^{(n)}(i_1, i_2, \cdots, i_n)\right| < \infty, \quad n = 1, 2, \cdots, L \tag{1.220b}$$

It follows from our considerations that the operator $\hat{V}$ can be realized, as shown in Figure 1.36.

In Figure 1.36, the linear part of the realizing system is a linear discrete subsystem with memory and transmittance, which can be expressed by the following vector:

$$\begin{bmatrix} 1 \\ z^{-1} \\ \vdots \\ z^{-M+1} \end{bmatrix} \tag{1.221a}$$

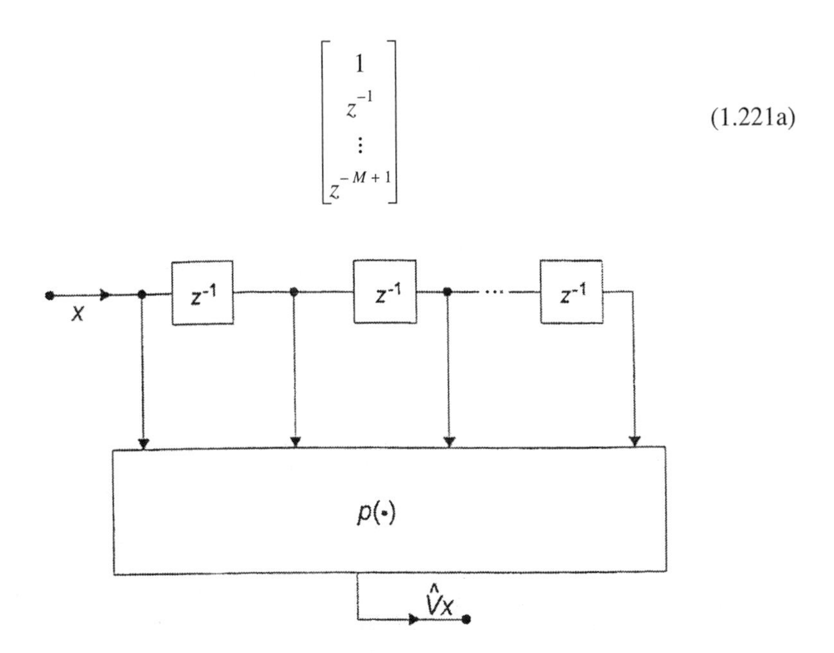

FIGURE 1.36 Realization of the approximating Volterra series operator $\hat{V}$ with z^{-1} meaning operation of shifting in discrete time by one sampling period.

where z stands for the z variable in the one-dimensional Z transform. Furthermore, in the above context, note that the relations

$$\underbrace{X(z)\cdot\begin{bmatrix} 1 \\ z^{-1} \\ \vdots \\ z^{-M+1} \end{bmatrix} = \begin{bmatrix} X(z) \\ X(z)z^{-1} \\ \vdots \\ X(z)z^{-M+1} \end{bmatrix}}_{z\text{-domain}} \Leftrightarrow \underbrace{\begin{bmatrix} x(k) \\ x(k-1) \\ \vdots \\ x(k-M+1) \end{bmatrix}}_{\text{discrete-time domain}} \tag{1.221b}$$

hold, which confirms the form of the linear part in Figure 1.36, with the input signal $x(k)$ shifted in time $(M - 1)$ times.

On the other hand, the nonlinear part of the realization presented in Figure 1.36 is a subsystem without memory. In this subsystem, only the operations of summation and multiplication are performed, according to the formula given by the polynomial p. This kind of mapping is known in the literature as the polynomial readout mapping.[1]

One well known model in the literature used to model linear systems (in the context of spectral analysis of stochastic processes) is the so-called moving average map[22] (MA-map), presented in Figure 1.37.

Observe the similarity of the MA-map structure presented in Figure 1.37 to the realization of the Volterra series approximator of Figure 1.36. The dashed-line

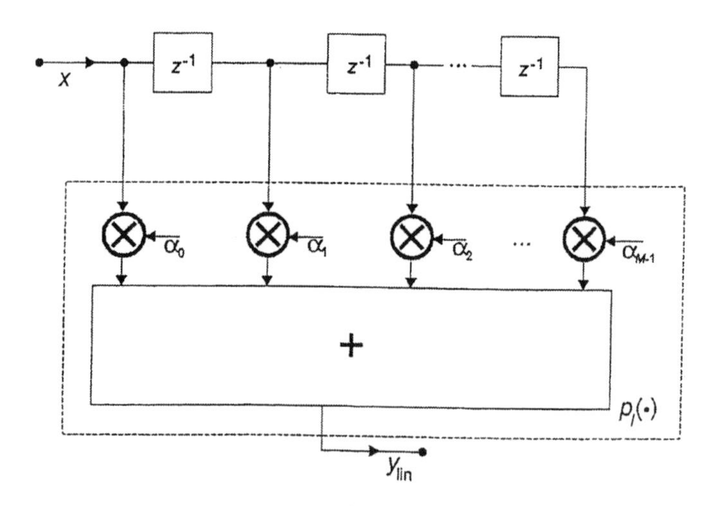

FIGURE 1.37 Structure of the moving average map.

rectangle of Figure 1.37 corresponds to the solid-line rectangle of Figure 1.36. The main difference between the structures presented lies in the fact that the polynomial $p(\cdot)$ in Figure 1.36 is quite general, but the polynomial $p_l(\cdot)$ of Figure 1.37, is a linear function of M variables, $x(k)$, $x(k-1), \cdots, x(k-M+1)$. The similarity of the structure of Figure 1.36 to the structure of the moving average map in Figure 1.37 resulted in naming the first a nonlinear moving average map[1] (NLMA).

The output signal in the structure of Figure 1.37 can be expressed in the following way:

$$\begin{aligned} y_{\text{lin}}(k) &= \alpha_0 x(k) + \alpha_1 x(k-1) + \alpha_2 x(k-2) + \cdots + \alpha_{M-1} x(k-M+1) \\ &= \sum_{i=0}^{M-1} \alpha_i x(k-i) = [x(k)\ x(k-1)\ \cdots\ x(k-M+1)] \begin{bmatrix} \alpha_0 \\ \alpha_1 \\ \vdots \\ \alpha_{M-1} \end{bmatrix} \\ &= p_l(x(k), x(k-1), \ldots, x(k-M+1)) \end{aligned} \tag{1.222}$$

Let us now divide the polynomial $p(\cdot)$ into two parts: a strictly linear one, $p_l(\cdot)$, and a strictly nonlinear one, $p_n(\cdot)$, such that we arrive at

$$\begin{aligned} & p(x(k), x(k-1), \cdots, x(k-M+1)) \\ &= p_l(x(k), x(k-1), \cdots, x(k-M+1)) \\ &\quad + p_n(x(k), x(k-1), \cdots, x(k-M+1)) \end{aligned} \tag{1.223a}$$

where

$$p_l(x(k), x(k-1), \cdots, x(k-M+1)) = \sum_{i=0}^{M-1} \alpha_i x(k-i) \tag{1.223b}$$

and

$$\begin{aligned} p_n(x(k), x(k-1), \cdots, x(k-M+1)) &= \alpha_0 + \sum_{i_1=0}^{M-1}\sum_{i_2=0}^{M-1} \alpha_{i_1 i_2} x(k-i_1)x(k-i_2) \\ &+ \cdots + \sum_{i_1=0}^{M-1}\sum_{i_2=0}^{M-1} \cdots \sum_{i_L}^{M-1} \alpha_{i_1 i_2 \ldots i_L} x(k-i_1)x(k-i_2)\cdots x(k-i_L) \end{aligned} \tag{1.223c}$$

Taking now into account Equations 1.223, and the structures presented in Figures 1.36 and 1.37, we conclude that the Volterra series approximator $\hat{V}x$ from Figure 1.36 can be divided into two mappings: strictly linear and strictly nonlinear moving average maps, as illustrated in Figure 1.38. In more detail, the structure of the strictly nonlinear moving average map is shown in Figure 1.39.

It is worth noting at this point that there is also an engineer's approach to the problem of nonlinear systems possessing finite memory that can be described by the Volterra series. The notion of a discrete-time nonlinear time-invariant causal system with finite memory, described by the Volterra series, has been introduced.[23] In other words, it has been defined such a system of which describing Volterra series contains only a finite number, say, M samples of the input signal. In this case, the Volterra series takes on the following form:

$$\begin{aligned} y(k) &= h^{(0)} + \sum_{i_1=0}^{M-1} h^{(1)}(i_1)x(k-i_1) + \sum_{i_1=0}^{M-1}\sum_{i_2=0}^{M-1} h^{(2)}(i_1, i_2)x(k-i_1)x(k-i_2) \\ &+ \sum_{i_1=0}^{M-1}\sum_{i_2=0}^{M-1}\sum_{i_3=0}^{M-1} h^{(3)}(i_1, i_2, i_3)x(k-i_1)x(k-i_2)x(k-i_3) + \cdots \end{aligned} \tag{1.224}$$

Note that the Volterra series given by Equation 1.224 is a series, that has an infinite number of components but contains only a finite number of samples of the input signal, $x(k)$, $x(k - 1)$, $\cdots$, $x(k - M + 1)$.

Observe that, if the series given by Equation 1.224 converges on some ball

$$B_{M_2} \stackrel{df}{=} \{x \in l^\infty(\mathbb{Z}) \mid \|x\| \le M_2\} \tag{1.225}$$

then it can be approximated on this ball by the series with a finite number of components, that is, by the series of the form 1.220a.

Note that from the engineer's point of view, the parameter M occurring in Equation 1.220a can be viewed as a measure of the memory length of a system

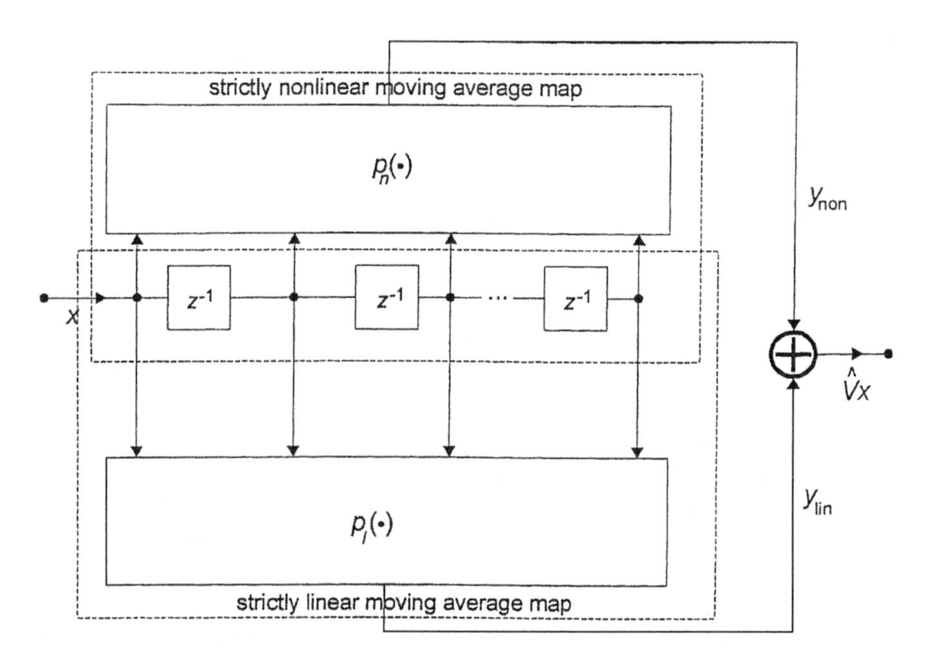

FIGURE 1.38 Division of the NLMA operator into two parts: strictly linear and strictly nonlinear.

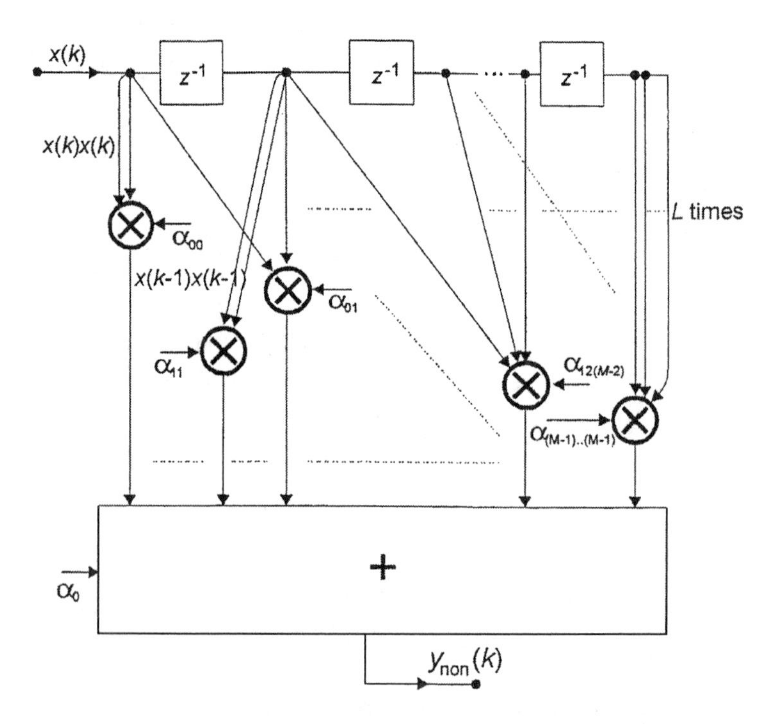

FIGURE 1.39 Structure of the strictly nonlinear moving average operator in detail.

considered.[23] Moreover, the parameter L in Equation 1.220a can be viewed as some measure of how strong the nonlinearity of a system is.

The work[23] is an example of a typical engineer's approach to the problem of modeling nonlinear discrete-time systems by means of description using the discrete Volterra series. This approach is characterized by the following two features:

1. Only a finite number, M, of input signal samples occurs in the Volterra series used for system description,
2. The series used is truncated at the component of the Lth order, being a L-fold sum of components of the form: $h^{(L)}(i_1, i_2, \cdots, i_L)\, x(k - i_1)x(k - i_2)\cdots x(k - i_L)$, (see Equation 1.220a).

Consider now time-invariant operators working on the sequences belonging to the space $l^\infty(\mathbb{Z}_+)$, that is, operators $N : l^\infty(\mathbb{Z}_+) \to l^\infty(\mathbb{Z}_+)$. We show in what follows that Theorems 1.1 and 1.2 are also valid for such operators. For this purpose, we consider the definition FMD2 of the fading memory to try to show that definition FMD1 follows from it for the extended operators N_f (see Equation 1.174) working on the extended sequences such as x_f (see Equation 1.173a). In other words, we proceed in the following way: assume that the operators $N : l^\infty(\mathbb{Z}_+) \to l^\infty(\mathbb{Z}_+)$ considered have the property of fading memory in the sense of the definition FMD2. Next, for these operators, we form the extended operators, according to definition 1.174. Furthermore, we assume that the extended operators work on the extended sequences such as $x_f \in l^\infty(\mathbb{Z})$ given by Equation 1.173a.

From the definition FMD2 for $k = 0$, we get

$$\sup_{0\le\tau\le 0} |x(\tau) - v(\tau)|\, w(0-\tau) < \delta \to |(Nx)(0) - (Nv)(0)| < \varepsilon \qquad (1.226a)$$

or, in another form

$$|x(0) - v(0)|\, w(0) < \delta \to |(Nx)(0) - (Nv)(0)| < \varepsilon \qquad (1.226b)$$

Taking into account the definitions of the extended operators and sequences given by Equations 1.174 and 1.173, respectively, we can rewrite the relation 1.226b in the following form:

$$|x_f(0) - v_f(0)|\, w(0) < \delta \to |(N_f x_f)(0) - (N_f v_f)(0)| < \varepsilon \qquad (1.227a)$$

Moreover, it is obvious from the definitions 1.173 and 1.174 that

$$\sup_{k<0} |x_f(k) - v_f(k)|\, w(-k) \equiv 0 < \delta \to |(N_f x_f)(0) - (N_f v_f)(0)| < \varepsilon \qquad (1.227b)$$

holds. Then, consideration of both relations 1.227a and 1.227b leads to

$$\sup_{k\le 0}|x_f(k)-v_f(k)|w(-k)<\delta \to |(N_f x_f)(0)-(N_f v_f)(0)|<\varepsilon \qquad (1.228)$$

Finally, the relation 1.228 only expresses definition FMD1 restricted, however, to the sequences $x_f(k)$.

Consider now another definition of the extended operator for $N: l^\infty(\mathbb{Z}_+)\to l^\infty(\mathbb{Z}_+)$. Let us name it N_g and define it in the following way:

$$(N_g x_f)(k) \overset{df}{=} (N x_f)(k) = \begin{cases}(Nx)(k) & k\ge 0\\ (Nx_f)(k) & k<0\end{cases} \qquad (1.229a)$$

and

$$(N_g x_{f\tau})(k) \overset{df}{=} (N x_{f\tau})(k) \qquad (1.229b)$$

where $x\in l^\infty(\mathbb{Z}_+)$, $x_f(k)$ is the extension of $x(k)$ given by Equation 1.173a, $x_{f\tau}(k) = x_f(k-\tau)$ is the sequence $x_f(k)$ time-shifted, and N is assumed to have a description, which is valid also for the extended sequences $x_f(k)$ and their time-shifted versions, belonging to $l^\infty(\mathbb{Z})$. Then, under the above assumption, the relations $(N_g x_f)(k) = (Nx_f)(k)$, $k<0$ in Equation 1.229a and $(N_g x_{f\tau})(k) = (Nx_{f\tau})(k)$ in Equation 1.229b make sense.

What do we mean under that the operator N has a description which is valid for the extended sequences x_f and their time-shifted versions $x_{f\tau}$? We explain this in the following example: Let N be given by $(Nx)(k) = x^2(k) + x(k-1)$ with $x(k)\in l^\infty(\mathbb{Z}_+)$ and $(Nx)(0) = x^2(0)$, as it would be calculated with $x(-1) = 0$. Then $(Nx_f)(k) = x^2(k) + x_f(k-1)$ is valid for any set $\{x_f(k), x_f(k-1)\}$ of any sequence $x_f(k)$. Furthermore, $(Nx_{f\tau})(k) = x_{f\tau}^2(k) + x_{f\tau}(k-1) = x^2{}_f(k-\tau) + x_f(k-\tau-1)$ is also valid for any $\tau\in\mathbb{Z}$.

Having defined the operator N_g, we shall check now whether this operator has the fading memory in the sense of definition FMD1, when the original operator $N: l^\infty(\mathbb{Z}_+)\to l^\infty(\mathbb{Z}_+)$ has, in the sense of definition FMD2. Consider again, quite formally, the definition FMD2. We rewrite this definition, introducing a new variable $\tau' = \tau - k$, which leads to

$$\sup_{-k\le\tau'\le 0}|x(k+\tau')-v(k+\tau')|w(-\tau')<\delta \to |(Nx)(k)-(Nv)(k)|<\varepsilon \qquad (1.230a)$$

Note that we can also rewrite relation 1.230a as follows:

$$\sup_{-k\le\tau'\le 0}|x(k+\tau')-v(k+\tau')|w(-\tau')<\delta \to |(Nx)(k+0)-(Nv)(k+0)|<\varepsilon \qquad (1.230b)$$

Further, relation 1.230b can be written with the use of the notion of the time-shifted extended sequences introduced just before. Then, we arrive at

$$\sup_{-k \le \tau' \le 0} |x_{f(-k)}(\tau') - v_{f(-k)}(\tau')| w(-\tau') < \delta \rightarrow |(Nx_{f(-k)})(0) - (Nv_{f(-k)})(0)| < \varepsilon \quad (1.231a)$$

because, having $x_f(k)$ and $v_f(k)$ given by Equations 1.173a and 1.173b, respectively, we get $x_{f(-k)}(\tau') = x_f(\tau' = k) = x(\tau' + k)$, and $v_{f(-k)}(\tau') = v_f(\tau' + k) = v(\tau' + k)$, $\tau' + k \geq 0$.

Note that relation 1.231a is not quite correctly written: instead of the operator N, we should use the operator N_g, because the operator N works only on the sequences belonging to $l^\infty(\mathbb{Z}_+)$, but the sequences $x_{f(-k)}$ and $v_{f(-k)}$ belong to the space $l^\infty(\mathbb{Z})$. The proper operator in relation 1.231a is N_g, working on the sequences belonging to the space $l^\infty(\mathbb{Z})$ according to the rule represented by the operator N. Hence, the relation 1.231a written correctly has the following form:

$$\sup_{-k \le \tau' \le 0} |x_{f(-k)}(\tau') - v_{f(-k)}(\tau')| w(-\tau') < \delta \rightarrow |(N_g x_{f(-k)})(0) - (N_g v_{f(-k)})(0)| < \varepsilon \quad (1.231b)$$

Take now into account any sequence belonging to some ball B of the space $l^\infty(\mathbb{Z})$ such that this sequence is identical with a certain sequence $x(\tau)$ for all $\tau \geq -k$, as shown in Figure 1.40c. Moreover, observe that when k increases, as in Figure 1.40d, the sequence $x_{f(-k)}(\tau)$ becomes identical with the same or another sequence of the ball B of the space $l^\infty(\mathbb{Z})$. This process can be continued into infinity, which allows us to conclude that each of the sequences $x(\tau) \in l^\infty(\mathbb{Z})$ can be viewed as a sequence $x_{f(-k)}(\tau)$ for $k = \infty$.

Because relation 1.231b holds for every k, so it also holds for $k = \infty$. This means, when taking into account the interpretation of the sequences $x_{f(-k)}$ for $k = \infty$ just given, that

$$\sup_{\tau \le 0} |x(\tau) - v(\tau)| w(-\tau) < \delta \rightarrow |(N_g x)(0) - (N_g v)(0)| < \varepsilon \quad (1.231c)$$

holds for all the sequences $x(\tau)$ of the ball B of the space $l^\infty(\mathbb{Z})$. This allows us to conclude that when the operator $N : l^\infty(\mathbb{Z}_+) \rightarrow l^\infty(\mathbb{Z}_+)$ possesses the fading memory in the sense of definition FMD2, then its extended operator N_g possesses the fading memory, in the sense of definition FMD1, too.

Comparing now the extended operators N_f and N_g, we see why the first cannot be used to reformulate Theorems 1.1 and 1.2 for the sequences belonging to the space $l^\infty(\mathbb{Z}_+)$. First of all, the operator N_f does not possess the fading memory for all the sequences of some ball $B \subset l^\infty(\mathbb{Z})$. It possesses the fading memory only for the extended sequences x_f (see relation 1.228). Moreover, the operator N_f is not a time-invariant operator.

We explain the latter in more detail in what follows. Let us clarify first what we mean by the property of time-invariance in the case of operators working on the sequences belonging to the space $l^\infty(\mathbb{Z}_+)$. Illustrated in Figures 1.41 and 1.42, note that the sequences (c) and (e) in Figure 1.41, and the sequences (a) and (c) in Figure 1.42, represent the sequences $(U_\tau Nx)(k)$ and $(NU_\tau x)(k)$, which should be equal to each other, according to definition 1.159 of the operator time-invariance property.

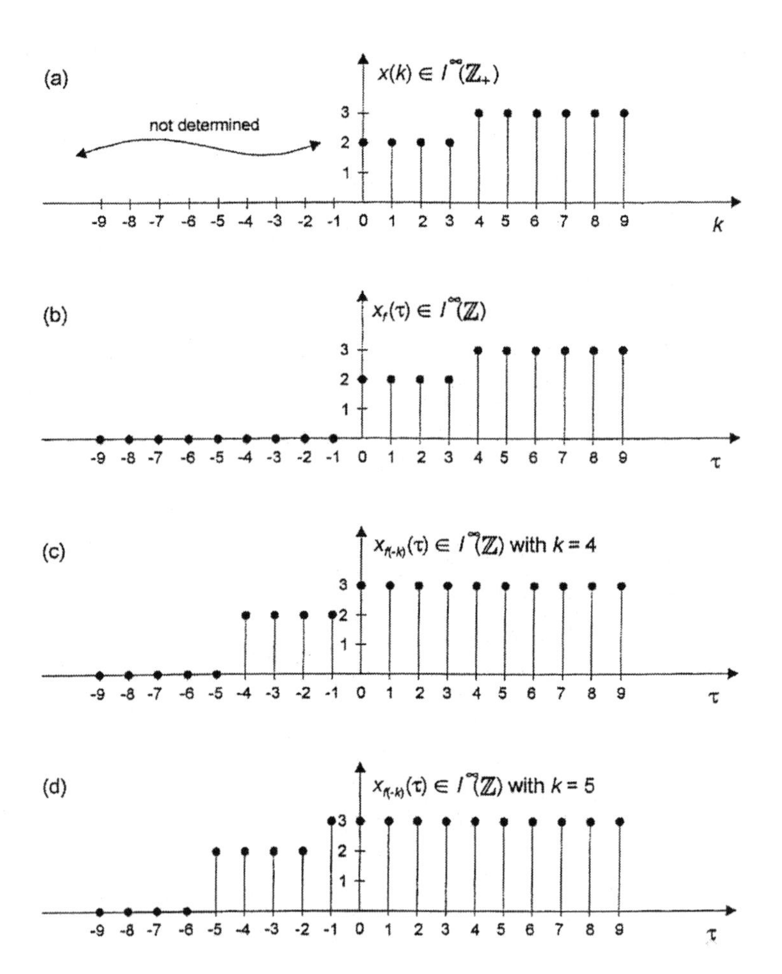

FIGURE 1.40 Illustration of construction of the sequences $x_{f(-k)}$: (a) the original sequence $x(k) \in l^\infty(\mathbb{Z}_+)$, (b) the extended sequence $x_f(\tau) \in l^\infty(\mathbb{Z})$, (c) the extended time-shifted sequence for $k = 4$, (d) the extended time-shifted sequence for $k = 5$.

In this context, observe that the sequences (c) and (e) of Figure 1.41 represent the same sequence. That is, operator N is time-invariant when the time-shifting is to the right. However, when the time-shifting is to the left, as in Figures 1.42a and c, the corresponding sequences are not equal to each other. So, then the time-invariance of the operator N does not hold. In conclusion, when we say that an operator $N : l^\infty(\mathbb{Z}_+) \to l^\infty(\mathbb{Z}_+)$ is time-invariant, this means the fulfillment of the time-invariance definition 1.159, with only positive values of τ.

Similarly, the operator N_f is time-invariant with respect to the time-shifting to the right, but not with respect to the time-shifting to the left. This is illustrated in Figures 1.43 and 1.44.

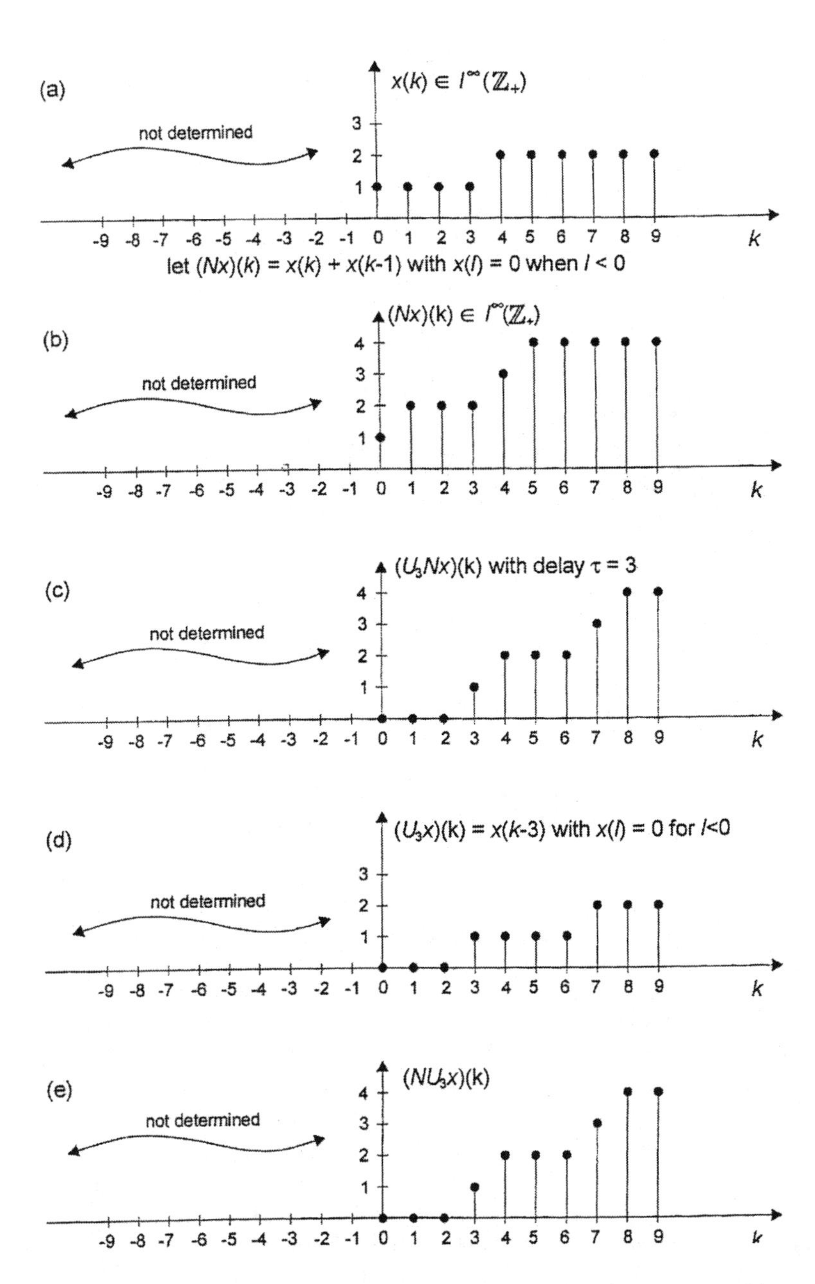

FIGURE 1.41 Illustration of the property of time-invariance of operators working on the sequences belonging to $l^\infty(\mathbb{Z}_+)$: (a) original input sequence, (b) output sequence $(Nx)(k)$, (c) sequence (b) time-shifted, (d) sequence (a) time-shifted, (e) output sequence for the input sequence (d).

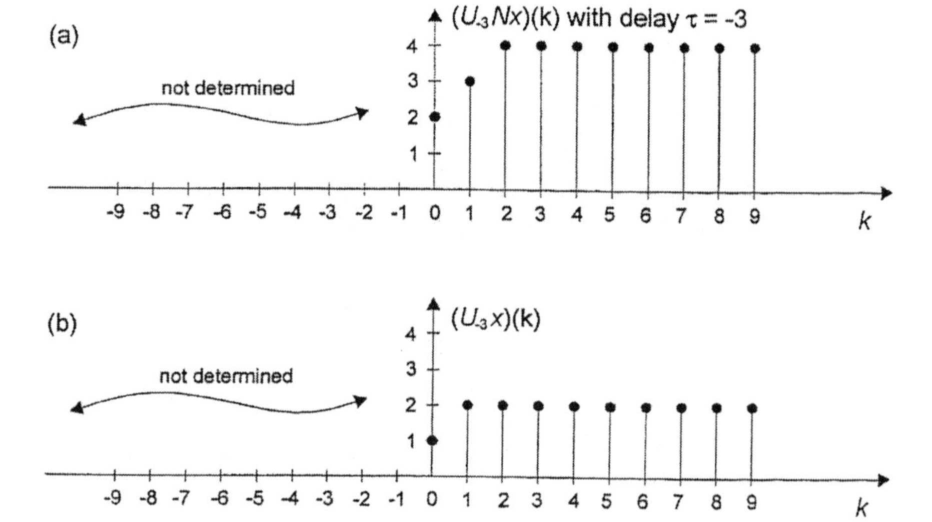

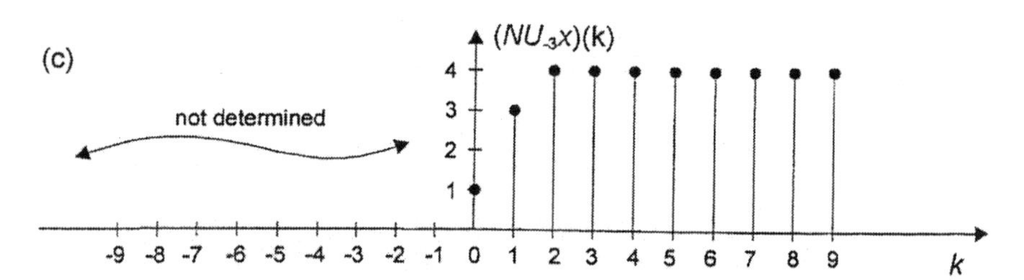

FIGURE 1.42 Checking the operator time-invariance property when time-shifting to the left is performed: (a) the sequence (b) of Figure 1.41 time-shifted to the left by $\tau = -3$, (b) the sequence (a) of Figure 1.41 time-shifted to the left by $\tau = -3$, (c) the output sequence for the input sequence (b) of this figure.

Note that the sequences (c) and (e) of Figure 1.43 represent the same sequences, but not the sequences (a) and (c) of Figure 1.44. Because the operator N_f represents a mapping from $l^\infty(\mathbb{Z})$ into itself, the time-invariance definition 1.159 should hold for any τ, both positive as well as negative ones. However, this is not the case here. The operator N_f cannot be regarded as a time-invariant one.

What is lacking in the operator N_f is present in the operator N_g. That is, the operator N_g behaves like the operator N_f for the time-shifting to the right, but differently for the time-shifting to the left. The time-shifting of the operator N_g to the left is illustrated in Figures 1.44d and e. It follows from Figures 1.44d and e that the sequences $(U_{-3}N_g x_f)(k)$ and $(N_g U_{-3} x_f)(k)$ are identical. This means that the operator N_g is a time-invariant mapping from $l^\infty(\mathbb{Z})$ into itself.

With this knowledge we are able to reformulate Theorems 1.1 and 1.2 for the $N : l^\infty(\mathbb{Z}_+) \to l^\infty(\mathbb{Z}_+)$ operators in the following way:

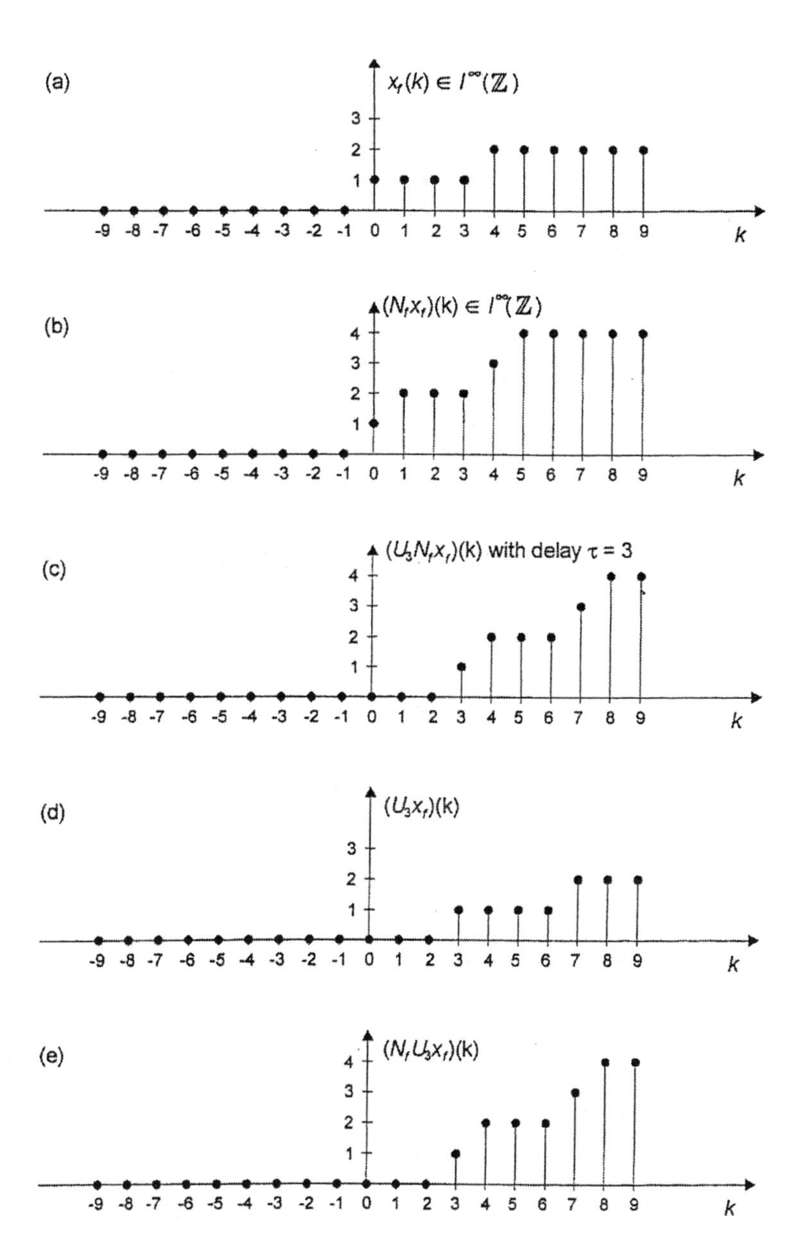

FIGURE 1.43 Checking time-invariance property of the operator N_f based on the operator N of Figure 1.41: (a) the extended sequence of that of Figure 1.41a, (b) the output sequence of the extended operator N_f, (c) the sequence (b) time-shifted, (d) the sequence (a) time-shifted, (e) the output sequence for the input sequence (d).

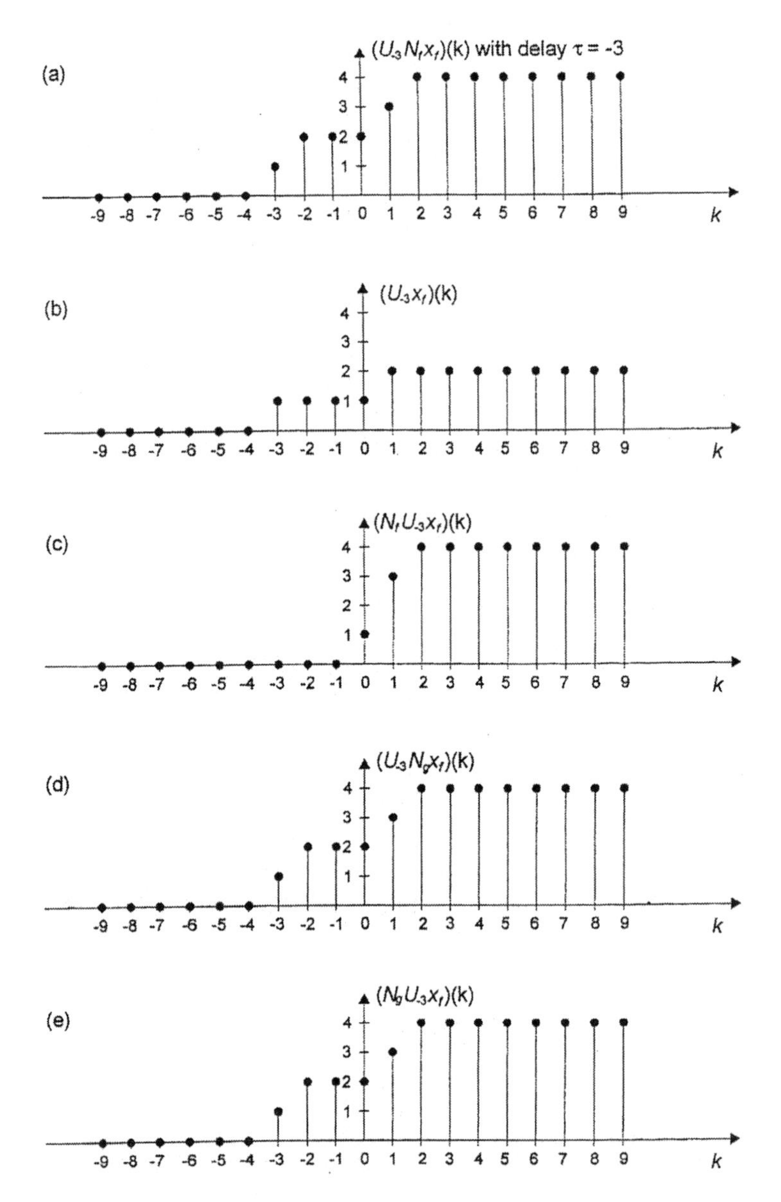

FIGURE 1.44 Checking time-invariance property of the operators N_f and N_g when time-shifting to the left is performed: (a) the sequence (b) of Figure 1.43 time-shifted to the left by $\tau = -3$, (b) the sequence (a) of Figure 1.43 time-shifted to the left by $\tau = -3$, (c) the output sequence of the operator N_f for the input sequence (b) of this figure, (d) the output sequence $(U_{-3}N_gx_f)(k)$, (e) the output sequence $(N_gU_{-3}x_f)(k)$.

Theorem 1.1 (for $x \in l^\infty(\mathbb{Z}_+)$)
Let $\varepsilon > 0$ and

$$B_+ \overset{df}{=} \{x \in l^\infty(\mathbb{Z}_+) \mid \|x\| \le M_3\} \tag{1.232a}$$

be a ball in $l^\infty(\mathbb{Z}_+)$ of radius M_3. If the $N : l^\infty(\mathbb{Z}_+) \to l^\infty(\mathbb{Z}_+)$ is a time-invariant operator possessing the fading memory in the sense of definition FMD2 on the ball B_+, then there exists such a Volterra series operator V given by Equations 1.1 and 1.11 with a finite number of components and of which nonlinear responses fulfill the stability condition (1.98) (with lower summation limits equal to zero) that the following inequality for extended operators and sequences

$$\|(N_g x_f)(k) - (V_g x_f)(k)\| \le \varepsilon, \quad k \in \mathbb{Z} \tag{1.232b}$$

or equivalently for $k \ge 0$

$$\|(Nx)(k) - (Vx)(k)\| \le \varepsilon, \text{ with } x(l) = 0, \text{ when } l < 0 \tag{1.232c}$$

holds for all $x \in B_+$.

Theorem 1.2 (for $x \in l^\infty(\mathbb{Z}_+)$)
Let B_+ be a ball in $l^\infty(\mathbb{Z}_+)$ of radius M_3 as given by definition 1.232a. If $N : l^\infty(\mathbb{Z}_+) \to l^\infty(\mathbb{Z}_+)$ is a time-invariant operator possessing the fading memory in the sense of definition FMD2 on the ball B_+, then, for any $\varepsilon > 0$, there exists such a polynomial p of M variables: $\mathbb{R}^M \to \mathbb{R}$ that the following:

$$\|(N_g x_f)(k) - p_g(x_f(k), x_f(k-1), \cdots, x_f(k-M+1))\| < \varepsilon \tag{1.233a}$$

or equivalently for $k \ge 0$

$$\|(Nx)(k) - p(x(k), x(k-1), \cdots, x(k-M+1))\| < \varepsilon, \text{ with } x(l) = 0, \text{ when } l < 0 \tag{1.233b}$$

holds for all $x \in B_+$ and with $y(k) = p(x(k), x(k-1), \cdots, x(k-M+1))$ considered as a mapping from $l^\infty(\mathbb{Z}_+)$ into itself.

Note that the operators V_g and p_g inequalities 1.232b and 1.233a, respectively, are the extended operators of V and p, and are defined by the defining Equations 1.229.

Concluding consideration of Theorems 1.1 and 1.2 (for operators $N : l^\infty(\mathbb{Z}_+) \to l^\infty(\mathbb{Z}_+)$), we denote similarly, the approximating Volterra series operator V in definition 1.232c by $\hat{V}$ and write the following:

$$
\begin{aligned}
(\hat{V}x)(k) &= p(x(k), x(k-1), \cdots, x(k-M+1)) \\
&= h^{(0)} + \sum_{i_1=0}^{M-1} h^{(1)}(i_1)x(k-i_1) + \sum_{i_1=0}^{M-1}\sum_{i_2=0}^{M-1} h^{(2)}(i_1, i_2)x(k-i_1)x(k-i_2) \\
&+ \cdots + \sum_{i_1=0}^{M-1}\sum_{i_2=0}^{M-1}\cdots\sum_{i_L=0}^{M-1} h^{(L)}(i_1, i_2, \cdots, i_L)x(k-i_1)x(k-i_2)\cdots x(k-i_L)
\end{aligned}
\tag{1.234a}
$$

with $x(l) = 0$, when $l < 0$ or, in other words, substituting $x(k - i_p)$, $p = 1, 2, \cdots$, L, by zero when $k - i_p$ eventually takes a negative value. This allows us to put Equation 1.234a into an alternative form:

$$
\begin{aligned}
(\hat{V}x)(k) &= h^{(0)} + \sum_{i_1=0}^{\min\{k, M-1\}} h^{(1)}(i_1)x(k-i_1) \\
&+ \sum_{i_1=0}^{\min\{k, M-1\}}\sum_{i_2=0}^{\min\{k, M-1\}} h^{(2)}(i_1, i_2)x(k-i_1)x(k-i_2) \\
&+ \cdots + \sum_{i_1=0}^{\min\{k, M-1\}}\sum_{i_2=0}^{\min\{k, M-1\}}\cdots\sum_{i_L=0}^{\min\{k, M-1\}} h^{(L)}(i_1, i_2, \cdots, i_L)x(k-i_1)x(k-i_2)\cdots x(k-i_L)
\end{aligned}
\tag{1.234b}
$$

The results obtained using the notion of fading memory can be related to those achieved by the use of the notion of approximately-finite memory. The theorem expressing this relation was given by Park and Sandberg.[4] For scalar sequences, it can be formulated in the following way:

Theorem 1.4

Let B_+ given by definition 1.232a be any closed ball in $l^\infty(\mathbb{Z}_+)$. Then a causal and time-invariant operator $N : l^\infty(\mathbb{Z}_+) \to l^\infty(\mathbb{Z}_+)$, has fading memory on B_+ if and only if it possesses approximately-finite memory in the sense of the definition AFM1 and the functional F_s associated with the operator N, defined as

$$
F_s x_k \overset{df}{=} (Nx)(k), \text{ where } x_k = \{x(k), x(k-1), \cdots, x(0), 0, 0, 0, \cdots\} \tag{1.235}
$$

is continuous on B_+ for each $k \in \{1, 2, 3, \cdots\}$.

Before going further, observe that the mapping N in Theorem 1.4 can be defined more generally as $N : l^\infty(\mathbb{Z}_+) \to C_R(\mathbb{Z}_+)$, where C_R means the collection of all $\mathbb{R}$-valued mappings defined on $\mathbb{Z}_+$; the definition of the mapping N given above occurs in the original formulation by Park and Sandberg.[4] Clarification of the definition of the functional F_s is also needed. If we denote by $c_s(\mathbb{Z}_+)$ the subset of $l^\infty(\mathbb{Z}_+)$, which consists of all the sequences with at most finitely many nonzero terms, then F_s can be assumed to be a mapping $F_s : c_s(\mathbb{Z}_+) \to \mathbb{R}$. Moreover, the functional F_s is associated uniquely with the causal time-invariant operator $N : l^{\infty}(\mathbb{Z}_+) \to C_R\,(\mathbb{Z}_+)$. The rule here is given by Equation 1.235.

Because of the importance of Theorem 1.4 for our considerations, we shall repeat now its proof, after Park and Sandberg.[4] For this purpose, let us denote by M_s a positive real number greater than the radius of B_+, that is $M_s > M_3$. Furthermore, assume that the operator $N : l^{\infty}(\mathbb{Z}_+) \to l^{\infty}(\mathbb{Z}_+)$ possesses the fading memory. Using relation 1.168 with $v(\tau) = (W_{k,a,}x)(\tau)$, we get

$$\sup_{0 \le \tau \le k} |x(\tau) - (W_{k,a}x)(\tau)| w(k-\tau) < \delta \to |(Nx)(k) - (NW_{k,a}x)(k)| < \varepsilon \quad (1.236a)$$

Of course inequaltiy 1.236a is not true for every value of the parameter a. Because this inequality should hold independently of the choice of $x \in B_+$, the only way to get this to hold is by choosing the parameter a properly. In this context, inequality 1.236a can be rewritten as

$$\sup_{0 \le \tau \le k-a} |x(\tau)| w(k-\tau) < \delta \to |(Nx)(k) - (NW_{k,a}x)(k)| < \varepsilon \qquad (1.236b)$$

because $x(\tau) = (W_{k,a,}x)(\tau)$, for $\tau = k, k-1, \cdots, k-a$. On the other hand, $|x(\tau)| < M_s$ for any τ. If we choose the parameter a such that $w(a) < \delta/M_s$ will hold, then

$$\sup_{0 \le \tau \le k-a} |x(\tau)| w(k-\tau) < M_s \cdot \frac{\delta}{M_s} = \delta \qquad (1.236c)$$

will also hold. This, of course, will mean that relation 1.236b is true for the values of a found. That is,

$$|(Nx)(k) - (NW_{k,a}x)(k)| < \varepsilon \qquad (1.236d)$$

holds for the parameter a fulfilling the inequality $w(a) < \delta/M_s$. Finally, we observe that inequality 1.236d is simply the definition AFM1 of the approximately-finite memory.

To complete the first part of the proof, we need to show the continuity of the associated functional F_s. For this purpose, we assume that sequences x and $v \in B_+$ and

$$\sup_{\tau \in \mathbb{Z}_+} |x(\tau) - v(\tau)| < \delta \qquad (1.237a)$$

holds. Furthermore, it follows from inequality 1.237a, and from the fact that $0 < w(k-\tau) \le 1$ for $k - \tau \in \mathbb{Z}_+$, that

$$\sup_{0 \le \tau \le k} |x(\tau) - v(\tau)| w(k-\tau) < \delta \qquad (1.237b)$$

also holds. Because of relation 1.168, we have

$$\sup_{\tau \in \mathbb{Z}_+} |x(\tau) - v(\tau)| < \delta \rightarrow |(Nx)(k) - (Nv)(k)| \tag{1.237c}$$

for each $k \in \{1, 2, 3, \cdots\}$. Finally, observe that relation 1.237c is the continuity definition for the functional F_s, because $F_s x_k = (Nx)(k)$.

To prove the converse part of Theorem 1.4, we need another result from the paper[4] by Park and Sandberg, which can be formulated for scalar sequences in the following way:

Theorem 1.5

Let denote $N : l^\infty(\mathbb{Z}_+) \rightarrow l^\infty(\mathbb{Z}_+)$ as any causal and time-invariant operator, and let B_+ given by 1.232a be any closed ball in $l^\infty(\mathbb{Z}_+)$. Then the statements given below are equivalent.

1. The operator N possesses approximately-finite memory on the ball B_+ in the sense of the definition AFM1 and the functional $N(\cdot)(k) : l^\infty(\mathbb{Z}_+) \rightarrow \mathbb{R}$ is continuous on the ball B_+ for each $k \in \mathbb{N} = \{1, 2, 3, \cdots\}$.
2. For any $\varepsilon > 0$, there exist such $m \in \mathbb{N}$, $a \in \mathbb{N}$, a real m-vector (consisting of m elements) $\mathbf{d}$, a real $m \times (a + 1)$ (having m rows and $(a + 1)$ columns) matrix $\mathbf{C}$, and a lattice map L: $\mathbb{R}^m \rightarrow \mathbb{R}$ that

$$|(Nx)(k) - L(\mathbf{d} + \mathbf{C} \cdot (\mathbf{P}_a x)(k))| < \varepsilon,\ k \in \mathbb{Z}_+ \tag{1.238a}$$

holds for all $x \in B_+$. Moreover, the map $\mathbf{P}_a$ in (1.238a) means picking a vector having $(a + 1)$ elements from a given sequence $x(k)$, according to the relation

$$(\mathbf{P}_a x)(k) = [x(k), x(k-1), \cdots, x(k-a)]^{\mathrm{T}} \tag{1.238b}$$

with T standing in relation 1.238b for the operation of transposition. Also, it is assumed in 1.238b that $x(l) = 0$ when $l < 0$, where l means $k - i$, $0 \le i \le a$. Furthermore, the lattice map L in relation 1.238a generates its output value $L\mathbf{z}$ from the elements of the vector $\mathbf{z} = [z_1, z_2, \cdots, z_m]^{\mathrm{T}}$ by a finite number of so-called lattice operations. The definition of the latter is as follows: let y and z be real numbers, then the lattice operations are given by

$$y \vee z = \max(y, z) \tag{1.238c}$$

and

$$y \wedge z = \min(y, z) \tag{1.238d}$$

Upon examination of Theorem 1.5, observe that something similar to picking from a sequence a vector of the kind shown in 1.238b occurred also in Equation

1.218. Using the notation of 1.238b, we could also write $p(x(k), x(k-1), \cdots, x(k-M+1))$ as $p(\mathbf{P}_{M-1}x)$. From the latter, it follows that the relation between the parameters a and M is $a = M - 1$.

The continuity of the functional $N(\cdot)(k) : l^{\infty}(\mathbb{Z}_{+}) \to \mathbb{R}$ for $k \in \mathbb{N}$, as used in Theorem 1.5, means the same as the continuity of the functional F_s from Theorem 1.4. This is because of definition 1.235 and the relation between the sequences x_k and $x(k)$, expressed by it.

The proof of Theorem 1.5 for vector-valued sequences can be found in the paper[4] by Park and Sandberg. Here, this proof is omitted.

Let us now start with the converse part of the proof of Theorem 1.4. For this purpose, we assume that the operator N possesses approximately-finite memory in the sense of the definition AFM1 and the functional F_s or, equivalently, the functional $N(\cdot)(k) : l^{\infty}(\mathbb{Z}_{+}) \to \mathbb{R}$ for $k \in \mathbb{N}$ is continuous on B_{+} for each k. Furthermore, take $\varepsilon > 0$ and any sequences $x(k)$ and $v(k)$ belonging to the ball B_{+}. Then using Theorem 1.5, we can say that there are $m \in \mathbb{N}$, $a \in \mathbb{N}$, a real m-vector $\mathbf{d}$, a real $m \times (a+1)$ matrix $\mathbf{C}$, and a lattice map L: $\mathbb{R}^{m} \to \mathbb{R}$ such that

$$\left|(Nx)(k) - (\hat{N}_{x})(k)\right| < \frac{\varepsilon}{3} \tag{1.239a}$$

and

$$\left|(Nv)(k) - (\hat{N}v)(k)\right| < \frac{\varepsilon}{3}, \; k \in \mathbb{Z}_{+} \tag{1.239b}$$

where the approximating operator $N : l^{\infty}(\mathbb{Z}_{+}) \to l^{\infty}(\mathbb{Z}_{+})$ for the sequences $x(k)$ and $v(k)$ is given by

$$(\hat{N}x)(k) \;=\; L(\mathbf{d} + \mathbf{C} \cdot (\mathbf{P}_{a}x)(k)) \tag{1.239c}$$

and

$$(\hat{N}v)(k) \;=\; L(\mathbf{d} + \mathbf{C} \cdot (\mathbf{P}_{a}v)(k)) \tag{1.239d}$$

respectively.

We recall now another fact in the paper[4] by Park and Sandberg regarding a certain property of the lattice map $L(\mathbf{d} + \mathbf{C}\cdot) : \mathbb{R}^{(a+1)} \to \mathbb{R}$; that is, its uniform continuity on the cube $\langle -M_s, M_s \rangle^{(a+1)}$. More precisely, if we choose δ_0 such that $\mathbf{z}_1$ and $\mathbf{z}_2$ lie in the cube $\langle -M_s, M_s \rangle^{(a+1)}$ and the length between them $|\mathbf{z}_1 - \mathbf{z}_2| \overset{df}{=} \max_{1 \le i \le a+1} |z_{1i} - z_{2i}| < \delta_0$, then

$$|\mathbf{z}_1 - \mathbf{z}_2| \le \delta_0 \to |L(\mathbf{d} + \mathbf{C}\mathbf{z}_1) - L(\mathbf{d} + \mathbf{C}\mathbf{z}_2)| < \frac{\varepsilon}{3} \tag{1.240}$$

Since relation 1.240 holds, consider any decreasing sequence $w : \mathbb{Z}_+ \rightarrow (0, 1>)$ such that $\lim_{k \to \infty} w(k) = 0$. Furthermore, take $\delta = \delta_0 w(a)$ and arbitrary x and v from the ball B_+, and arbitrary $k \in \mathbb{Z}_+$, too. Note then

$$|x(\tau) - v(\tau)| w(k - \tau) < \delta, \tau = 0, 1, \cdots, k \tag{1.241a}$$

implies

$$|(\mathbf{P}_a x)(k) - (\mathbf{P}_a v)(k)| < \delta_0 \tag{1.241b}$$

The relation 1.241 holds because it follows from the definition of the absolute value of a vector and from the inequality 1.241a that

$$|(\mathbf{P}_a x)(k) - (\mathbf{P}_a v)(k)| = \max_{0 \le i \le a} |x(k - i) - v(k - i)| < \frac{\delta}{w(i_{\max})} = \frac{\delta_0 w(a)}{w(i_{\max})} \tag{1.242a}$$

where i_{max} stands for this index for which max $|x(k - i) - v(k - i)|$ occurs. Furthermore, because the index i_{max} can take a value from the range $0 \le i_{max} \le a$, and $w(i) > w(i + 1)$ holds for each $i \in \mathbb{Z}_+$, we can write

$$\frac{w(a)}{w(i_{\max})} \le 1 \tag{1.242b}$$

Finally, taking into account inequality 1.242b in 1.242a, we get inequality 1.241b. The inequality 1.242a holds for indices i that fulfil the inequality $k - i < 0$. For these indices, however, $x(k - i) = v(k - i) = 0$, which indicates that i_{max} cannot come from the range for which $k - i < 0$ holds.

Identifying $\mathbf{z}_1$ with $\mathbf{P}_a x$ and $\mathbf{z}_2$ with $\mathbf{P}_a v$ and taking into account the definitions 1.239c and 1.239d, we conclude from relation 1.240 that 1.241b implies

$$|(\hat{N} x)(k) - (\hat{N} v)(k)| < \frac{\varepsilon}{3} \tag{1.243a}$$

To summarize, observe that because the inequalities 1.239a and 1.239b hold for all $k \in \mathbb{Z}_+$, and the relation 1.243a holds as well, we can conclude that $\sup_{0 \le \tau \le k} |x(\tau) - v(\tau)| w(k - \tau) < \delta$ implies

$$\begin{aligned} |(Nx)(k) - (Nv)(k)| &\le |(Nx)(k) - (\hat{N} x)(k)| \\ &+ |(\hat{N} x)(k) - (\hat{N} v)(k)| + |(\hat{N} v)(k) - (Nv)(k)| < \varepsilon \end{aligned} \tag{1.243b}$$

The latter is the definition FMD2 of fading memory. This completes the proof of Theorem 1.4.

The important conclusion that follows from Theorem 1.4, is that the definition FMD2 of the fading memory and the definition AFM1 of the approximately-finite memory are equivalent only when some continuity condition is added to the latter.

Also worth noting is that the fading memory property for mappings $N: l^\infty(\mathbb{Z}_+) \to l^\infty(\mathbb{Z}_+)$ is equivalent to the continuity of the associated functionals F_s, defined by 1.235, with respect to the norm $\sup_{k \geq 0} |x(k)| w(k)$. To show this, assume that an operator $N: l^\infty(\mathbb{Z}_+) \to l^\infty(\mathbb{Z}_+)$ possesses the fading memory property according to the definition FMD2. Then we can write

$$\sup_{0 \leq \tau \leq k} |x(\tau) - v(\tau)| w(k - \tau) < \delta \to |(Nx)(k) - (Nv)(k)| < \varepsilon \tag{1.244a}$$

To proceed further, note that the order of elements in the sequence x_k is reversed compared to the order of elements in the sequence $x(k)$ for the indices from 0 to k, that is, $x_k(0) = x(k)$, $x_k(1) = x(k-1)$, $\cdots$, $x_k(k) = x(0)$. Keeping this in mind, and taking into account definition 1.235, we can rewrite relation 1.244a in the following form:

$$\sup_{\tau \in \mathbb{Z}_+} |x_k(\tau) - v_k(\tau)| w(\tau) < \delta \to |F_s x_k - F_s v_k| < \varepsilon \tag{1.244b}$$

With the definition of the norm

$$\|x\|_{\omega 1} \overset{df}{=} \sup_{\tau \in \mathbb{Z}_+} |x(i)| w(i) \tag{1.244c}$$

we get from relation 1.244b

$$\|x_k - v_k\|_{\omega 1} < \delta \to |F_s x_k - F_s v_k| < \varepsilon \tag{1.244d}$$

Finally, we conclude that relation 1.244d is the definition of the uniform continuity of the functional F_s on the set $c_s(\mathbb{Z}_+) \cap B_+$.

We can also proceed conversely. Starting from relation 1.244d, with the corresponding interpretations of the sequences x_k and x, and keeping in mind definition 1.235, we arrive at relation 1.244a, that is, the fading memory definition FMD2.

Knowing the relation between the definitions of the fading memory and of the approximately-finite memory, we can formulate equivalently Theorems 1.1 and 1.2 (for $x \in l^\infty(\mathbb{Z}_+)$) using the phrase "possessing the approximately-finite memory in the sense of the definition AFM1 together with the continuity property of the functional $N(\cdot)(k): l^\infty(\mathbb{Z}_+) \to \mathbb{R}$ for each $k \in \mathbb{N}$" instead of the phrase "possessing the fading memory in the sense of the definition FMD2." With this comment, we conclude Section 1.9. The next section will be devoted to the Volterra series representations for special input sequences, binary sequences.

1.10 DISCRETE VOLTERRA SERIES FOR BINARY SIGNALS

Binary signals are most often used in digital systems. If the systems considered are nonlinear and possess the fading memory property, their behavior can be described by the Volterra series approximator, as shown in the previous section. With the specific form of binary signals, however, the form of the approximator, given by Equation 1.220a, can be further reduced. This section is devoted to this task.

Assume that the input sequence $x(k)$ in Equation 1.220a is a binary sequence; that is, its elements have the following form:

$$x(k) = a + bx_B(k) \tag{1.245}$$

where a and b are constants, and $x_B(k)$ takes on the value 0 or 1. Thus, $x(k)$ takes on only two distinct values: a, for $x_B(k) = 0$, or $a + b$, for $x_B(k) = 1$.

Equation 1.245) is a general form expressing all the possible forms of binary signals, as illustrated in Figure 1.45 by means of three characteristic examples.

To proceed further, we observe now that the multiplication of any element of the sequence $x(k)$ by itself produces an element of the general form expressed by Equation 1.245.

That is, we get

$$x(k) \cdot x(k) = (a + bx_B(k))^2 = -a(a+b) + (2a+b)(a + bx_B(k)) \tag{1.246a}$$

because $x_B(k) \cdot x_B(k) = x_B(k)$. Furthermore, we can rewrite Equation 1.246a as

$$x(k) \cdot x(k) = a_1 + b_1 x(k) \tag{1.246b}$$

where the constants a_1 and b_1 are given by

$$a_1 = 2a + b \tag{1.246c}$$

and

$$b_1 = -a(a+b) \tag{1.246d}$$

Using the notion of symmetric impulse responses (see the defining Equation 1.15), we can rewrite Equation 1.220a for the Volterra series approximator $\hat{V}$ as

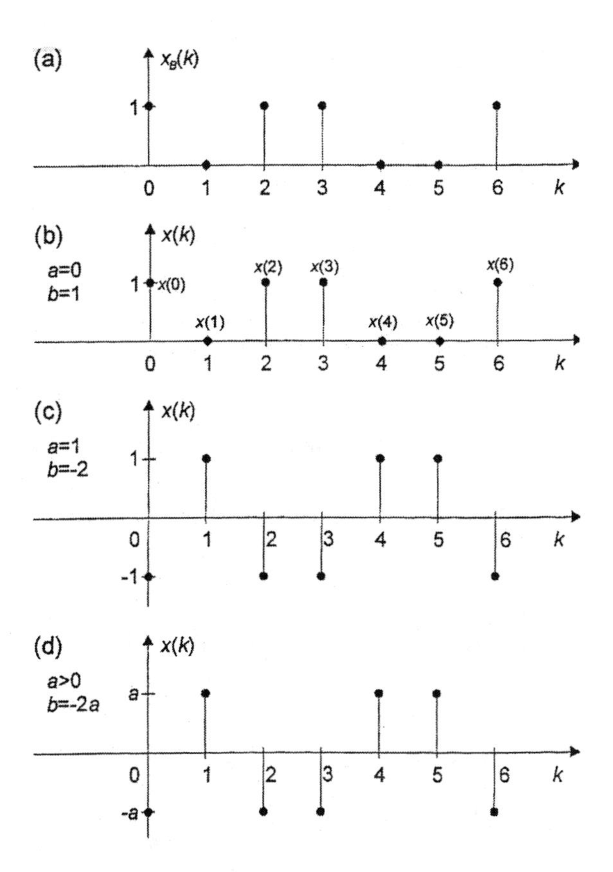

FIGURE 1.45 Illustration of the formula 1.245 describing any binary signal, (a) assumed $x_B(k)$, (b) $x(k)$ for $a = 0$ and $b = 1$, (c) $x(k)$ for $a = 1$ and $b = -2$, (d) $x(k)$ for $a > 0$ and $b = -2a$.

$$(\hat{V}x)(k) = h_{sym}^{(0)} + \sum_{i_1=0}^{M-1} h_{sym}^{(1)}(i_1)x(k-i_1) + \sum_{i_1=0}^{M-1}\sum_{i_2=0}^{M-1} h_{sym}^{(2)}(i_1, i_2)x(k-i_1)x(k-i_2)$$
$$+ \cdots + \sum_{i_1=0}^{M-1}\sum_{i_2=0}^{M-1}\cdots\sum_{i_L=0}^{M-1} h_{sym}^{(L)}(i_1, i_2, \cdots, i_L)x(k-i_1)x(k-i_2)\cdots x(k-i_L) \tag{1.247a}$$

Moreover, the Volterra series approximator $\hat{V}$ can be equivalently expressed by means of the triangular impulse responses (see the defining Equations 1.19). Then it takes on the following form:

$$(\hat{V}x)(k) = h_{tri}^{(0)} + \sum_{i_1=0}^{M-1} h_{tri}^{(1)}(i_1)x(k-i_1) + \sum_{i_1=0}^{M-1}\sum_{i_2=0}^{M-1} h_{tri}^{(2)}(i_1, i_2)x(k-i_1)x(k-i_2)$$
$$+\cdots+\sum_{i_1=0}^{M-1}\sum_{i_2=0}^{M-1}\cdots\sum_{i_L=0}^{M-1} h_{tri}^{(L)}(i_1, i_2, \cdots, i_L)x(k-i_1)x(k-i_2)\cdots x(k-i_L) \tag{1.247b}$$

Observe that the representation 1.247b takes into account only distinct multiples of signal samples at different time instants; in other words, the components such as $x(k - i)x(k - j) = x(k - j)x(k - i)$, $x(k - i)x(k - j)\, x(k - l) = x(k - j)\, x(k - i)\, x(k - l) = \cdots = x(k - l)\, x(k - j)\, x(k - i)$ and so on, occur only once in the expression. This is so because, for example, when $h^{(2)}{}_{tri}(2, 1) \neq 0$, then $h^{(2)}{}_{tri}(1, 2) = 0$ follows immediately from the definition of triangular impulse responses. Consequently, this means that $x(k - 2)x(k - 1) = x(k - 1)\, x(k - 2)$ occurs only once in Equation 1.247b. Similarly, $x(k - 2)x(k - 1)x(k - 3) = x(k - 1)x(k - 2)x(k - 3)$ occurs only once, and so on, which allows us to rewrite Equation 1.247b as

$$(\hat{V}x)(k) = h_{tri}^{(0)} + \sum_{i_1=0}^{M-1} h_{tri}^{(1)}(i_1)x(k-i_1) + \sum_{i_1=i_2=0}^{M-1} h_{tri}^{(2)}(i_1, i_2)x(k-i_1)x(k-i_2)$$
$$+\cdots+\sum_{i_1=i_2=\cdots=i_L=0}^{M-1} h_{tri}^{(L)}(i_1, i_2, \cdots, i_L)x(k-i_1)x(k-i_2)\cdots x(k-i_L) \tag{1.247c}$$

where the n-fold summation symbol $\sum_{i_1=i_2=\cdots=i_L=0}^{M-1}$ is over all the terms containing those triangular impulse responses that are not zero by definition.

We obtain further reduction of the number of components in the Volterra series approximator given by Equation 1.247c, by the use of Equation 1.246b. Using the relation 1.246b, we reduce these components, which have the same values of the indices, for example, $i_1 = i_2 = 0$, $i_1 = i_2 = 1$, and so on. To illustrate, consider a term $h_{tri}^{(3)}(2, 2, 1)x(k - 2)\, x(k - 2)x(k - 1) = h_{tri}^{(3)}(2, 2, 1)x^2(k - 2)\, x(k - 1)$ as an example. To reduce the quadratic term $x^2(k - 2)$ in the above expression, we use the general relation 1.246b, which gives $x^2(k - 2) = a_1 + b_1x(k - 2)$. Substituting the latter into the previous expression, we arrive at $h_{tri}^{(3)}(2, 2, 1)(a_1 + b_1x(k - 2))\, x(k - 1) = a_1h_{tri}^{(3)}(2, 2, 1)x(k - 1) + b_1h_{tri}^{(3)}(2, 2, 1)x(k - 2)\, x(k - 1)$. The two components achieved cannot be further reduced. The first contributes to the components that have the following form of a constant multiplied by $x(k - 1)$. The second component contributes to the components possessing the form of a constant multiplied by $x(k - 2)\, x(k - 1)$.

From the above procedure, we see that the successive applications of the Equation 1.246b in Equation 1.247c lead to getting an equivalent form of the Volterra series approximator for binary signals; that is, to

$$(\hat{V}x)(k) = d_0^{(0)} + \sum_{i_1=0}^{M-1} d_{i_1}^{(1)} x(k-i_1) + \sum_{\substack{i_1=0,\, i_2=0 \\ i_1 \neq i_2}}^{M-1} d_{i_1 i_2}^{(2)} x(k-i_1)x(k-i_2) + \cdots$$
$$+ d_{012\cdots(M-1)}^{(M)} x(k)x(k-1)\cdots x(k-M+1) \tag{1.247d}$$

for $L \geq M$. In Equation 1.247d, $d_0^{(0)}$, $d_{i_1}^{(1)}$, $d_{i_1 i_2}^{(2)}$, $\cdots$, $d_{012\cdots(M-1)}^{(M)}$, are the resulting coefficients, taking into account all the contributions to the corresponding combinations of signal samples $x^0(k) = 1$, $x(k - i_1)$, $x(k - i_1)\, x(k - i_2)$, $\cdots$, $x(k)x(k - 1)\cdots x(k - M + 1)$, respectively. It is clear that these coefficients depend upon the values of the impulse responses and upon the values of the coefficients a and b. The summation symbol in Equation 1.247d is slightly modified in comparison to that in Equation 1.247c. Namely, the components with the same values of indices are excluded in (1.247d). Finally, we stress that, in the combinations $x(k - i_1)\, x(k - i_2)$, $\cdots$, $x(k - i_n)$, $n = 1, 2, \cdots, M$, in Equation 1.247d, only the signal samples $x(k)$, $x(k - 1)$, $\cdots$, $x(k - M + 1)$ to the power of one or zero occur.

The representation given by Equation 1.247d holds only for $L \geq M$ |·| For L< M, that is, when the value of the measure of system nonlinearity strength is less than the value of the system memory length measure, the longest possible combination in 1.247c, as for example, $x(k)x(k - 1)\cdots x(k - L + 1)$, contains a lesser number of signal samples than the longest combination in Equation 1.247d. In other words, the series 1.247c for binary signals in the case of $L < M$ must have a lesser number of components than that given by Equation 1.247d. It has then the form

$$(\hat{V}x)(k) = d_0^{(0)} + \sum_{i_1=0}^{M-1} d_{i_1}^{(1)} x(k-i_1) + \sum_{\substack{i_1=0,\, i_2=0 \\ i_1 \neq i_2}}^{M-1} d_{i_1 i_2}^{(2)} x(k-i_1)x(k-i_2) + \cdots$$
$$+ \sum_{\substack{i_1=0,\, \cdots,\, i_L=0 \\ i_1 \neq i_2 \neq \cdots \neq i_L}}^{M-1} d_{i_1 i_2 \cdots i_L}^{(L)} x(k-i_1)x(k-i_2)\cdots x(k-i_L) \tag{1.247e}$$

for $L < M$.

Comparison of the expressions 1.247d and 1.247e shows that the length of the system memory, M, decides about the number of components to be taken into account when $L > M$. Conversely, when the system nonlinearity is relatively small in comparison to the system memory length, that is, when $L < M$, this nonlinearity, not the system memory, decides about the number of the series components to be taken into account.

A similar expansion to that in Equation 1.247d has been derived using other arguments, not related to the topic of the Volterra series (see Reference 54).

Example 1.5

Let us illustrate in this example the expansion given by Equation 1.247d. For this purpose, assume that $L = 4$ and $M = 3$. Then, we get from Equation 1.247d

$$(\hat{V}x)(k) = d_0^{(0)} + \sum_{i_1=0}^{2} d_{i_1}^{(1)} x(k-i_1) + \sum_{\substack{i_1=0,\, i_2=0 \\ i_1 \neq i_2}}^{2} d_{i_1 i_2}^{(2)} x(k-i_1)x(k-i_2)$$

$$+ d_{012(120)}^{(3)} x(k)x(k-1)x(k-2) + d_0^{(0)} = [d_0^{(1)} x(k) + d_1^{(1)} x(k-1) + d_2^{(1)} x(k-2)]$$

$$+ [d_{01}^{(2)} x(k)x(k-1) + d_{02}^{(2)} x(k)x(k-2) + d_{12}^{(2)} x(k-1)x(k-2)] + d_{012}^{(3)} x(k)x(k-1)x(k-2)$$

This expansion is presented in Figure 1.46.

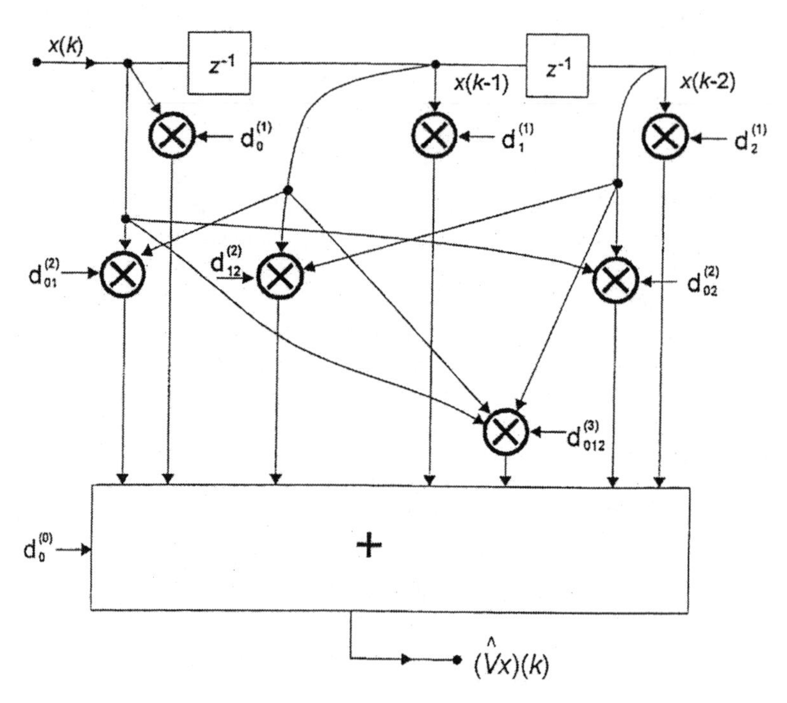

FIGURE 1.46 Illustration of the Volterra series approximator for binary signals for $L=4$ and $M=3$.

Example 1.6

Here we illustrate the expansion given by Equation 1.247e. For this purpose, we assume $L = 2$ and $M = 3$. Then, we get from Equation 1.247e

$$(\hat{V}x)(k) = d_0^{(0)} + \sum_{i_1=0}^{2} d_{i_1}^{(1)} x(k-i_1) + \sum_{\substack{i_1=0,\, i_2=0 \\ i_1 \neq i_2}}^{2} d_{i_1 i_2}^{(2)} x(k-i_1)x(k-i_2)$$

$$= d_0^{(0)} + [d_0^{(1)} x(k) + d_1^{(1)} x(k-1) + d_2^{(1)} x(k-2)]$$

$$+ [d_{01}^{(2)} x(k)x(k-1) + d_{02}^{(2)} x(k)x(k-2) + d_{12}^{(2)} x(k-1)x(k-2)]$$

A graphical illustration of the expansion is presented in Figure 1.47.

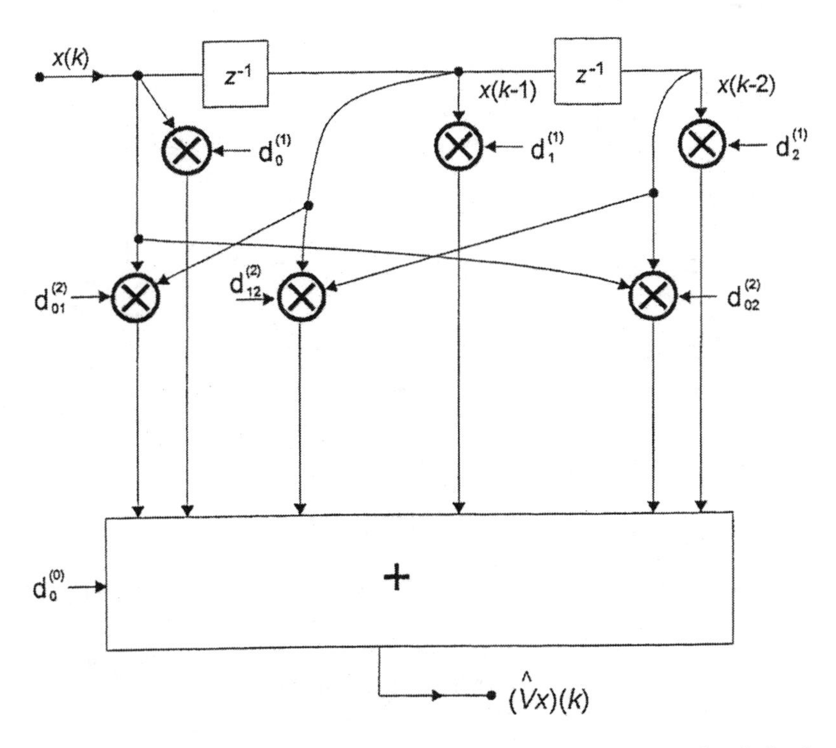

FIGURE 1.47 Illustration of the Volterra series approximator for binary signals for $L = 2$ and $M = 3$.

Comparison of Examples 1.5 and 1.6 shows a general rule that, depending upon the relation between the measure of system memory length, M, and the measure of system nonlinearity strength, L, a decisive about the number of components in the Volterra series approximator is either the first or the second of the parameters. In Example 1.5, where $L > M$ holds, there occur no components of higher order than three. In other words, the approximator highest order component is that of order three, which equals the system memory length $M = 3$. Conversely, in Example 1.6, we have $M > L$. L is less than M, and as before, the lesser parameter decides on the highest order component in the Volterra series approximator. Because $L = 2$, the highest order components in Example 1.6 are the components of the second order, $d_{01}^{(2)}\, x(k)\, x(k-1)$, $d_{02}^{(2)}\, x(k)x(k-2)$, and $d_{12}^{(2)}\, x(k-1)x(k-2)$.

Comparison of Examples 1.5 and 1.6 leads also to another observation. For this purpose, we need the notion of a linear finite impulse response (FIR) system (filter).[24] The response $y(k)$ to the input signal $x(k)$ of such a system (filter) is given by

$$y(k) \overset{df}{=} \sum_{i=0}^{M-1} b(i)x(k-i) \tag{1.248}$$

where $b(i)$ are the samples of the system (filter) impulse response. Equation 1.248 is illustrated in Figure 1.48.

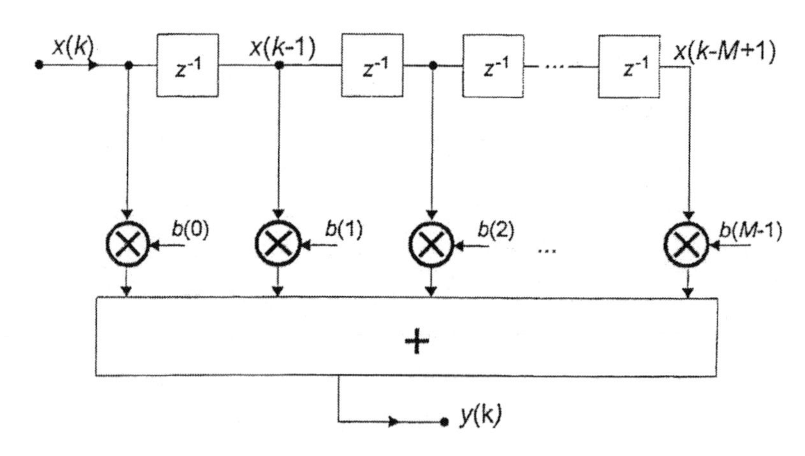

FIGURE 1.48 Structure of a linear FIR system (filter).

Comparison of the structure of a linear FIR filter in Figure 1.48 with the structure of the moving average map in Figure 1.37 shows that, in fact, both the structures are identical when the samples of the impulse response are identified with the coefficients α, $i = 0, 1, \cdots, M - 1$ in the following way:

$$b(i) = \alpha_i, \quad i = 0, 1, \cdots, M-1 \tag{1.249}$$

We prefer now, however, to use the notion of the FIR filter rather than the notion of the moving average map. The latter is used to model linear systems in the context of spectral analysis of stochastic processes. On the other hand, the finite impulse response filter name underlines that property which we want to exploit now, that is, the finite memory property expressed by an impulse response of a finite length.

The linear parts of the structures in Figures 1.46 and 1.47 represent a FIR filter of the same length because, in both cases, we have assumed $M = 3$. We conclude that the system memory length fully determines the length of the impulse response of a system linear part, and this is independent of the system nonlinearity strength expressed by parameter L. On the other hand, note that the parameter L affects the coefficients $d_i^{(1)} = 0, 1, \cdots, M - 1$, which we identify here with the samples of the impulse response of a FIR filter. That is, we write

$$d_i^{(1)} = b(i), \quad i = 0, 1, \cdots, M-1 \tag{1.250}$$

The coefficients $b(i)$ are equal to the corresponding samples $h_{tri}^{(1)}(i)$, that is, $b(i) = h_{tri}^{(1)}(i) = h^{(1)}(i)$ only in the case of a linear approximator, when $L = 1$. In the case of a nonlinear approximator, when $L > 1$, the coefficients $d_i^{(1)} = 0, 1, \cdots, M$

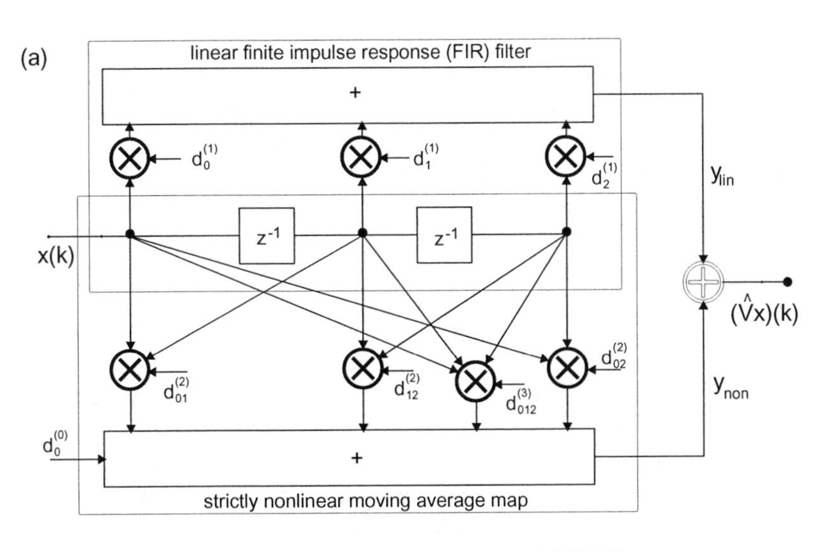

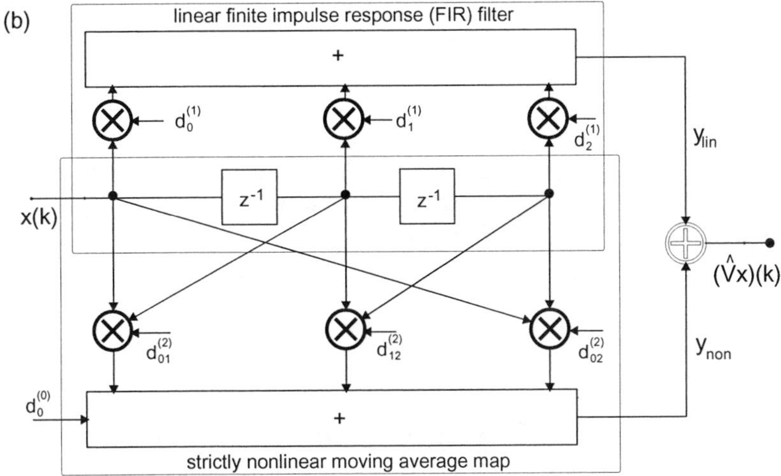

FIGURE 1.49 (a) The structure of Figure 1.46 redrawn with the strictly linear part (FIR filter) separated, (b) the structure of Figure 1.47 redrawn with the strictly linear part (FIR filter) separated.

− 1 depend also upon the form of a binary signal and upon the values of the system impulse responses of higher orders. We illustrate this by means of an example. Note that, as we have just shown, the term

$$h_{tri}^{(3)}(2, 2, 1)x^2(k-2)x(k-1)$$

is equal to

$$a_1 h_{tri}^{(3)}(2, 2, 1)x(k-1) + b_1 h_{tri}^{(3)}(2, 2, 1)x(k-2)x(k-1)$$

From the above, it is evident that the coefficient $a_1 h_{tri}^{(3)}(2, 2, 1)$ contributes to the value of the coefficient $d_1^{(1)}$. Moreover, the coefficient $b_1 h_{tri}^{(3)}(2, 2, 1)$ contributes to the value of the coefficient $d_{12}^{(2)}$. Hence, we conclude that linear FIR filters shown in Figure 1.49 for the structures derived in Examples 1.5 and 1.6 are linear because they fulfill the defining Equation 1.248. However, the filter's coefficients depend upon the remaining structure, that is, upon the nonlinear part and upon the form of the binary input signal. The values of the coefficients $d_0^{(1)}$, $d_1^{(1)}$, and $d_2^{(1)}$ are different in Figure 1.49a and Figure 1.49b because of the different nonlinearity strength assumed in the examples 1.5 and 1.6. Also different are the nonlinear moving average maps in Figure 1.49a and Figure 1.49b, that is, the values of the corresponding coefficients and the number of these coefficients.

The form of the Volterra series approximator for binary input signals is a particular case of the general formula (see Equations 1.247a or 1.247b). One could, of course, try to find simplified versions of these formulas for input signals possessing more than two amplitude levels. This would be, however, cumbersome and unsatisfactory in most cases. We propose, in the cases of signals with more than two amplitude levels, to use the general formula given by Equation 1.220a or alternatively, by Equations 1.247a or 1.247b.

1.11 ASSOCIATED EXPANSIONS

Origins of what is now called the Volterra series in the technical literature lie in the works[25, 26] of Vito Volterra, who investigated functional expansion of a type in which the discrete-time variant can be expressed as

$$\begin{aligned} h^{(0)} &+ \sum_{i_1=i_a}^{i_b} h^{(1)}(i_1)x(i_1) + \sum_{i_1=i_a}^{i_b}\sum_{i_2=i_a}^{i_b} h^{(2)}(i_1, i_2)x(i_1)x(i_2) \\ &+ \sum_{i_1=i_a}^{i_b}\sum_{i_2=i_a}^{i_b}\sum_{i_3=i_a}^{i_b} h^{(3)}(i_1, i_2, i_3)x(i_1)x(i_2)x(i_3) + \cdots \end{aligned} \tag{1.251}$$

where $h^{(o)}$ is a constant and $h^{(n)}(i_1, i_2, \cdots, i_i)$, $i = 1, 2, \cdots$, and $x(i_1)$, $x(i_2)$, $\cdots$, with $i_a \le i_i \le i_b$, represent samples of the corresponding functions $h^{(n)}$ and x.

Observe that the expression 1.251 is a typical functional, that is a mapping into the set of real numbers $\mathbb{R}$. More precisely, identifying the function x_i in expression 1.251 with the input signal, a sequence from the space $l^\infty(i_a \le i \le i_b)$ with the norm

$$\|x\|_{i_a i_b} \overset{df}{=} \sup_{i_a \le i \le i_b} |x(i)| = \max_{i_a \le i \le i_b} |x(i)| \tag{1.252}$$

we can interpret expression 1.251 as a mapping of the type: $l^\infty(i_a \le i \le i_b) \to \mathbb{R}$.

Introducing a parameter k into expression 1.251 by assuming for instance that x depends upon this parameter; that is, we have now to work with $x(k, i_i)$, $i = 1, 2,$

…, and assuming then that each $x(k, i_i)$ can be written in the form $x(k, i_i) = x(k - i_i)$ and $0 \le i_i \le k$, we get finally from expression 1.251 the expansion 1.12

$$y(k) = h^{(0)} + \sum_{i_1=0}^{k} h^{(1)}(i_1)x(k-i_1) + \sum_{i_1=0}^{k}\sum_{i_2=0}^{k} h^{(2)}(i_1, i_2)x(k-i_1)x(k-i_2)$$
$$+\sum_{i_1=0}^{k}\sum_{i_2=0}^{k}\sum_{i_3=0}^{k} h^{(3)}(i_1, i_2, i_3)x(k-i_1)x(k-i_2)x(k-i_3) + \cdots \tag{1.12}$$

where $y(k)$ means the resulting sequence. We arrived thereby at what has been called the Volterra series in Section 1.1. This is no longer a functional but an operator, that is, a mapping of sequences $x(k)$ into sequences $y(k)$.

Using similar arguments as above, we are able to get from expression 1.251 all the other forms of the Volterra series presented in Section 1.1. For example, letting $-\infty \le i_i \le \infty$ and substituting $x(k - i_i)$ into expression 1.251 instead of $x(i_i)$, $i = 1, 2, \cdots$, we get the expansion given by Equations 1.1 and 1.3. On the other hand, when we introduce the parameter k by substituting $h^{(0)}(k)$ in place of $h^{(0)}$, $h^{(1)}(k, i_1)$ in place of $h^{(1)}(i_1)$, $h^{(2)}(k, i_1, i_2)$ in place of $h^{(2)}(i_1, i_2)$, and so on, in expression 1.251, we get, as a result, the general form of the Volterra series for time-dependent systems, as given by Equations 1.1 and 1.2.

Consider once again what has been called the discrete Volterra series for time-independent systems in Section 1.1, that is,

$$y(k) = h^{(0)} + \sum_{i_1=-\infty}^{\infty} h^{(1)}(i_1)x(k-i_1) + \sum_{i_1=-\infty}^{\infty}\sum_{i_2=-\infty}^{\infty} h^{(2)}(i_1, i_2)x(k-i_1)x(k-i_2)$$
$$+\sum_{i_1=-\infty}^{\infty}\sum_{i_2=-\infty}^{\infty}\sum_{i_3=-\infty}^{\infty} h^{(3)}(i_1, i_2, i_3)x(k-i_1)x(k-i_2)x(k-i_3) + \cdots$$

(1.1) and (1.3)

An analog version of the above series contains, of course, integral symbols in place of summation ones. Different authors who have applied the series in their investigations of nonlinear analog systems have often used quite different names. For instance, Schetzen[27] uses the Volterra series term in the context of nonlinear time-invariant systems with memory, and Chua and Ng[28] use Volterra functional series term for these systems.

Furthermore, for the series without the constant component $h^{(0)}$, Kuo[29] uses the name Volterra series time-domain representation, Saleh[12] uses Volterra series expansion, and Bedrosian and Rice[30] simply use Volterra series. Boyd, Chua, and Desoer in Reference 31 use three equivalent names: Volterra series operator, Volterra series, and Volterra series expansion, without giving any limits of integration under the integration symbols. Note that the first name used underscores the fact that the Volterra series represents a mapping which is an operator, not a functional. The third of the names mentioned underscores the fact that the Volterra series is an expansion.

As we already know from Section 1.9, this expansion, truncated or not, can represent an approximation of a nonlinear system response.

In this context, the notion of a functional expansion (series) can be understood as a mapping, transforming functions into reals, that is, as a functional. On the other hand, it can be also understood as a mapping transforming a set of functions into another set of functions (which is a main subject of functional analysis), that is, as an operator. As mentioned, the above notion has been used in both meanings in the technical literature, which sometimes led to misunderstandings. In this book, we use the Volterra series term for all the forms of series described in Section 1.1, as is assumed in most papers on these subjects. Because of our restriction here to consideration of nonlinear discrete-time systems, our series is called the discrete Volterra series.

We shall now explain what the associated Volterra series, or associated Volterra series expansions, are. These expansions have been introduced to the literature by Sandberg.[17, 32] For nonlinear discrete-time systems, these Volterra series expansions have the following form[3]:

$$y(k) = \sum_{n=1}^{\infty} \sum_{i_1=-\infty}^{k} \sum_{i_2=-\infty}^{k} \cdots \sum_{i_n=-\infty}^{k} h^{(n)}(k-i_1, k-i_2, \cdots, k-i_n)x(i_1)x(i_2)\cdots x(i_n) \quad (1.253a)$$

where $x, y \in l^\infty(\mathbb{Z})$, $k \in \mathbb{Z}$. When needed, the constant component $h^{(0)}$ can be added to the expansion 1.253a.

The expansion 1.253a can be written in an equivalent form with $k - i_i$, $i = 1, 2, \ldots$, standing in the expansion as arguments of the input signal x. To get this representation from expansion 1.253a, one has to introduce new variables $i_i' = k - i_i$, $i = 1, 2, \cdots$, in 1.253a. This leads, after changing the summation limits, because of $0 \le i_i' \le \infty$, and after dropping the prime symbol at each i_i', to

$$y(k) = \sum_{n=1}^{\infty} \sum_{i_1=0}^{\infty} \sum_{i_2=0}^{\infty} \cdots \sum_{i_n=0}^{\infty} h^{(n)}(i_1, i_2, \cdots, i_n)x(k-i_1)x(k-i_2)\cdots x(k-i_n) \quad (1.253b)$$

Both expansions 1.253a and 1.253b are equivalent to the following expansions:

$$y(k) = \sum_{n=1}^{\infty} \sum_{i_1=-\infty}^{\infty} \sum_{i_2=-\infty}^{\infty} \cdots \sum_{i_n=-\infty}^{\infty} h^{(n)}(k-i_1, k-i_2, \cdots, k-i_n)x(i_1)x(i_2)\cdots x(i_n) \quad (1.253c)$$

and

$$y(k) = \sum_{n=1}^{\infty} \sum_{i_1=-\infty}^{\infty} \sum_{i_2=-\infty}^{\infty} \cdots \sum_{i_n=-\infty}^{\infty} h^{(n)}(i_1, i_2, \cdots, i_n)x(k-i_1)x(k-i_2)\cdots x(k-i_n) \quad (1.253d)$$

for causal systems defined in Section 1.1: that is, for systems of which nonlinear impulse responses fulfill: $h^{(n)}(i_1, i_2, \cdots, i_n) \equiv 0$, $n = 1, 2, \ldots$, when even one of the arguments $i_i < 0$, $i = 1, 2, \cdots, n$. It follows from this property that $h^{(n)}(k - i_1, k - i_2, \cdots, k - i_n)$ equals zero if even one $i_i > k$. Therefore, we can let each i_i in expansion 1.253a go to infinity without changing the value of $y(k)$. Similarly, because $h^{(n)}(i_1, i_2, \cdots, i_n)$ equals zero if even one $i_i < 0$, we can let each i_i in expansion 1.253b go to minus infinity without changing the value of $y(k)$.

Note also that expansion 1.253d is identical with that represented by Equations 1.1 and 1.3, when dropping the constant component in the latter.

In Section 1.1, we have defined the Volterra series using the notion of a nonlinear system. We have considered such systems in which response to an input can be represented in the form of a series such as that assumed in Equations 1.1 and 1.2 or in Equations 1.1 and 1.3. Furthermore, note that we have called these series possessing the infinite summation limits, the discrete Volterra series. We see now that, when omitting the component independent of an input signal, they are nothing other than the associated series in the case of considering causal systems. In the case of nonlinear causal systems, our basic definitions of the discrete Volterra series from Section 1.1 are conceptually identical with the definition of the Volterra series expansions of Sandberg.

Of course, the Volterra series may not exist for a given system. That a given system possesses a representation in the form of a Volterra series or, in Sandberg's terminology, of an associated Volterra series expansion, must be proven in each case. Furthermore, the associated Volterra series expansion represents a natural description of a system response only when input signals are defined on the whole time-axis, from $-\infty$ to $+\infty$, as, for instance, is the case with signals taken from the space $l^\infty(\mathbb{Z})$. However, in many cases in which input signals are defined only for nonnegative times, $k \geq 0$, the associated Volterra series expansions are taken as the input-output representations. The advantage of this is that, as in the linear case, the occurrence of the summation limits $-\infty$ and $+\infty$ enables simplification of some calculations. For example, the two-sided Z transform can be then applied to such representations, (see Section 1.3). However, such use of the associated Volterra series has, as a result, the modification of the original model of a system. This fact is illustrated in Figure 1.50.

As shown in Figure 1.50, the set of allowable input signals to a system is an inherent part of any model of this system. In Figure 1.50a, an original model of a nonlinear causal and time-invariant system is presented. This model consists of the set of input signals defined only for nonnegative times, $k \geq 0$. Assume that an input-output representation in the form of a Volterra series exists for the above model, and assumes the form given by Equation 1.12, which, after the change of variables $k - i_i = i_i'$, $0 \leq i_i' \leq k$, and dropping the prime at each i_i', can be also expressed as:

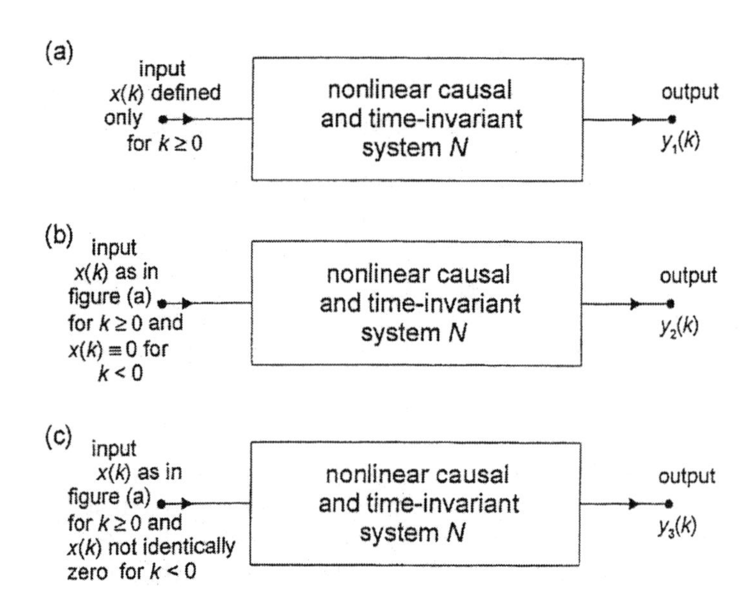

FIGURE 1.50 (a) An original model of a nonlinear causal and time-invariant system N for which input signals are defined only for nonnegative times, $k \geq 0$, (b) a modified model with input signals as in (a) for $k \geq 0$ and identically zero for $k < 0$, (c) another modified model with input signals as in (a) for $k \geq 0$ and not identically zero for $k < 0$.

$$y_1(k) = h^{(0)} + \sum_{i_1=0}^{k} h^{(1)}(k-i_1)x(i_1) + \sum_{i_1=0}^{k}\sum_{i_2=0}^{k} h^{(2)}(k-i_1, k-i_2)x(i_1)x(i_2) + \sum_{i_1=0}^{k}\sum_{i_2=0}^{k}\sum_{i_3=0}^{k} h^{(3)}(k-i_1, k-i_2, k-i_3)x(i_1)x(i_2)x(i_3) + \cdots \tag{1.254a}$$

The fact that the original model of a system possesses the Volterra series representation does not mean that any modified model, such as that in Figure 1.50b, also possesses such a representation. This is so because any modified model is another model, even when it regards only the allowable domain of inputs different from those in the original model. Assume, however, that the modified model in Figure 1.50b possesses the Volterra series representation. It can then be expressed by Equation 1.11, or, equivalently, by Equation 1.253a with the component $h^{(0)}$ added to it. Further, because of $x(k) \equiv 0$ for every $k < 0$, the latter simplifies to Equation 1.254a. The output response of the system in Figure 1.50a is equal to the output response of the system in Figure 1.50b, that is,

$$y_2(k) = y_1(k) \tag{1.254b}$$

In other words, the associated model of Figure 1.50b leads to the same result as the original one, when the Volterra series input-output representation does exist for it. We stress once again that the existence of the Volterra series input-output representation for the associated model must be checked.

Consider now another associated model, presented in Figure 1.50c. If the Volterra series input-output representation exists for this model, it has the form expressed by Equation 1.11, or, equivalently, by Equation 1.253a with the component $h^{(0)}$ added to it. Because the output of the original system is given by Equation 1.254a, the difference between the outputs $y_3(k)$ and $y_1(k)$ can be expressed as

$$y_3(k)-y_1(k)=\sum_{n=1}^{\infty}\sum_{i_1=-\infty}^{-1}\sum_{i_2=-\infty}^{-1}\cdots\sum_{i_n=-\infty}^{-1}h^{(n)}(k-i_1,k-i_2,\cdots,k-i_n)x(i_1)x(i_2)\cdots x(i_n) \tag{1.254c}$$

where the input signal samples $x(i_1)$, $x(i_2)$, $\cdots$, $x(i_n)$, occur for negative arguments $i_i < 0$ $i = 1, 2, \cdots, n$; these are the samples of the input signal as defined in Figure 1.50c.

Thus, in this case, using the associated model presented in Figure 1.50c instead of the original model of Figure 1.50a, we must answer the question whether the difference between the outputs given by Equation 1.254c is essential for the model replacement. In this context, a theorem presented for the first time by Sandberg[17] is very important. We present here its discrete-time version for scalar-valued inputs and outputs. It is a simplified version in comparison with the original theorem presented by Sandberg[17] for continuous-time systems; this theorem is formulated in what follows.

From Section 1.7, we already know what the linear normed spaces are. When these spaces are also complete as metric spaces, then they are called Banach spaces.[13] We draw the reader's attention to the fact that the spaces $l^\infty(\mathbb{Z})$ and $l^\infty(\mathbb{Z}_+)$ defined and used in the previous sections are just such spaces; that is, they are Banach spaces.

It is important to be aware of the above fact because the authors of some papers use the notion of Banach space in their formulations. The others, however, exploit the notion of linear normed space possessing the metric $d(x, v) = \|x - v\|$. As we know, both notions can be used equivalently.

Finally, we sketch in what follows the proof of the fact that the linear normed spaces $l^\infty(\mathbb{Z})$ and $l^\infty(\mathbb{Z}_+)$ are complete. For this purpose, we take into account any Cauchy sequence $\{x_n\}$ (for the definition of the Cauchy sequence, see Section 1.9). That is, we take such a sequence for which

$$d(x_n, x_m) = \sup_{k\in K}|x_n(k)-x_m(k)| \rightarrow 0,\ n, m\rightarrow\infty \tag{1.255}$$

holds, where $K = \mathbb{Z}$ or $K = \mathbb{Z}_+$. Whence the relation $|x_n(k) - x_m(k)| \rightarrow 0,\ n, m \rightarrow \infty$ holds for any given k. That is, the sequence $\{x_n(k)\}$ for a given k, being a sequence of the following real numbers $\{x_0(k),\ x_1(k),\ x_2(k),\ \cdots\}$, is a Cauchy sequence in the space of reals. At this point, we recall a very well-known fact from the functional analysis that the space of real numbers with the metric $d(x, v) = |x - v|$ is complete.[13] Further, it follows that the sequence $\{x_n(k)\}$ for a given k has a limit, say, $x^*(k)$, for every k.

Let us now check whether x^* is an element of the space $l^\infty(K)$, $K = \mathbb{Z}$ or $K = \mathbb{Z}_+$, and whether it is a limit of x_n and x_m, that is, $\lim_{n\to\infty} x_m = \lim_{m\to\infty} x_m = x^*$. In this context, we use the observation just made, which allows us to write

$$|x_n(k) - x^*(k)| \to 0,\ n \to \infty \tag{1.256a}$$

and similarly, for $x_m(k)$

$$|x_m(k) - x^*(k)| \to 0,\ m \to \infty \tag{1.256b}$$

for every k. Because of this, it is possible to choose such n and m for any $\varepsilon > 0$, so that the following inequalities

$$\sup_{k\in K} |x_n(k) - x^*(k)| < \varepsilon \tag{1.257a}$$

and

$$\sup_{k\in K} |x_m(k) - x^*(k)| < \varepsilon \tag{1.257b}$$

hold as well. That is, x^* is a limit of x_n and x_m.

To show that x^* belongs to the space $l^\infty(K)$, $K = \mathbb{Z}$ or $K = \mathbb{Z}_+$, take into account the obvious equality

$$x^* = x^* - x_n + x_n \tag{1.258a}$$

that holds for every k. Proceeding further, see that this equality allows us to write

$$\sup_{k\in K} |x^*(k)| \le \sup_{k\in K} |x^*(k) - x_n(k)| + \sup_{k\in K} |x_n(k)| \tag{1.258b}$$

Using then in 1.258b the inequality 1.257a and the fact that $x_n \in l^\infty(K)$, that is, $\sup_{k\in K} |x_n(k)| < \infty$, we get from inequality 1.258b

$$\sup_{k\in K} |x^*| \le \varepsilon + \|x\| < \infty \tag{1.258c}$$

We conclude from inequality 1.258c that x^* is an element of the space $l^\infty(K)$, where K means $\mathbb{Z}$ or $\mathbb{Z}_+$.

Returning to the formulation of the theorem, recall that the partial responses of the Volterra series for the original model and the associated model in Figure 1.50 have the following form

$$y_O^{(n)} = \sum_{i_1=0}^{k} \cdots \sum_{i_n=0}^{k} h^{(n)}(k - i_1, \cdots, k - i_n)x(i_1)\cdots x(i_n),\ n = 1, 2, 3, \cdots \tag{1.259a}$$

and

$$y_A^{(n)} = \sum_{i_1=-\infty}^{k} \cdots \sum_{i_n=-\infty}^{k} h^{(n)}(k-i_1,\cdots,k-i_n)x(i_1)\cdots x(i_n),\ n = 1,2,3,\cdots \quad (1.259b)$$

respectively.

It is clear that each of the partial responses given by Equations 1.259a and 1.259b describes a mapping. In the case of 1.259a, these mappings (for $n = 1, 2, 3,\ldots$) can be considered as the mappings from the Banach space $l^\infty(\mathbb{Z}_+)$, or in short, from the space $l^\infty(\mathbb{Z}_+)$ into itself. Similarly, the mappings represented by Equation 1.259b can be considered as the mappings from the Banach space $l^\infty(\mathbb{Z})$, or in short, from the space $l^\infty(\mathbb{Z})$ into itself.

It can be proven, in analogy to the continuous-time case,[17] that the mappings given by Equation 1.259a belong to the set $P_1(m)$, and the mappings given by 1.259b belong to the set $M_1(m)$; for the definitions of the sets $P_1(m)$ and $M_1(m)$, see Section 1.8.

After clarification of some notions, let us now formulate the theorem.

Theorem 1.6

Let the mappings $y_O{}^{(n)}: l^\infty(\mathbb{Z}_+) \to l^\infty(\mathbb{Z}_+)$ and $y_A{}^{(n)}: l^\infty(\mathbb{Z}) \to l^\infty(\mathbb{Z})$ for $n =$ $1, 2, 3, \ldots$ be given by Equations 1.259a and 1.259b, respectively. Furthermore, let there be positive constants σ_1, σ_2 and δ such that $\delta<\sigma_2^{-1}$ and $\left\|y_O^{(n)}x\right\|_0 \le \sigma_1(\sigma_2\|x\|_0)^n$ for each n and $x \in l^\infty(\mathbb{Z}_+)$, where the norm $\|x\|_{k_0=0} = \sup_{k \ge k_0=0}|x(k)|$ and $\|x\|_0<\delta$. Moreover, let B_δ denote the open ball in $l^\infty(\mathbb{Z})$ of radius δ centered at the zero element of $l^\infty(\mathbb{Z})$, that is, B_δ is the set $\{x \in l^\infty(\mathbb{Z}_+) \mid \|x\| < \delta\}$, where $\|x\|$ is the usual norm $\|x\| = \sup_{k\in\mathbb{Z}}|x(k)|$. Then the series $\sum_{n=1}^{\infty} y_A^{(n)}x$ converges in $l^\infty(\mathbb{Z})$ uniformly with respect to $x \in B_\delta$, and the mapping $Hx = \sum_{n=1}^{\infty} y_A^{(n)}x : B_\delta \to l^\infty(\mathbb{Z})$ has the following two properties:

1. $\left|(Hx)(k) - (HQ_{k_0}x)(k)\right| \to 0$ uniformly with respect to $x \in B_\delta$ as $(k - k_0) \to \infty$
2. $(Hx - Hv)(k)$ goes to the zero element of $l^\infty(\mathbb{Z})$ as $k \to \infty$ for x and v belonging to B_δ whenever $(x - v)(k)$ goes to the zero element of $l^\infty(\mathbb{Z})$

Commenting on Theorem 1.6, we note that the zero element of the space $l^\infty(\mathbb{Z})$ of bounded scalar-valued sequences is, as we already know from Section 1.7, the sequence $\theta(k) = \{\cdots, 0, 0, 0, 0, \ldots\}$. Furthermore, recall that the truncation operator Q_{k_0} occurring in the first property has been already defined. See the defining Equation 1.189 in Section 1.8. However, one new notion still occurs in the above theorem, that needs some explanation, namely, the notion of uniform convergence of a sequence or a series.

To explain what the uniform convergence of the series $y_A(k, x) = \sum_{n=1}^{\infty}(y_A^{(n)}x)(k)$, with respect to $x \in B_\delta$ means, denote the partial series in the above series in the following way:

$$y_{1A}(k, x) = (y_A^{(1)}x)(k) \tag{1.260a}$$

$$y_{2A}(k, x) = \sum_{n=1}^{2} (y_A^{(n)}x)(k) \tag{1.260b}$$

$$y_{3A}(k, x) = \sum_{n=1}^{3} (y_A^{(n)}x)(k) \tag{1.260c}$$

and so on. Then, choose any $\varepsilon > 0$ and such n_0 that the inequality

$$\sup_{k \in K} |y_{nA}(k, x) - y_A(k, x)| < \varepsilon \tag{1.261a}$$

holds for all $n \geq n_0$. In this context, the uniform convergence of the series considered means that the inequality 1.261a holds for all $n \geq n_0$ independently of the choice of the sequence $x \in B_\delta$.

Similarly, the uniform convergence in the property 1 means that choosing any $\varepsilon > 0$, we can find such k' that, for all $(k - k_0) \geq (k' - k_0)$, the inequality

$$|(Hx)(k) - (HQ_{k_0}x)(k)| < \varepsilon \tag{1.261b}$$

holds independently of the choice of $x \in B_\delta$.

The proof of Theorem 1.6 is omitted in this book. We shall present another result of Sandberg regarding associated expansions in a more general framework for vector-valued sequences. For the latter, we shall also present the proof of a corresponding theorem.

It follows immediately from Theorem 1.6 that, if the original model of Figure 1.50a possesses the Volterra series representation in which partial responses fulfill some requirements, then the associated expansion $Hx = \sum_{n=1}^{\infty} y_A^{(n)}x$ does exist, with the input sequence x in the associated model of Figure 1.50c equal to the input sequence x in the original of Figure 1.50a for $k \geq 0$, and having such values that $\|x\| < \delta$ holds also for $k < 0$. Moreover, the representation $(HQ_{k_0}x)(k)$ is equal to the Volterra series representation in the original model of Figure 1.50a when $k_0 = 0$ is assumed. Furthermore, it follows from the property 1 that, for accordingly high values of k, the difference between $(Hx)(k)$ and $(HQ_0x)(k)$ is so small that the associated series can be practically identified with the series of the original system in Figure 1.50a. In other words, coming back to our question regarding the difference given by Equation 1.254c, this difference is not essential for accordingly high values of k.

By the way, note that if the partial responses in the Volterra series of the original system in Figure 1.50a fulfill the requirements indicated in Theorem 1.6, this series then converges absolutely. This follows from the fact that the series consisting of the normed partial responses converges,

$$\sum_{n=1}^{\infty} \left\|(y_O^{(n)}x)(k)\right\|_0 < \infty \tag{1.262a}$$

The latter holds because we can write the following relations:

$$\left\|\sum_{n=1}^{\infty} y_O^{(n)}x\right\|_0 \le \sum_{n=1}^{\infty} \left\|y_O^{(n)}x\right\|_0 \le \sum_{n=1}^{\infty} \sigma_1(\sigma_2\|x\|_0)^n \tag{1.262b}$$

Continuing further, we write

$$\sum_{n=1}^{\infty} \sigma_1(\sigma_2\|x\|_0)^n \le \sum_{n=1}^{\infty} \sigma_1(\sigma_2\delta)^n \tag{1.262c}$$

Finally, the geometric series on the right-hand side of relation 1.262c converges because $\sigma_2\delta < 1$. We get

$$\sum_{n=1}^{\infty} \sigma_1(\sigma_2\delta)^n = \sigma_1\frac{\sigma_2\delta}{1-\sigma_2\delta} < \infty \tag{1.262d}$$

which ends the proof of the absolute convergence of the series $\sum_{n=1}^{\infty} y_O^{(n)}x$.

Worth noting at this point is that the technique presented in inequalities 1.262a to d can also be used for estimation of an error made by truncation of higher terms in the Volterra series. To illustrate this point, let us consider the model of a system presented in Figure 1.50a and assume that the exact representation for it exists in the form of an infinite Volterra series $\sum_{n=1}^{\infty} y_O^{(n)}x$. Furthermore, assume that a truncated representation obtained from the infinite Volterra series by keeping in it only, say, L components, is used instead of that exact one. Then the error made by such a series replacement can be expressed as

$$\sum_{n=1}^{\infty} y_O^{(n)}x - \sum_{n=1}^{L} y_O^{(n)}x = \sum_{n=L+1}^{\infty} y_O^{(n)}x \tag{1.263a}$$

Using the same technique as that presented in inequalities 1.262a to d, to the right-hand side expression in Equation 1.263a, we get

$$\left\|\sum_{n=L+1}^{\infty} y_O^{(n)}x\right\|_0 \le \sum_{n=L+1}^{\infty} \left\|y_O^{(n)}x\right\|_0 \le \sum_{n=L+1}^{\infty} \sigma_1(\sigma_2\|x\|_0)^n \tag{1.263b}$$

And, continuing, we write

$$\sum_{n=L+1}^{\infty} \sigma_1(\sigma_2\|x\|_0)^n \le \sum_{n=L+1}^{\infty} \sigma_1(\sigma_2\delta)^n = \sigma_1\frac{(\sigma_2\delta)^{L+1}}{1-\sigma_2\delta} \le er \qquad (1.263c)$$

where *er* means a maximal admissible value assumed for the error estimate.

Solving inequality 1.263c for *L*, after performing some simple operations, we arrive at

$$L \ge \frac{\ln\left[\frac{er(1-\sigma_2\delta)}{\sigma_1}\right]}{\ln(\sigma_2\delta)} - 1 \qquad (1.263d)$$

where ln means a logarithm with the base equal to *e*.

Note that inequality 1.263d determines the number of components in the truncated Volterra series that are necessary to guarantee the truncation error does not to exceed the value of *er*.

Consider now other results[3] of Sandberg published in 1992 that can be seen as a continuation of the considerations presented in Reference 17 for continuous-time systems. The paper[3] develops further ideas given in Reference 17 with respect to the accuracy of truncations often made in practical applications using the associated Volterra series. The problem of a uniform approximation with doubly finite Volterra series is formulated and solved in Reference 3.

We now present the results of Sandberg's paper[3]. In particular, the results regarding nonlinear systems, which can be modeled by the following system equations,

$$\mathbf{w} = A(\mathbf{x}) + C(\mathbf{z}) \qquad (1.264a)$$

$$\mathbf{y} = D(\mathbf{x}) + B(\mathbf{z}) \qquad (1.264b)$$

$$\mathbf{z} = N(\mathbf{w}) \qquad (1.264c)$$

where $\mathbf{x}$ is a vector-valued input signal (sequence) to a system and $\mathbf{y}$ is a vector-valued output signal (sequence) of a system. Furthermore, it is assumed that the input sequences (that is the elements of $\mathbf{x}(k)$) belong to a subset of the (scalar) space $l^\infty(\mathbb{Z}_+)$ and the output vector-valued sequence $\mathbf{y}(k)$ belongs to the linear space E of $\mathbb{R}^m$-valued functions defined on $\mathbb{Z}_+$, where $m \in \mathbb{N}$. The first two equations, 1.264a and 1.264b, describe the linear part of a system, but Equation 1.264c describes its nonlinear part. The mappings *A*, *C*, *D*, and *B* are linear ones, mapping the space *E* into itself. The behavior of all the nonlinear elements of a system is described collectively by means of the mapping *N*. And the vector-valued sequence $\mathbf{w}(k)$ of the space *E* represents the input to this mapping. Similarly, the vector $\mathbf{z}$ consists of output sequences of the mapping *N*. Because it is assumed that the output vector-valued sequences also belong to the space *E*, the nonlinear mapping *N* is a mapping from one subset of *E* into another.

With regard to the same dimension *m* assumed for all the vectors occurring in Equations 1.264, we draw the reader's attention to the fact that the above assumption can have such a consequence on the system model that some elements of the vectors

w, **x**, **z**, and **y** (sequences) eventually will have no effect on the system in some situations. Then they will be simply ignored as redundant. Finally, it is also assumed in the system model that any nonzero initial conditions are taken into account in the system input vector-valued sequence **x**.

The model given by Equations 1.264, used widely in the literature for modeling nonlinear systems, is presented in graphical form in Figure 1.51.

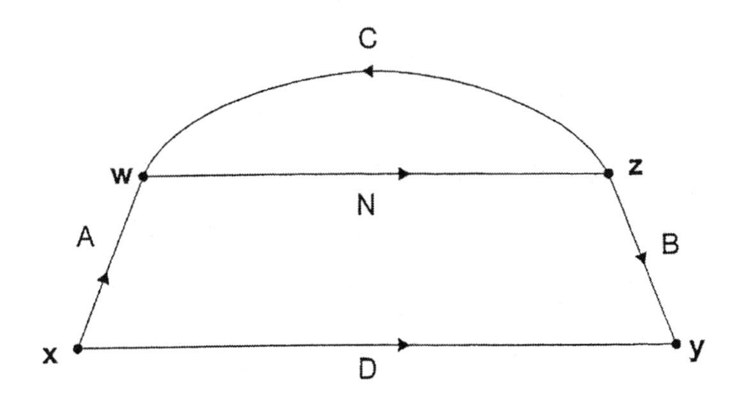

FIGURE 1.51 Graphical representation of nonlinear system modeling Equations 1.264.

Now having introduced the nonlinear system description presented in Figure 1.51, we draw the reader's attention to the fact that the vector-valued sequences, are understood here both as the vectors of which elements are sequences and as the sequences of which elements are vectors. For example,

$$\mathbf{x}(k) = \begin{bmatrix} x_1(0), x_1(1), x_1(2), \ldots \\ \vdots \\ x_m(0), x_m(1), x_m(2), \ldots \end{bmatrix} \tag{1.265a}$$

and

$$\mathbf{x}(k) = \left\{ \begin{bmatrix} x_1(0) \\ \vdots \\ x_m(0) \end{bmatrix}, \begin{bmatrix} x_1(1) \\ \vdots \\ x_m(1) \end{bmatrix}, \begin{bmatrix} x_1(2) \\ \vdots \\ x_m(2) \end{bmatrix}, \ldots \right\} \tag{1.265b}$$

mean in principle have the same meaning.

A large class of nonlinear systems can be described by the model represented by Equations 1.264 and the signal flow graph of Figure 1.51. For example, nonlinear systems governed by a system of difference equations of the form $\mathbf{y}(k + 1) = \mathbf{f}(\mathbf{y}(k), \mathbf{x}(k), k)$, $k \in \mathbb{Z}_+$, $\mathbf{y}(0) = \mathbf{y}_0$ where, **f** means a vector consisting of nonlinear functions, can be put into the above form. To illustrate this, we present the following example:

Example 1.7

Assume we have a system governed by the following difference equation:

$$y(k+1) = \frac{1}{3}y(k) + (y(k))^2 + (y(k))^3 + x(k)$$

where $k \in \mathbb{Z}_+$, and the initial condition $y(0) = 0$. Comparing the above equation with the general form $\mathbf{y}(k + 1) = \mathbf{f}(\mathbf{y}(k), \mathbf{x}(k), k)$, we see that the dimension of our equation is $m = 1$, and the function $f(y(k), x(k), k) = \frac{1}{3} y(k) + (y(k))^2 + (y(k))^3 + x(k)$. This function is obviously nonlinear because of the occurrence of quadratic and cubic terms. Moreover, we identify $x(k)$ with the input sequence and $y(k)$ with the output sequence, respectively, in the underlying model of Figure 1.51.

Note that the difference equation in this example can be rewritten in an equivalent form as a system of the following equations:

$$w(k) = y(k)$$

$$z(k) = (w(k))^2 + (w(k))^3$$

$$y(k+1) = \frac{1}{3}y(k) + x(k) + z(k)$$

Furthermore, using the Z transform, the third equation in the above system can be solved for $y(k)$. The final result in the time-domain will then have the following form:

$$y(k) = \sum_{i=0}^{k} g(k-i)(x(i) + z(i)) = \sum_{i=0}^{k} g(k-i)x(i) + \sum_{i=0}^{k} g(k-i)z(i)$$

with the linear impulse response $g(k)$ given by

$$g(0) = 0$$

and

$$g(k) = \left(\frac{1}{3}\right)^{k-1} \text{ for } k \geq 1$$

Moreover, because $w(k) = y(k)$ in the first of the modelling equations in our example, we can write

$$w(k) = \sum_{i=0}^{k} g(k-i)x(i) + \sum_{i=0}^{k} g(k-i)z(i)$$

as well. Thus, we see that all the signals in this example are scalar-valued sequences, and the linear mappings A, C, D, and B have the form of the discrete-time convolution with the impulse response $g(k)$ as given above, that is,

$$(As)(k) = (Cs)(k) = (Ds)(k) = (Bs)(k) = \sum_{i=0}^{k} g(k-i)s(i)$$

where $s(k)$ means a scalar-valued sequence. The nonlinear mapping N in the example considered has the form of a sum of quadratic and cubic terms. Finally, these observations are illustrated in Figure 1.52.

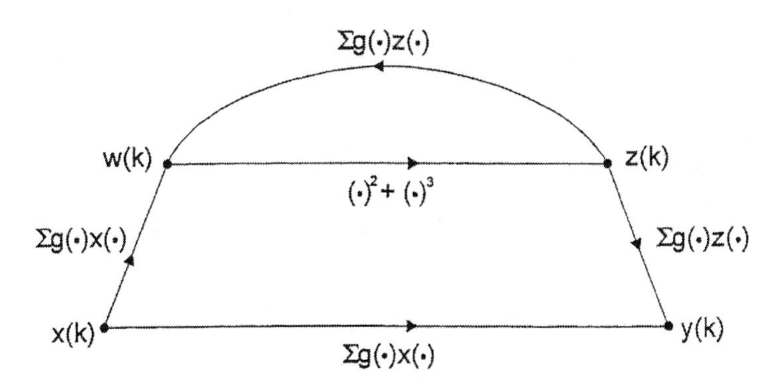

FIGURE 1.52 Illustration of modeling equations obtained in Example 1.7.

The following three assumptions are made with regard to the model given by Equations 1.264:

1. The mapping A possesses the discrete-time convolution representation

$$(A\mathbf{x})(k) = \sum_{i=0}^{k} A(k-i)\cdot\mathbf{x}(i) = \sum_{i=0}^{k}\begin{bmatrix} a_{11}(k-i) & a_{12}(k-i) & \dots & a_{1m}(k-i) \\ a_{21}(k-i) & a_{22}(k-i) & \dots & a_{2m}(k-i) \\ \vdots & \vdots & \vdots & \vdots \\ a_{m1}(k-i) & a_{m2}(k-i) & \dots & a_{mm}(k-i) \end{bmatrix} . \begin{bmatrix} x_1(i) \\ x_2(i) \\ \vdots \\ x_m(i) \end{bmatrix}$$

$$= \begin{bmatrix} \sum_{i=0}^{k} [a_{11}a_{12}\cdots a_{1m}]_{(k-i)}\cdot \begin{bmatrix} x_1 \\ x_2 \\ \vdots \\ x_m \end{bmatrix}_{(i)} \\ \sum_{i=0}^{k} [a_{21}a_{22}\cdots a_{2m}]_{(k-i)}\cdot \begin{bmatrix} x_1 \\ x_2 \\ \vdots \\ x_m \end{bmatrix}_{(i)} \\ \vdots \\ \sum_{i=0}^{k} [a_{m1}a_{m2}\cdots a_{mm}]_{(k-i)}\cdot \begin{bmatrix} x_1 \\ x_2 \\ \vdots \\ x_m \end{bmatrix}_{(i)} \end{bmatrix} \tag{1.266a}$$

that is, A is a real $m \times m$ matrix-valued map defined on $\mathbb{Z}_+$ such that the elements of the matrix $\mathbf{A}$, dependent upon the discrete-time, a_{rs}, $r = 1, \cdots, m$, $s = 1, \ldots, m$, fulfill the condition

$$\sup_k \sum_{i=0}^{k} |a_{rs}(i)| < \infty, k \in \mathbb{Z}_+ \tag{1.266b}$$

Moreover, the same representation possess the remaining linear mappings C, D, and B, that is,

$$(C\mathbf{x})(k) = \sum_{i=0}^{k} \mathbf{C}(k-i) \cdot \mathbf{x}(i) \tag{1.266c}$$

$$(D\mathbf{x})(k) = \sum_{i=0}^{k} \mathbf{D}(k-i) \cdot \mathbf{x}(i) \tag{1.266d}$$

$$(B\mathbf{x})(k) = \sum_{i=0}^{k} \mathbf{B}(k-i) \cdot \mathbf{x}(i) \tag{1.266e}$$

where all the matrices **C**, **D**, and **B** have order $m \times m$, and the elements of these matrices fulfill a condition similar to that given by condition 1.266b.

The discrete-time convolution representations 1.266a, 1.266c, 1.266d, and 1.266e represent linear multiple-input and multiple-output systems, where both the input vector **x** and the output vector $A\mathbf{x}$, $C\mathbf{x}$, $D\mathbf{x}$, or $B\mathbf{x}$ have the same dimension m. For details of representation of such systems, see Section 1.6, where the more general case of nonlinear multiple-input and multiple-output systems has been considered.

2. Let now $l^\infty(\mathbb{Z}_+)$ mean the normed linear space of vector-valued sequences defined on $\mathbb{Z}_+$ with the norm given by Equations 1.157. Moreover, let B_{p+} mean the open ball in the above space of radius ρ, centered at the zero element of this space,

$$\boldsymbol{\theta} = \left\{ \begin{matrix} 1 \\ 2 \\ \vdots \\ m \end{matrix} \begin{bmatrix} 0 \\ 0 \\ \vdots \\ 0 \end{bmatrix}, \begin{bmatrix} 0 \\ 0 \\ \vdots \\ 0 \end{bmatrix}, \begin{bmatrix} 0 \\ 0 \\ \vdots \\ 0 \end{bmatrix}, \dots \right\}$$

We assume then that there exists such a positive real number δ that the mapping N is defined on $B_{\delta+}$ by the following equation:

$$(N\mathbf{w})(k) = \boldsymbol{\eta}(\mathbf{w}(k)), k \in \mathbb{Z}_+ \tag{1.267a}$$

where the mapping $\boldsymbol{\eta}(\mathbf{w})$ represents a vector of dimension m, of which elements are given by

$$\eta_i(\mathbf{w}) = \sum_{j=1}^{j_0} \lambda_{ij}(w_i) \tag{1.267b}$$

for some positive integer j_0, and for all $i = 1, \ldots, m$. Moreover, each function $\lambda_{ij}:(-\delta, \delta) \to \mathbb{R}$ in Equation 1.267b has an extension into the complex plane, that is, $\lambda_{ij}(z_i)$, where z_i is a complex number, is properly defined. Furthermore, it is assumed that each function $\lambda_{ij}(z_i)$ is an analytic function, taking the disk $|z_i| < \delta$ into the complex plane and having the property that $\lambda_{ij}(0) = 0$. At this point, we also recall how the theory of complex functions defines analytic functions. Thus, a complex function given on some disk D is analytic if it is differentiable in each point of the disk D.

3. The assumptions 1 and 2 hold, and for each $\mathbf{w} \in l^\infty(\mathbb{Z}_+)$, where $l^\infty(\mathbb{Z}_+)$ now means the normed linear space of vector-valued sequences defined on $\mathbb{Z}_+$, there exists a unique vector-valued sequence $\mathbf{z} \in l^\infty(\mathbb{Z}_+)$ such that Equation

$$\mathbf{w}(k) = \mathbf{z}(k) - \sum_{i=0}^{k} \mathbf{C}(k-i) \cdot \mathbf{J} \cdot \mathbf{z}(i) \tag{1.268a}$$

holds. The matrix $\mathbf{C}$ in Equation 1.268a is defined by Equation 1.266c, and the matrix $\mathbf{J}$ of order $m \times m$ is a Jacobian matrix given by

$$\mathbf{J} = \frac{\partial N(\mathbf{w})}{\partial \mathbf{w}} = \begin{bmatrix} \frac{\partial \eta_1}{\partial w_1} & \frac{\partial \eta_1}{\partial w_2} & \cdots & \frac{\partial \eta_1}{\partial w_m} \\ \frac{\partial \eta_2}{\partial w_1} & \frac{\partial \eta_2}{\partial w_2} & \cdots & \frac{\partial \eta_2}{\partial w_m} \\ \vdots & \vdots & \vdots & \vdots \\ \frac{\partial \eta_m}{\partial w_1} & \frac{\partial \eta_m}{\partial w_2} & \cdots & \frac{\partial \eta_m}{\partial w_m} \end{bmatrix} \tag{1.268b}$$

The Jacobian matrix in Equation 1.268a is calculated at the point (of a vector space) $w_1 = w_2 = w_1 = \cdots = w_m = 0$.

For readers not familiar with the Jacobian matrix, we observe from Equation 1.268b that this matrix can be treated as a derivative of the vector $N(\mathbf{w})$ with respect to the vector $\mathbf{w}$. Because this derivative must take into account all the components of both the vectors $N(\mathbf{w})$ and $\mathbf{w}$, it assumes the form of a matrix of scalar derivatives, as given in Equation 1.268b.

Finally, note from Equation 1.268b that the components of the vector $N(\mathbf{w})$ are $\eta_i(\mathbf{w})$, $i = 1, 2, \cdots, m$, that is, $N(\mathbf{w})$ has the form

$$N(\mathbf{w}) = \begin{bmatrix} \eta_1(\mathbf{w}) \\ \eta_2(\mathbf{w}) \\ \vdots \\ \eta_m(\mathbf{w}) \end{bmatrix} \tag{1.268c}$$

We now introduce an auxiliary vector **p**

$$\mathbf{p} = \begin{bmatrix} p_1 \\ p_2 \\ \vdots \\ p_m \end{bmatrix}_{(i)} = \begin{bmatrix} \begin{bmatrix} \frac{\partial \eta_1}{\partial w_1} & \frac{\partial \eta_1}{\partial w_2} & \cdots & \frac{\partial \eta_1}{\partial w_m} \end{bmatrix} \cdot \begin{bmatrix} z_1 \\ z_2 \\ \vdots \\ z_m \end{bmatrix}_{(i)} \\ \begin{bmatrix} \frac{\partial \eta_2}{\partial w_1} & \frac{\partial \eta_2}{\partial w_2} & \cdots & \frac{\partial \eta_2}{\partial w_m} \end{bmatrix} \cdot \begin{bmatrix} z_1 \\ z_2 \\ \vdots \\ z_m \end{bmatrix}_{(i)} \\ \vdots \\ \begin{bmatrix} \frac{\partial \eta_m}{\partial w_1} & \frac{\partial \eta_m}{\partial w_2} & \cdots & \frac{\partial \eta_m}{\partial w_m} \end{bmatrix} \cdot \begin{bmatrix} z_1 \\ z_2 \\ \vdots \\ z_m \end{bmatrix}_{(i)} \end{bmatrix} \tag{1.269a}$$

Using then the vector **p** and the notation as in Equation 1.266a, we can rewrite Equation 1.268a in explicit form, as

$$\begin{bmatrix} w_1 \\ w_2 \\ \vdots \\ w_m \end{bmatrix}_{(k)} = \begin{bmatrix} z_1 \\ z_2 \\ \vdots \\ z_m \end{bmatrix}_{(k)} - \begin{bmatrix} \sum_{i=0}^{k} [c_{11}c_{12}\cdots c_{1m}]_{(k-i)} \cdot \begin{bmatrix} p_1 \\ p_2 \\ \vdots \\ p_m \end{bmatrix}_{(i)} \\ \sum_{i=0}^{k} [c_{21}c_{22}\cdots c_{2m}]_{(k-i)} \cdot \begin{bmatrix} p_1 \\ p_2 \\ \vdots \\ p_m \end{bmatrix}_{(i)} \\ \vdots \\ \sum_{i=0}^{k} [c_{m1}c_{m2}\cdots c_{mm}]_{(k-i)} \cdot \begin{bmatrix} p_1 \\ p_2 \\ \vdots \\ p_m \end{bmatrix}_{(i)} \end{bmatrix} \tag{1.269b}$$

Let us now solve Equation 1.268a for $\mathbf{z}(k)$. We start with Equation 1.269b and transform it into the Z-domain using the corresponding transform relations developed in Section 1.3. We get then

$$\begin{bmatrix} W_1(z) \\ W_2(z) \\ \vdots \\ W_m(z) \end{bmatrix} = \begin{bmatrix} Z_1(z) \\ Z_2(z) \\ \vdots \\ Z_m(z) \end{bmatrix} - \begin{bmatrix} C_{11}(z)P_1(z) + C_{12}(z)P_2(z) + \cdots + C_{1m}(z)P_m(z) \\ C_{21}(z)P_1(z) + C_{22}(z)P_2(z) + \cdots + C_{2m}(z)P_m(z) \\ \vdots \\ C_{m1}(z)P_1(z) + C_{m2}(z)P_2(z) + \cdots + C_{mm}(z)P_m(z) \end{bmatrix} \quad (1.270a)$$

where $W_1(z), \cdots, Z_1(z), \cdots, C_{11}(z), \cdots, P_1(z), \cdots$ are the Z transforms of $w_1(k), \cdots, z_1(k), \cdots, c_{11}(k), \cdots, p_1(k), \cdots$ respectively.

Moreover, transforming Equation 1.269a into the Z-domain, we obtain

$$\begin{bmatrix} P_1(z) \\ P_2(z) \\ \vdots \\ P_m(z) \end{bmatrix} = \mathbf{J} \cdot \begin{bmatrix} Z_1(z) \\ Z_2(z) \\ \vdots \\ Z_m(z) \end{bmatrix} \quad (1.270b)$$

where $P_1(z), \cdots, P_m(z)$ are the Z transforms of $p_1(k), \cdots, p_m(k)$, respectively. Substituting then Equation 1.270b into Equation 1.270a, and rewriting in a more compact form gives

$$\mathbf{W}(z) = \mathbf{Z}(z) - \mathbf{C}(z) \cdot \mathbf{P}(z) = \mathbf{Z}(z) - \mathbf{C}(z) \cdot \mathbf{J} \cdot \mathbf{Z}(z) \quad (1.271a)$$

Finally, solving Equation 1.271a for $\mathbf{Z}(z)$, we arrive at

$$\mathbf{Z}(z) = [\mathbf{I}_m - \mathbf{C}(z) \cdot \mathbf{J}]^{-1}\mathbf{W}(z) \quad (1.271b)$$

where $\mathbf{I}_m$ is the identity matrix of order $m \times m$. That is, $\mathbf{I}_m$ is given by

$$\mathbf{I}_m = \begin{matrix} & 1\ 2\ 3 \ldots m \\ \begin{matrix} 1 \\ 2 \\ 3 \\ \vdots \\ m \end{matrix} & \begin{bmatrix} 1 & 0 & 0 & \ldots & 0 \\ 0 & 1 & 0 & \ldots & 0 \\ 0 & 0 & 1 & \ldots & 0 \\ \vdots & \vdots & \vdots & \vdots & \vdots \\ 0 & 0 & 0 & \ldots & 1 \end{bmatrix} \end{matrix} \quad (1.271c)$$

Moreover, the -1 symbol in Equation 1.271b means the matrix inverse.

Applying the inverse Z transform defined by Equation 1.38, to Equation 1.271b, we get

$$\mathbf{z}(k) = Z^{-1}\{[\mathbf{I}_m - \mathbf{C}(z)\cdot\mathbf{J}]^{-1}\mathbf{W}(z)\} \tag{1.272a}$$

where Z^{-1} denotes the operation of transforming from the Z-domain into the discrete-time domain defined by Equation 1.38. We see from the above equation that, if the "input vector" $\mathbf{w}(k)$ is bounded, then the "output vector" $\mathbf{z}(k)$ is bounded as well, when the inverse matrix

$$[\mathbf{I}_m - \mathbf{C}(z)\cdot\mathbf{J}]^{-1} \tag{1.272b}$$

does exist. In other words, a hypothetical system with the "input" vector-valued sequence $\mathbf{w}(k)$ and "output" vector-valued sequence $\mathbf{z}(k)$ is bounded-input bounded-output stable.

Of course, it follows from the derivations presented above that assumption 3 is equivalent to the existence of the inverse matrix given by 1.272b. Moreover, the Jacobian matrix simplifies when the relation 1.267b in assumption 2 holds. Then we can write

$$\frac{\partial\eta_i}{\partial w_s} = \frac{\partial}{\partial w_s}\left(\sum_{j=1}^{j_0}\lambda_{ij}(w_i)\right) = \begin{cases} 0 & \text{for} \quad s \neq i \\ \displaystyle\sum_{j=1}^{j_0}\frac{\partial\lambda_{ij}(w_s)}{\partial w_s} & \text{for} \quad s = i \end{cases} \tag{1.273a}$$

which allows us to simplify Equation 1.268b to

$$\mathbf{J} = \begin{bmatrix} \frac{\partial\eta_1}{\partial w_1} & 0 & \dots & 0 \\ 0 & \frac{\partial\eta_2}{\partial w_2} & \dots & 0 \\ \vdots & \vdots & \vdots & \vdots \\ 0 & 0 & \dots & \frac{\partial\eta_m}{\partial w_m} \end{bmatrix} \tag{1.273b}$$

In other words, when the mapping N can be expressed by m single-input single-output nonlinear operations as in Equation 1.267b, then the corresponding Jacobian matrix is a diagonal matrix.

Let us now define Sandberg's "conversion" operators[3] S_t (truncating) and S_e (expanding). So the map $S_t: l^\infty(\mathbb{Z}) \to l^\infty(\mathbb{Z}_+)$ is such that it truncates a vector-valued sequence defined on $\mathbb{Z}$ into a vector-valued sequence defined on $\mathbb{Z}_+$. That is,

$$(S_t\mathbf{x})(k) = \mathbf{x}(k) \quad \text{for} \quad k \in \mathbb{Z}_+ \tag{1.274}$$

where $\mathbf{x}(k) \in l^\infty(\mathbb{Z})$. By the way, note that the operator S_t is identical with Sandberg's operator R_{k_0} defined in Section 1.8, when assuming $k_0 = 0$ in the latter and redefining to allow for vector-valued sequences to be incorporated.

Similarly, the map $S_e : l^\infty(\mathbb{Z}_+) \to l^\infty(\mathbb{Z})$ is such that it expands a vector-valued sequence defined on $\mathbb{Z}_+$ into a vector-valued sequence defined on $\mathbb{Z}$ according to the following relation:

$$(S_e\mathbf{x})(k) = \begin{cases} \mathbf{x}(k) & \text{for} \quad k \in \mathbb{Z}_+ \\ \boldsymbol{\theta}(k) & \text{for} \quad k \in (\mathbb{Z} - \mathbb{Z}_+) \end{cases} \tag{1.275}$$

where $\mathbf{x}(k) \in l^\infty(\mathbb{Z}_+)$ and $\boldsymbol{\theta}\,(k) \in l^\infty(\mathbb{Z})$.

In Section 1.6, the multiple-input and multiple-output systems have been considered. For such systems, the Volterra series representation has been presented, using the impulse responses in the form of matrices of order $M \times N^n$, where M meant the number of outputs and N meant the number of inputs. Now we adopt this kind of system description to our model of Figure 1.51. First, we recall what was assumed about the model of Figure 1.51: all the vectors shown in this figure have the same dimension m, including the input vector $\mathbf{x}$ and the output vector $\mathbf{y}$. Therefore, corresponding matrices $\mathbf{h}^{(n)}$, $n = 1, 2, 3, \cdots$, are of order $m \times m^n$ in our case of considering the model of Figure 1.51. Continuing, we denote by $H^{(n)}$, $n = 1, 2, 3, \cdots$, the set of all real $m \times m^n$ matrix-valued maps defined on $\mathbb{Z}_+{}^n$ such that each component of the related matrix $\mathbf{h}^{(n)}$ satisfies the condition

$$\sup_k \sum_{i_1=0}^{k} \sum_{i_2=0}^{k} \cdots \sum_{i_n=0}^{k} \left| h_{rs}^{(n)}(i_1, i_2, \cdots, i_n) \right| < \infty \tag{1.276}$$

where $r = 1, \cdots, m$ and $s = 1, \cdots, m^n$.

To be able to formulate Sandberg's theorem regarding the existence of strict-sense representations, published for the first time in Reference 3, we need yet to define a class of maps denoted by Sandberg as V_{h_n}. So assuming $\mathbf{h}^{(n)} \in H^{(n)}$, we define the map $V_{h_n} : l^\infty(\mathbb{Z}) \to l^\infty(\mathbb{Z})$ as

$$(V_{h_n}\mathbf{x})(k) \stackrel{df}{=} \sum_{i_1=-\infty}^{k} \sum_{i_2=-\infty}^{k} \cdots \sum_{i_n=-\infty}^{k} \mathbf{h}^{(n)}(k-i_1, k-i_2, \cdots, k-i_n) \cdot (\mathbf{x}(i_1) \otimes \mathbf{x}(i_2) \otimes \cdots \otimes \mathbf{x}(i_n)) \tag{1.277}$$

where the symbol $\otimes$ stands for the Kronecker product of matrices (see Section 1.6).

In what follows, when the symbol of the absolute value $|\cdot|$ is used with regard to a vector, it is taken in the sense of definition 1.157a. Similarly, when the norm symbol $\|\cdot\|$ regards a vector-valued sequence, it is taken in the sense of definition 1.157b.

Theorem 1.7 (Sandberg's theorem regarding existence of strict-sense representations)

Consider a nonlinear system belonging to a class of systems that can be modeled by Equations 1.264, and assume that the assumptions 1, 2, and 3 are met. Then there exist positive numbers ρ, σ_1, and σ_2, as well as elements $\mathbf{h}^{(1)}$, $\mathbf{h}^{(2)}$, $\cdots$ of the sets $H^{(1)}$, $H^{(2)}$, $\cdots$, respectively, and an open neighborhood $U \subset B_{\delta+}$ of the zero element $\boldsymbol{\theta}$, such that

(1) $\sigma_2\rho < 1$ holds.
(2) For each $\mathbf{x} \in B_{\delta+}$, there exists a unique $\mathbf{w}$, $\mathbf{z}$, and $\mathbf{y}$ of U, $l^\infty(\mathbb{Z}_+)$, and $l^\infty(\mathbb{Z}_+)$, respectively, satisfying Equations 1.264.
(3) The output can be expressed in the form of the series that follows.

$$\mathbf{y} = \sum_{n=1}^{\infty} S_t V_{h_n}(S_e \mathbf{x}), \ \mathbf{x} \in B_{\delta_+} \tag{1.278a}$$

of which components satisfy

$$\|S_t V_{h_n}(S_e \mathbf{x})\| \leq \sigma_1(\sigma_2 \|\mathbf{x}\|)^n \tag{1.278b}$$

for $\mathbf{x} \in l^\infty(\mathbb{Z}_+)$.

Comments regarding the proof of Theorem 1.7 can be found in Reference 3. Here, this proof is omitted.

Consider now the main result of Reference 3; that is, a theorem regarding extended representations. The term *extended representations*, or *extended expansions*, is used in Reference 3 as a synonym for the terms *associated representations*, or *associated expansions*. We formulate the theorem as follows.

Theorem 1.8 (Sandberg's theorem regarding existence of extended (associated) representations)

Let all the assumptions of Theorem 1.7 are valid and the positive numbers ρ, σ_1, and σ_2, and the elements $\mathbf{h}^{(1)}$, $\mathbf{h}^{(2)}$, $\cdots$ of the sets $H^{(1)}$, $H^{(2)}$, $\cdots$, respectively, as defined there. Then, the following holds:

(1) The series

$$S(\mathbf{x}) = \sum_{n=1}^{\infty} V_{h_h}(\mathbf{x}) \tag{1.279a}$$

converges uniformly with respect to $\mathbf{x} \in B_\delta \subset l^\infty(\mathbb{Z})$, where B_δ stands for the open ball in $l^\infty(\mathbb{Z})$ of radius ρ centered at the zero element of $l^\infty(\mathbb{Z})$. That is, for given $\varepsilon > 0$ the inequality

$$\sup_{k \in \mathbb{Z}} \left| \sum_{n=1}^{n'} V_{h_h}(\mathbf{x}) - \sum_{n=1}^{\infty} V_{h_h}(\mathbf{x}) \right| < \varepsilon \tag{1.279b}$$

holds for all $n' \geq n_0$ independently of the choice of $\mathbf{x} \in B_\delta$. Moreover, the series given by Equation 1.279a is called the extended or associated representation.

(2) If $(\mathbf{x}(k) - \mathbf{v}(k)) \to \boldsymbol{\theta}_m$ as $k \to \infty$, where $\mathbf{x}(k)$, $\mathbf{v}(k) \in B_\delta$, and $\boldsymbol{\theta}_m$ is the zero vector of order m,

$$\boldsymbol{\theta}_m = \begin{matrix} 1 \\ 2 \\ \vdots \\ m \end{matrix} \begin{bmatrix} 0 \\ 0 \\ \vdots \\ 0 \end{bmatrix}$$

then

$$(S\mathbf{x})(k) - (S\mathbf{v})(k) \to \boldsymbol{\theta}_m \quad \text{as } k \to \infty \tag{1.279c}$$

holds.

(3) For $q, n \in \{1, 2, \cdots\}$, let the mapping $T_q : H^{(n)} \to H^{(n)}$ be given by

$$(T_q \mathbf{h}^{(n)})(i_1, i_2, \ldots, i_n) = \begin{cases} \boldsymbol{\theta}_{m \times m^n} & \text{for } \max(i_1, i_2, \ldots, i_n) > q \\ \mathbf{h}^{(n)}(i_1, i_2, \ldots, i_n) & \text{otherwise} \end{cases} \tag{1.279d}$$

where $\boldsymbol{\theta}_{m \times m^n}$ denotes the zero matrix of order $m \times m^n$,

$$\boldsymbol{\theta}_{m \times m^n} = \begin{matrix} \\ 1 \\ 2 \\ \vdots \\ m \end{matrix} \begin{matrix} 1 & 2 & \ldots & m^n \\ \end{matrix} \begin{bmatrix} 0 & 0 & \ldots & 0 \\ 0 & 0 & \ldots & 0 \\ \vdots & \vdots & \vdots & \vdots \\ 0 & 0 & \ldots & 0 \end{bmatrix}$$

Then, for any positive $\varepsilon > 0$, there exist positive integers p and q such that the inequality

$$\left\| S(\mathbf{x}) - \sum_{n=1}^{p} V_{T_q h_n}(\mathbf{x}) \right\| < \varepsilon \tag{1.279e}$$

holds for all $\mathbf{x} \in B_\delta$.

By the way, note that the operator T_q defined by Equation 1.279d has nothing to do with the Sandberg's operator $T_{k_1k_2}$ defined in Section 1.8. They are quite different operators.

(4) For given $\varepsilon > 0$ and a positive integer p, there exist such a $\rho_1 \in (0, \rho)$ and a positive integer q that the following inequality:

$$\left\| S(\mathbf{x}) - \sum_{n=1}^{p} V_{T_q h_n}(\mathbf{x}) \right\| < \varepsilon \tag{1.279f}$$

holds for all $\mathbf{x} \in B_\delta$, where B_δ denotes the open ball in $l^\infty(\mathbb{Z})$ of radius ρ_1 centered at the zero element of $l^\infty(\mathbb{Z})$.

Because of the importance of Theorem 1.8, we present here its proof along the lines given by Sandberg in Reference 3. We show first that the following lemma holds.

Lemma 1.4. Let the assumptions of Theorem 1.8 be valid. Then, for each n and any $k_1 \in \mathbb{Z}$ and any real number $\gamma > 0$, there exists an integer $k_2 \in \{\cdots, 0, 1, \cdots, k_1 - k_1\}$ such that the following inequality

$$\left| V_{h_n}(\mathbf{x})(k) - V_{h_n}(\mathbf{x}_{k_2})(k) \right| \le \gamma \|\mathbf{x}\|^n,\ k \le k_1 \tag{1.280}$$

holds for all $\mathbf{x} \in l^\infty(\mathbb{Z})$. Moreover, it is assumed in inequality 1.280 that the vector-valued sequence $\mathbf{x}_{k_2}$ is given by $\mathbf{x}_{k_2}(k) = \mathbf{x}(k)$ for $k \ge k_2$ and $\mathbf{x}_{k_2}(k) = \mathbf{\theta}_m$ for each $k < k_2$.

Regarding inequality 1.280, we see that it resembles inequality 1.190 defining the so-called decaying memory (SDM1) property of an operator, which has been formulated in Section 1.8. Furthermore, extending inequality 1.190 to allow for vector-valued sequences, we get nothing other than inequality 1.280 after additionally making the following identifications: $N \leftrightarrow V_{h_n}$, $\varepsilon \leftrightarrow \gamma$, $m \leftrightarrow n$, and after extending the definition of the truncation operator Q_τ, given by Equation 1.189, to vector-valued sequences. Then, of course, the vector-valued sequence $\mathbf{x}_{k_2}(k)$ will be expressed with the use of the operator $Q_{\tau = k_2}$ by $\mathbf{x}_{k_2}(k) = (Q_{k_2}\mathbf{x})(k)$.

Note that we get as a byproduct of the above observation the evidence of possessing the property of decaying memory in the sense of Sandberg's first definition (SDM1) by the maps V_{h_n}.

To prove Lemma 1.4, observe first that using the defining Equation 1.277 for $V_{h_n}(\mathbf{x})$, the symbolic notation for components in the Kronecker product given by Equation 1.121, and the definition of the vector-valued sequence $\mathbf{x}_{k_2}(k)$, the left-hand side of inequality 1.280 can be rewritten as

$$\begin{array}{c} \text{1st} \\ \text{component} \\ \\ \vdots \\ \\ m\text{th} \\ \text{component} \end{array} \left| \begin{array}{c} \sum_{i_1=-\infty}^{k} \cdots \sum_{i_n=-\infty}^{k} \sum_{s=1}^{m^n} h_{1s}^{(n)}(k-i_1, \ldots, k-i_n)(x^n)_s - \\ - \sum_{i_1=k_2}^{k} \cdots \sum_{i_n=k_2}^{k} \sum_{s=1}^{m^n} h_{1s}^{(n)}(k-i_1, \ldots, k-i_n)(x^n)_s \\ \vdots \\ \sum_{i_1=-\infty}^{k} \cdots \sum_{i_n=-\infty}^{k} \sum_{s=1}^{m^n} h_{ms}^{(n)}(k-i_1, \ldots, k-i_n)(x^n)_s - \\ - \sum_{i_1=k_2}^{k} \cdots \sum_{i_n=k_2}^{k} \sum_{s=1}^{m^n} h_{ms}^{(n)}(k-i_1, \ldots, k-i_n)(x^n)_s \end{array} \right| \tag{1.281a}$$

For a given k, let the maximal (in the sense of absolute value) component of the vector $V_{h_n}(\mathbf{x})(k) - V_{h_n}(\mathbf{x}_{k_2})(k)$ given in the explicit form in expression 1.281a, occur for, say, $r = r_0$. According to 1.281a, this component has the form

$$\sum_{i_1=-\infty}^{k} \cdots \sum_{i_n=-\infty}^{k} \sum_{s=1}^{m^n} h_{r_0 s}^{(n)}(k-i_1, \ldots, k-i_n)(x^n)_s - \sum_{i_1=k_2}^{k} \cdots \sum_{i_n=k_2}^{k} \sum_{s=1}^{m^n} h_{r_0 s}^{(n)}(k-i_1, \ldots, k-i_n)(x^n)_s \tag{1.281b}$$

Let us now take the absolute value of the above component and introduce new variables $k - i_1 = i_1'$, $\cdots$, $k - i_n = i_n'$. To perform this operation correctly, we also need to show the arguments in $(x^n)_s$. Hence, we get

$$\left| \sum_{i'_1=0}^{\infty} \cdots \sum_{i'_n=0}^{\infty} \sum_{s=1}^{m^n} h_{r_0 s}^{(n)}(i'_1, \ldots, i'_n)(x(k-i'_1)\cdots x(k-i'_n))_s \right. \tag{1.281c}$$
$$\left. - \sum_{i'_1=0}^{k-k_2} \cdots \sum_{i'_n=0}^{k-k_2} \sum_{s=1}^{m^n} h_{r_0 s}^{(n)}(i'_1, \ldots, i'_n)(x(k-i'_1)\cdots x(k-i'_n))_s \right|$$

Observe that some components in the expression under the absolute value in expression 1.281c have the same magnitude. They differ only in signs. In each case, one of them is preceded by a plus sign, and the second by, a minus sign. When added to each other, they give, in effect, zero. After grouping these components in pairs, we get the following expressions:

$$\sum_{s=1}^{m^n} (h_{r_0 s}^{(n)}(i'_1, \ldots, i'_n) - h_{r_0 s}^{(n)}(i'_1, \ldots, i'_n))(x(k-i'_1)\cdots x(k-i'_n))_s \tag{1.281d}$$

where $i_1', \cdots, i_n' \in \{0, \cdots, k - k_2\}$.

Let us now consider each of the components under the summation symbols in expression 1.281c and use the triangle inequality

$$|a+b+c\ldots| \le |a| + |b| + |c| + \cdots$$

where a, b, c, ... stand for the grouped components in the form

$$\sum_{s=1}^{m^n} h_{r_0 s}^{(n)}(i'_1, \ldots, i'_n)(x(k-i'_1)\cdots x(k-i'_n))_s \tag{1.281e}$$

or in the form given by expression 1.281d. Furthermore, in the cases where $|a|$, $|b|$, $|c|$, $\cdots$ are equal to zero, let us use additionally the following obvious inequalities:

$$|0| = |a-a| \le |a| - |a|,\ 101 = |\mathrm{b}-\mathrm{b}| \le |\mathrm{b}| - |\mathrm{b}|, |0| = |\mathrm{c}-\mathrm{c}| \le |\mathrm{c}| - |\mathrm{c}|$$

It follows then that the absolute value of the expression given by 1.281c is equal to or less than the value of following expression:

$$\begin{aligned} &\sum_{i'_1=0}^{\infty} \cdots \sum_{i'_n=0}^{\infty} \left| \sum_{s=1}^{m^n} h_{r_0 s}^{(n)}(i'_1, \ldots, i'_n)(x(k-i'_1)\cdots x(k-i'_n))_s \right| \\ &- \sum_{i'_1=0}^{k-k_2} \cdots \sum_{i'_n=0}^{k-k_2} \left| \sum_{s=1}^{m^n} h_{r_0 s}^{(n)}(i'_1, \ldots, i'_n)(x(k-i'_1)\cdots x(k-i'_n))_s \right| \end{aligned} \tag{1.281f}$$

Obviously, the expressions $x(k-i'_1)\cdots x(k-i'_n)$ occurring in expression 1.281f are bounded, and we have

$$|x(k-i'_1)\cdots x(k-i'_n)| \le \|\mathbf{x}\|^n \tag{1.281g}$$

for any set of k, $i'_{1,i}$, $\cdots$, i'_n. Moreover, using the triangle inequality again and expression 1.281g, we get for $\sum_{s=1}^{m^n} h_{r_0 s}^{(n)}(\cdot)(x(\cdot)\cdots x(\cdot))_s$

$$\begin{aligned} &\left| \sum_{s=1}^{m^n} h_{r_0 s}^{(n)}(i'_1, \ldots, i'_n)(x(k-i'_1)\cdots x(k-i'_n))_s \right| \\ &\le \sum_{s=1}^{m^n} \left| h_{r_0 s}^{(n)}(i'_1, \ldots, i'_n) \right| (|x(k-i'_1)\cdots x(k-i'_n)|)_s \\ &\le \sum_{s=1}^{m^n} \left| h_{r_0 s}^{(n)}(i'_1, \ldots, i'_n) \right| (\|\mathbf{x}\|)^n = (\|\mathbf{x}\|)^n \sum_{s=1}^{m^n} \left| h_{r_0 s}^{(n)}(i'_1, \ldots, i'_n) \right| \end{aligned} \tag{1.281h}$$

Applying then inequality 1.281h to expression 1.281f, dropping the prime at each i'_i, and coming back to expressions 1.281c and 1.281b, we write

$$\left|\sum_{i_1=-\infty}^{k}\cdots\sum_{i_n=-\infty}^{k}\sum_{s=1}^{m^n}h_{r_0s}^{(n)}(k-i_1,\ldots,k-i_n)(x^n)_s\right. \tag{1.281i}$$
$$\left.-\sum_{i_1=k_2}^{k}\cdots\sum_{i_n=k_2}^{k}\sum_{s=1}^{m^n}h_{r_0s}^{(n)}(k-i_1,\ldots,k-i_n)(x^n)_s\right|$$
$$\leq\left[\sum_{i_1=0}^{\infty}\cdots\sum_{i_n=0}^{\infty}\sum_{s=1}^{m^n}\left|h_{r_0s}^{(n)}(i_1,\ldots,i_n)\right|-\sum_{i_1=0}^{k-k_2}\cdots\sum_{i_n=0}^{k-k_2}\sum_{s=1}^{m^n}\left|h_{r_0s}^{(n)}(i_1,\ldots,i_n)\right|\right]\cdot(\|\mathbf{x}\|)^n$$

Recall that it has been assumed $k \geq k_1 > k_2$ in the formulation of Lemma 1.4. Hence, we have $k \geq k_1$ in inequality 1.281i. This implies the following inequality:

$$\sum_{i_1=0}^{k_1-k_2}\cdots\sum_{i_n=0}^{k_1-k_2}\sum_{s=1}^{m^n}\left|h_{r_0s}^{(n)}(i_1,\ldots,i_n)\right|\leq\sum_{i_1=0}^{k-k_2}\cdots\sum_{i_n=0}^{k-k_2}\sum_{s=1}^{m^n}\left|h_{r_0s}^{(n)}(i_1,\ldots,i_n)\right|$$

Using the above inequality in 1.281i gives, finally

$$\left|\sum_{i_1=-\infty}^{k}\cdots\sum_{i_n=-\infty}^{k}\sum_{s=1}^{m^n}h_{r_0s}^{(n)}(k-i_1,\ldots,k-i_n)(x^n)_s\right. \tag{1.281j}$$
$$\left.-\sum_{i_1=k_2}^{k}\cdots\sum_{i_n=k_2}^{k}\sum_{s=1}^{m^n}h_{r_0s}^{(n)}(k-i_1,\ldots,k-i_n)(x^n)_s\right|$$
$$\leq\left[\sum_{i_1=0}^{\infty}\cdots\sum_{i_n=0}^{\infty}\sum_{s=1}^{m^n}\left|h_{r_0s}^{(n)}(i_1,\ldots,i_n)\right|-\sum_{i_1=0}^{k-k_2}\cdots\sum_{i_n=0}^{k-k_2}\sum_{s=1}^{m^n}\left|h_{r_0s}^{(n)}(i_1,\ldots,i_n)\right|\right]\cdot(\|\mathbf{x}\|)^n$$
$$\leq\left[\sum_{i_1=0}^{\infty}\cdots\sum_{i_n=0}^{\infty}\sum_{s=1}^{m^n}\left|h_{r_0s}^{(n)}(i_1,\ldots,i_n)\right|-\sum_{i_1=0}^{k_1-k_2}\cdots\sum_{i_n=0}^{k_1-k_2}\sum_{s=1}^{m^n}\left|h_{r_0s}^{(n)}(i_1,\ldots,i_n)\right|\right]\cdot(\|\mathbf{x}\|)^n$$

Inequality 1.281j ends the proof of Lemma 1.4 because the value of the expression in brackets on the extreme right-hand side of inequality 1.281j can be made arbitrarily small by the choice of k_2 sufficiently small. That is, letting $k_1 - k_2$ to be nearer to the value of the upper summation limit ∞ occurring in sums of the first component in brackets in 1.281j has as an effect the decrease of the difference between the first and the second component in brackets in 1.281j. This difference can be made arbitrarily small. Moreover, note that the bound on the extreme right-hand side of 1.281j is independent of k. However, it can happen that this bound takes on the maximal value for some other index r'_0, different from that assumed initially for a given k. Note that this does not create any problem because we take the difference for the index r'_0 and make this difference as small as needed by choosing k_2 sufficiently small. In conclusion, the inequality 1.280 is satisfied in each case. Moreover, it can be rewritten in the following form:

$$\left\|V_{h_n}(\mathbf{x})(k)-V_{h_n}(\mathbf{x}_{k_2})(k)\right\|\leq\gamma\|\mathbf{x}\|^n \tag{1.281k}$$

because inequality 1.280 holds for any $k_1 \in \mathbb{Z}$.

Before going further, let us first show that the series $\sum_{n=1}^{\infty} S_t V_{h_n}(S_e \mathbf{x})$ from Theorem 1.7 converges uniformly. To do this, we take any positive number $\varepsilon > 0$ and ask under what conditions the following inequality

$$\left\| \sum_{n=1}^{n'} S_t V_{h_n}(S_e \mathbf{x}) - \sum_{n=1}^{\infty} S_t V_{h_n}(S_e \mathbf{x}) \right\| < \varepsilon \tag{1.282a}$$

holds. In inequality 1.282a $\mathbf{x} \in B_{\delta+} \subset l^{\infty}(\mathbb{Z}_+)$.

According to point 3 of Theorem 1.7, the series 1.278a converges. So inequality 1.282a must hold for all n' up to some n_0. Furthermore, note that inequality 1.282a can be rewritten as

$$\left\| \sum_{n=n'+1}^{\infty} S_t V_{h_n}(S_e \mathbf{x}) \right\| < \varepsilon \tag{1.282b}$$

On the other hand, observe that, using inequality 1.278b and the triangle inequality, we can write

$$\left\| \sum_{n=n'+1}^{\infty} S_t V_{h_n}(S_e \mathbf{x}) \right\| \le \sum_{n=n'+1}^{\infty} \sigma_1 (\sigma_2 \|\mathbf{x}\|)^n \tag{1.282c}$$

Under the assumptions of Theorem 1.7, $\|\mathbf{x}\| < \rho$ and $\sigma_2 \rho < 1$, inequality 1.282c can be put into the form

$$\left\| \sum_{n=n'+1}^{\infty} S_t V_{h_n}(S_e \mathbf{x}) \right\| \le \sum_{n=n'+1}^{\infty} \sigma_1 (\sigma_2 \rho)^n = \frac{\sigma_1 (\sigma_2 \rho)^{n'+1}}{1 - \sigma_2 \rho} \tag{1.282d}$$

where the expression on the extreme right side of relation 1.282d follows from the formula for the infinite convergent geometric series.

Note that, given any $\varepsilon > 0$, n' can be so chosen that the inequality

$$\frac{\sigma_1 (\sigma_2 \rho)^{n'+1}}{1 - \sigma_2 \rho} < \varepsilon \tag{1.282e}$$

holds. This is possible because, as assumed, $\sigma_2 \rho < 1$. Furthermore, observe that inequality 1.282a holds independently of the choice of $\mathbf{x} \in B_{\delta+}$. That is, the series 1.278a converges uniformly with respect to the vector-valued input sequence $\mathbf{x}$.

Let us now consider the term $\| V_{h_n}(\mathbf{x})(k) \|$, where $\mathbf{x} \in B_{\delta} \subset l^{\infty}(\mathbb{Z})$. Observe that this term can be written in the following equivalent form:

$$\left\| V_{h_n}(\mathbf{x})(k) \right\| = \left\| V_{h_n}(\mathbf{x})(k) - V_{h_n}(\mathbf{x}_{k_2})(k) + V_{h_n}(\mathbf{x}_{k_2})(k) \right\| \tag{1.283a}$$

where $\mathbf{x}_{k_2}$ is defined as in inequality 1.280. Obviously, $\|\mathbf{x}_{k_2}\| \le \|\mathbf{x}\|$, holds.

To proceed further with Equation 1.283a, we need some bound on the expression $V_{h_n}(\mathbf{x}_{k_2})(k)$. To arrive at such a bound, consider first the vector-valued sequence $\mathbf{x}_{k_2}(k) \in l^\infty(\mathbb{Z})$ occurring in $V_{h_n}(\mathbf{x}_{k_2})(k)$. This sequence has the form shown below.

$$\begin{array}{l} \text{discrete} \\ \text{time} \rightarrow \end{array} \quad \cdots \; k_2-2 \quad k_2-1 \qquad k_2 \qquad\qquad k_2+1 \qquad \cdots$$

$$\mathbf{x}_{k_2}(k) = \left\{ \ldots, \begin{bmatrix} 0 \\ 0 \\ \vdots \\ 0 \end{bmatrix}, \begin{bmatrix} 0 \\ 0 \\ \vdots \\ 0 \end{bmatrix}, \begin{bmatrix} x_{(k_2)1}(k_2) \\ x_{(k_2)2}(k_2) \\ \vdots \\ x_{(k_2)m}(k_2) \end{bmatrix}, \begin{bmatrix} x_{(k_2)1}(k_2+1) \\ x_{(k_2)2}(k_2+1) \\ \vdots \\ x_{(k_2)m}(k_2+1) \end{bmatrix}, \ldots \right\}$$

Note that, without loss of generality, we can assume k_2 to be negative in $\mathbf{x}_{k_2}(k)$. Hence, when we shift the sequence $\mathbf{x}_{k_2}(k)$ by k_2 discrete-time units on the discrete-time axis, this means shifting to the right such that all the nonzero elements of the sequence occur for times $k \ge 0$, and all the elements being identically zero vectors occur for times $k < 0$. We then get the following vector-valued sequence:

$$\begin{array}{l} \text{discrete} \\ \text{time} \rightarrow \end{array} \quad \cdots \quad -2 \quad -1 \qquad 0 \qquad\qquad 1 \qquad \cdots$$

$$\mathbf{x}_s(k) = \mathbf{x}_{k_2}(k+k_2) = \left\{ \ldots, \begin{bmatrix} 0 \\ 0 \\ \vdots \\ 0 \end{bmatrix}, \begin{bmatrix} 0 \\ 0 \\ \vdots \\ 0 \end{bmatrix}, \begin{bmatrix} x_{(k_2)1}(k_2) \\ x_{(k_2)2}(k_2) \\ \vdots \\ x_{(k_2)m}(k_2) \end{bmatrix}, \begin{bmatrix} x_{(k_2)1}(k_2+1) \\ x_{(k_2)2}(k_2+1) \\ \vdots \\ x_{(k_2)m}(k_2+1) \end{bmatrix}, \ldots \right\}$$

Using Equation 1.277, we now write the expression $(V_{h_n}\mathbf{x}_s)(k)$ for the sequence $\mathbf{x}_s(k) = \mathbf{x}_{k_2}(k + k_2)$. That is,

$$\begin{aligned} (V_{h_n}\mathbf{x}_s)(k) &= (V_{h_n}\mathbf{x}_{k_2})(k+k_2) \\ &= \sum_{i_1=-\infty}^{k} \cdots \sum_{i_n=-\infty}^{k} \mathbf{h}^{(n)}(k-i_1, \ldots, k-i_n) \cdot (\mathbf{x}_s(i_1) \otimes \cdots \otimes \mathbf{x}_s(i_n)) \\ &= \sum_{i_1=-\infty}^{k} \cdots \sum_{i_n=-\infty}^{k} \mathbf{h}^{(n)}(k-i_1, \ldots, k-i_n) \cdot (\mathbf{x}_{k_2}(i_1+k_2) \otimes \cdots \otimes \mathbf{x}_{k_2}(i_n+k_2)), \; k \in \mathbb{Z} \end{aligned} \tag{1.283b}$$

Note that, because of the form of the sequence $\mathbf{x}_s(k)$, the expression 1.283b simplifies for nonnegative times to

$$(V_{h_n}\mathbf{x}_s)(k) = (V_{h_n}\mathbf{x}_{k_2})(k+k_2)$$

$$= \sum_{i_1=0}^{k} \cdots \sum_{i_n=0}^{k} \mathbf{h}^{(n)}(k-i_1, \ldots, k-i_n) \cdot (\mathbf{x}_s(i_1) \otimes \ldots \otimes \mathbf{x}_s(i_n))$$

$$= \sum_{i_1=0}^{k} \cdots \sum_{i_n=0}^{k} \mathbf{h}^{(n)}(k-i_1, \ldots, k-i_n) \cdot (\mathbf{x}_{k_2}(i_1+k_2) \otimes \ldots \otimes \mathbf{x}_{k_2}(i_n+k_2)), k \in \mathbb{Z}_+ \tag{1.283c}$$

Observe at this point that the expression for $S_t V_{h_n}(\mathbf{x}_s)(k)$, has the same form and the same value for each $k \geq 0$ as the expression given by Equation 1.283c. Moreover, we get $(V_{h_n}\mathbf{x}_s)(k) = \mathbf{\theta}_m$ for each $k < 0$ from expression 1.283b. The above results allow us to write

$$\|(V_{h_n}\mathbf{x}_s)(k)\| = \|S_t V_{h_n}(\mathbf{x}_s)(k)\| \leq \sigma_1(\sigma_2\|\mathbf{x}_s\|)^n \leq \sigma_1(\sigma_2\|\mathbf{x}\|)^n \tag{1.283d}$$

where $V_{h_n}(\mathbf{x}_{k_2})(k) \in l^\infty(\mathbb{Z})$ and $S_t V_{hn}(\mathbf{x}_s)(k) \in l^\infty(\mathbb{Z}_+)$. The bounds on the right-hand side of relation 1.283d follow from inequality1.278b and the fact that $\|\mathbf{x}_s\| = \|\mathbf{x}_{k_2}\| \leq \|\mathbf{x}\|$.

Performing the time-shifting back to the left by $-k_2$ discrete-time units on the vector-valued sequence $(V_{h_n}\mathbf{x}_s)$, given by Equation 1.283b, we get

$$(V_{h_n}\mathbf{x}_{k_2})(k+k_2-k_2)$$

$$= \sum_{i_1=-\infty}^{k-k_2} \cdots \sum_{i_n=-\infty}^{k-k_2} \mathbf{h}^{(n)}(k-k_2-i_1, \ldots, k-k_2-i_n) \cdot (\mathbf{x}_{k_2}(i_1+k_2) \otimes \ldots \otimes \mathbf{x}_{k_2}(i_n+k_2)) \tag{1.283e}$$

Furthermore, introducing new variables $i'_1 = i_1 + k_2, \cdots, i'_n = i_n + k_2$ in Equation 1.283e, we obtain

$$(V_{h_n}\mathbf{x}_{k_2})(k) = \sum_{i'_1=-\infty}^{k} \cdots \sum_{i'_n=-\infty}^{k} \mathbf{h}^{(n)}(k-i'_1, \ldots, k-i'_n) \cdot (\mathbf{x}_{k_2}(i'_1) \otimes \ldots \otimes \mathbf{x}_{k_2}(i'_n)) \tag{1.283f}$$

And after dropping the prime at each i'_i in Equation 1.283f, we arrive at

$$(V_{h_n}\mathbf{x}_{k_2})(k) = \sum_{i_1=-\infty}^{k} \cdots \sum_{i_n=-\infty}^{k} \mathbf{h}^{(n)}(k-i_1, \ldots, k-i_n) \cdot (\mathbf{x}_{k_2}(i_1) \otimes \ldots \otimes \mathbf{x}_{k_2}(i_n)) \tag{1.283g}$$

The expression 1.283g is identical to the expression that can be obtained for $(V_{h_n}\mathbf{x}_{k_2})(k)$ from the defining Equation 1.277.

Observe that the norm of any vector-valued sequence belonging to the space $l^\infty(\mathbb{Z})$ does not change under the operation of time shifting. In particular, the above rule applies to the sequence $V_{h_n}(\mathbf{x}_{k_2})(k) \in l^\infty(\mathbb{Z})$. With this fact and the result given by inequality 1.283d taken into account, we can write the following:

$$\|(V_{h_n}\mathbf{x}_{k_2})(k)\| \leq \sigma_1(\sigma_2\|\mathbf{x}\|)^n \tag{1.283h}$$

Recall at this point that $(V_{h_n}\mathbf{x}_{k_2})(k)$ and $V_{h_n}(\mathbf{x}_{k_2})(k)$ have exactly the same meaning according to the notational convention assumed for writing mappings (operators) in Section 1.7.

Applying the triangle inequality, and inequalities 1.281k and 1.283h in 1.283a, we get

$$\|(V_{h_n}\mathbf{x})(k)\| \leq \|V_{h_n}(\mathbf{x})(k) - V_{h_n}(\mathbf{x}_{k_2})(k)\| + \|V_{h_n}(\mathbf{x}_{k_2})(k)\| \leq \gamma\|\mathbf{x}\|^n + \sigma_1(\sigma_2\|\mathbf{x}\|)^n \tag{1.283i}$$

On the other hand, we know from the preceding considerations that γ in inequality 1.283i can be made arbitrarily small by letting $k_2 \to -\infty$. In so doing, inequality 1.283i takes on the following form:

$$\|(V_{h_n}\mathbf{x})(k)\| \leq \sigma_1(\sigma_2\|\mathbf{x}\|)^n \tag{1.283j}$$

Inequality 1.283j allows us to prove point 1 of Theorem 1.8. That is, the series $S(\mathbf{x})$ given by Equation 1.279a converges because

$$\left\|\sum_{n=1}^{\infty} V_{h_n}(\mathbf{x})(k)\right\| \leq \sum_{n=1}^{\infty} \|V_{h_n}(\mathbf{x})(k)\| \tag{1.284}$$

$$\leq \sum_{n=1}^{\infty} \sigma_1(\sigma_2\|\mathbf{x}\|)^n \leq \sum_{n=1}^{\infty} \sigma_1(\sigma_2\rho)^n = \frac{\sigma_1\sigma_2\rho}{1-\sigma_2\rho} < \infty$$

holds, when $\sigma_2\rho < 1$, as assumed.

Because the proof of the uniform convergence of the series $S(\mathbf{x})$ given by Equation 1.279a is the same as the proof presented in inequalities 1.282a to 1.282e for the series $\sum_{n=1}^{\infty} S_t V_{h_n}(S_e\mathbf{x})$, it is omitted here. This proof relies on the property given by inequality 1.283j, the counterpart of the property 1.278b for sequences belonging to the space $l^\infty(\mathbb{Z}_+)$.

To prove point 2 of Theorem 1.8, consider the absolute value of the difference $|(S\mathbf{x})(k) - (S\mathbf{v})(k)|$, which, using 1.279a, can be rewritten as

$$|(S\mathbf{x})(k) - (S\mathbf{v})(k)| = \left|\sum_{n=1}^{\infty} (V_{h_n}(\mathbf{x})(k) - V_{h_n}(\mathbf{v})(k))\right| \tag{1.285a}$$

Furthermore, by applying the triangle inequality to the components on the right-hand side of Equation 1.285a, we can write

$$\left|\sum_{n=1}^{\infty}(V_{h_n}(\mathbf{x})(k)-V_{h_n}(\mathbf{v})(k))\right| \le \left|\sum_{n=1}^{p}(V_{h_n}(\mathbf{x})(k)-V_{h_n}(\mathbf{v})(k))\right| + \left|\sum_{n=p+1}^{\infty}(V_{h_n}(\mathbf{x})(k)-V_{h_n}(\mathbf{v})(k))\right| \quad (1.285b)$$

for any $p \in \mathbb{N}$.

Given any $\varepsilon_1 > 0$, assume now that

$$|\mathbf{x}(k)-\mathbf{v}(k)| < \varepsilon_1 \quad (1.286)$$

holds for all $k \ge k_3$. Because we can go in inequality 1.286 with $\varepsilon_1 \to 0$, this means that $\mathbf{x}(k) \to \mathbf{v}(k)$ for $k \to \infty$. To apply inequality 1.286 in 1.285b, we need some other results. The property of uniform continuity can help, so we concentrate on proving this property for operators V_{h_n}. Using Equations 1.277 and 1.121 for symbolic notation of Kronecker products, the difference $|V_{h_n}(\mathbf{x})(k) - V_{h_n}(\mathbf{v})(k)|$ can be expressed as follows:

$$|V_{h_n}(\mathbf{x})(k)-V_{h_n}(\mathbf{v})(k)| = \left\| \begin{bmatrix} \left[V_{h_n}(\mathbf{x})(k)-V_{h_n}(\mathbf{v})(k)\right]_1 \\ \left[V_{h_n}(\mathbf{x})(k)-V_{h_n}(\mathbf{v})(k)\right]_2 \\ \vdots \\ \left[V_{h_n}(\mathbf{x})(k)-V_{h_n}(\mathbf{v})(k)\right]_m \end{bmatrix} \right\| \quad (1.287a)$$

where the rth, $r = 1, \cdots, m$, component of the vector is given by

$$[V_{h_n}(\mathbf{x})(k)-V_{h_n}(\mathbf{v})(k)]_r = \sum_{i_1=-\infty}^{k} \cdots \sum_{i_n=-\infty}^{k} \sum_{s=1}^{m^n} h_{rs}^{(n)}(k-i_1, \ldots, k-i_n)((x^n)_s-(v^n)_s) \quad (1.287b)$$

Note that the application of the algebraic equality

$$a^n - b^n = (a-b)(a^{n-1}+a^{n-2}b+\ldots+ab^{n-2}+b^{n-1}),\ a, b \in \mathbb{R} \quad (1.288a)$$

to the expression $((x^n)_s - (v^n)_s)$ in Equation 1.287b allows us to write it in the form

$$((x^n)_s-(v^n)_s) = ((x-v)(x^{n-1}+x^{n-2}v\ldots+xv^{n-2}+v^{n-1}))_s \quad (1.288b)$$

Substituting then Equation 1.288b into 1.287b, taking the absolute value and applying the corresponding bounds to the components of the sequences **x** and **v**, we get

$$\left|[V_{h_n}(\mathbf{x})(k) - V_{h_n}(\mathbf{v})(k)]_r\right| \tag{1.289a}$$

$$\leq \|\mathbf{x}-\mathbf{v}\|\Bigg((\|\mathbf{x}\|^{n-1} + \|\mathbf{x}\|^{n-2}\|\mathbf{v}\| + \cdots + \|\mathbf{x}\|\,\|\mathbf{v}\|^{n-2} + \|\mathbf{v}\|^{n-1})$$

$$\bullet \sum_{i_1=-\infty}^{k} \cdots \sum_{i_n=-\infty}^{k} \sum_{s=1}^{m^n} \left|h_{rs}^{(n)}(k-i_1, \ldots, k-i_n)\right| \Bigg)$$

We obtain further simplification of inequality 1.289a using the procedure of introducing new variables $k - i_1 = i'_1$ and then dropping the prime at each $i',\ldots,_1$ and assuming that $\mathbf{x}, \mathbf{v} \in B_\delta$, that is, $\|\mathbf{x}\|$, $\|\mathbf{v}\|$ are less than ρ. Then, we arrive at

$$\left|[V_{h_n}(\mathbf{x})(k) - V_{h_n}(\mathbf{v})(k)]_r\right| \leq \|\mathbf{x}-\mathbf{v}\|(n\rho^{n-1}) \cdot \sum_{i_1=0}^{k} \cdots \sum_{i_n=0}^{\infty} \sum_{s=1}^{\infty} \left|h_{rs}^{(n)}(i_1, \ldots, i_n)\right| \tag{1.289b}$$

Furthermore, observe that the expression on the right-hand side of inequality 1.289b does not depend upon k. Moreover, denote by β_n the following expression:

$$\beta_n = \max_{1\leq r\leq m}\left(\sum_{s=1}^{m^n} \sum_{i_1=0}^{\infty} \cdots \sum_{i_n=0}^{\infty} \left|h_{rs}^{(n)}(i_1, \ldots, i_n)\right| \right) \tag{1.289c}$$

Applying expression 1.289c to inequality 1.289b, and then coming back to Equation 1.287a allows us to write for the latter

$$\left|V_{h_n}(\mathbf{x})(k) - V_{h_n}(\mathbf{v})(k)\right| \leq \left\|V_{h_n}(\mathbf{x}) - V_{h_n}(\mathbf{v})\right\| \leq \|\mathbf{x}-\mathbf{v}\|(n\rho^{n-1})\beta_n \tag{1.289d}$$

From inequality 1.289d, it is evident that $V_{h_n}(\mathbf{x})$ is uniformly continuous. This is so because, according to the uniform continuity definition, for any ε_2 such that $|V_{h_n}(\mathbf{x})(k) - V_{h_n}(\mathbf{v})(k)| < \varepsilon_2$ we can choose such ε_1 that $\|\mathbf{x} - \mathbf{v}\| < \varepsilon_1$, and the latter inequality implies the first inequality for all **x** and **v** belonging to B_ρ (and satisfying $\|\mathbf{x} - \mathbf{v}\| < \varepsilon_1$). In fact, choosing $\|V_{h_n}(\mathbf{x})(k) - V_{h_n}(\mathbf{v})(k)\| < \varepsilon_2$, we get from inequality 1.289d $\varepsilon_1 = \varepsilon_2(n\rho^{n-1}\beta_n)^{-1}$ as a bound on $\|\mathbf{x} - \mathbf{v}\|$. Moreover, it follows then from 1.289d that $\|\mathbf{x} - \mathbf{v}\| < \varepsilon_1$ implies $|V_{h_n}(\mathbf{x})(k) - V_{h_n}(\mathbf{v})(k)| < \varepsilon_2$.

With the uniform continuity of the operator V_{hn} already proven, we now return to inequality 1.285b and rewrite it in the following way:

$$\left|\sum_{n=1}^{\infty}(V_{h_n}(\mathbf{x})(k) - V_{h_n}(\mathbf{v})(k))\right| \tag{1.290}$$
$$\leq \left|\sum_{n=1}^{p}(V_{h_n}(\mathbf{x})(k) - V_{h_n}(\mathbf{v})(k))\right| + \sum_{n=p+1}^{\infty}\left|V_{h_n}(\mathbf{x})(k)\right| + \sum_{n=p+1}^{\infty}\left|V_{h_n}(\mathbf{v})(k)\right|$$

Given any $\varepsilon_2 > 0$, we choose then such p that

$$\left|\sum_{n=1}^{\infty} V_{h_n}(\mathbf{x})(k) - \sum_{n=1}^{p} V_{h_n}(\mathbf{x})(k)\right| = \left|\sum_{n=p+1}^{\infty} V_{h_n}(\mathbf{x})(k)\right| \leq \frac{\sigma_1(\sigma_2\rho)^{p+1}}{1-\sigma_2\rho} \leq \frac{\varepsilon_2}{4} \tag{1.291a}$$

and accordingly,

$$\left|\sum_{n=1}^{\infty} V_{h_n}(\mathbf{v})(k) - \sum_{n=1}^{p} V_{h_n}(\mathbf{v})(k)\right| = \left|\sum_{n=p+1}^{\infty} V_{h_n}(\mathbf{v})(k)\right| \leq \frac{\sigma_1(\sigma_2\rho)^{p+1}}{1-\sigma_2\rho} \leq \frac{\varepsilon_2}{4} \tag{1.291b}$$

The validity of inequalities 1.291a and 1.291b follows, of course, from the fact already proven that the series $\sum_{n=1}^{\infty} V_{h_n}(\mathbf{x})$ and $\sum_{n=1}^{\infty} V_{h_n}(\mathbf{v})$ converge for the vector-valued sequences belonging to B_ρ.

Let us now consider the first component on the right-hand side of inequality 1.290. Note that this component can be rewritten, without changing its value, in the following way:

$$\left|\sum_{n=1}^{\infty} V_{h_n}(\mathbf{x})(k) - V_{h_n}(\mathbf{v})(k)\right| \tag{1.292a}$$
$$= \left|\sum_{n=1}^{p}(V_{h_n}(\mathbf{x})(k) - V_{h_n}(\mathbf{x}_{k_2})(k) - V_{h_n}(\mathbf{v})(k)\right.$$
$$\left.+ V_{h_n}(\mathbf{v}_{k_2})(k) + V_{h_n}(\mathbf{x}_{k_2})(k) - V_{h_n}(\mathbf{v}_{k_2})(k))\right|$$
$$\leq \sum_{n=1}^{p}\left|V_{h_n}(\mathbf{x})(k) + -V_{h_n}(\mathbf{x}_{k_2})(k) - V_{h_n}(\mathbf{v})(k)\right.$$
$$\left.+ V_{h_n}(\mathbf{v}_{k_2})(k) + V_{h_n}(\mathbf{x}_{k_2})(k) - V_{h_n}(\mathbf{v}_{k_2})(k)\right|$$

where the vector-valued sequences $\mathbf{x}_{k_2}$ and $\mathbf{v}_{k_2}$ are defined as $\mathbf{x}_{k_2}(k) = \mathbf{x}(k)$ and $\mathbf{v}_{k_2}(k) = \mathbf{v}(k)$ for $k \geq k_2$, and as $\mathbf{x}_{k_2} = \boldsymbol{\theta}_m$ and $\mathbf{v}_{k_2} = \boldsymbol{\theta}_m$ for each $k < k_2$. The inequality on the extreme right-hand side of inequality 1.292a follows, of course, from the application of the triangle inequality to the preceding expression in 1.292a.

Now we shall use the uniform continuity property of the operators V_{h_n}, $n = 1, 2, \cdots$, proven at the beginning of the proof of point 2 of Theorem 1.8. It follows from this property that having a positive number equal to $\varepsilon_2/(4p)$, and taking into account the fact that $\mathbf{x}(k) \to \mathbf{v}(k)$, when $k \to \infty$, we can choose such k_3 that

$$\left\| V_{h_n}(\mathbf{x}_{k_3})(k) - V_{h_n}(\mathbf{v}_{k_3})(k) \right\| < \frac{\varepsilon_2}{4p} \tag{1.292b}$$

holds for each V_{h_n}, $n = 1, 2, \cdots, p$, and for all $k \geq k_3$ such that

$$\left\| \mathbf{x}_{k_3}(k) - \mathbf{v}_{k_3}(k) \right\| < \varepsilon_1 \tag{1.292c}$$

With regard to inequality 1.292c, see also inequality 1.286.

The next step is searching for such a $k_2 = k_3$ that the following inequalities

$$\left| V_{h_n}(\mathbf{x})(k) - V_{h_n}(\mathbf{x}_{k_2})(k) \right| \leq \gamma \|\mathbf{x}\|^n \leq \gamma \rho^n \leq \frac{\varepsilon_2}{8p} \tag{1.292d}$$

and

$$\left| V_{h_n}(\mathbf{v})(k) - V_{h_n}(\mathbf{v}_{k_2})(k) \right| \leq \gamma \|\mathbf{v}\|^n \leq \gamma \rho^n \leq \frac{\varepsilon_2}{8p} \tag{1.292e}$$

are satisfied for all $k \geq k_1 > k_2$, and for each $n = 1, 2, \cdots, p$, when $\mathbf{x}, \mathbf{v} \in B_\delta$. The satisfaction of inequalities 1.292d and 1.292e is guaranteed by inequality 1.280. That is, choosing the correspondingly small value of k_2 causes the value of γ in 1.292d and 1.292e to be as small as needed. If the value of k_2 needed for satisfaction of inequalities 1.292d and 1.292e is greater than or equal to the value of k_3 determined by inequalities 1.292b and 1.292c, we then simply choose $k_2 = k_3$. On the other hand, when the value of k_2 needed for satisfaction of inequalities 1.292d and 1.292e is less than the value of k_3 determined by inequalities 1.292b and 1.292c, we must change the value of k_1 influencing γ in inequalities 1.292d and 1.292e. In this context, observe from inequality 1.281j that the increase of k_1 causes the decrease of γ in inequality 1.280, that is, the decrease of γ in inequalities 1.292d and 1.292e as well. We increase k_1 until the satisfaction of inequalities 1.292d and 1.292e is obtained for $k_2 \geq k_3$, and then do the same as in the first case. That is, we put $k_2 = k_3$. Consequently, we obtain satisfaction of all the inequalities 1.292b, 1.292d, and 1.292e at the same time for all $k \geq k_1 > k_3 = k_2$.

To proceed further, let us apply the triangle inequality to the extreme right-hand side of inequality 1.292a. This gives

$$\begin{aligned} \left| \sum_{n=1}^{p} \left(V_{h_n}(\mathbf{x})(k) - V_{h_n}(\mathbf{v})(k) \right) \right| &\leq \sum_{n=1}^{p} \left| V_{h_n}(\mathbf{x})(k) - V_{h_n}(\mathbf{x}_{k_2})(k) \right| \\ &+ \sum_{n=1}^{p} \left| V_{h_n}(\mathbf{v})(k) - V_{h_n}(\mathbf{v}_{k_2})(k) \right| + \sum_{n=1}^{p} \left| V_{h_n}(\mathbf{x}_{k_2})(k) - V_{h_n}(\mathbf{v}_{k_2})(k) \right| \end{aligned} \tag{1.292f}$$

Then, using inequalities 1.292d, 1.292e, and 1.292b in inequality 1.292f leads to

$$\left|\sum_{n=1}^{p}(V_{h_n}(\mathbf{x})(k)-V_{h_n}(\mathbf{v})(k))\right| < p\cdot\frac{\varepsilon_2}{8p}+p\cdot\frac{\varepsilon_2}{8p}+p\cdot\frac{\varepsilon_2}{4p} = \frac{\varepsilon_2}{8}+\frac{\varepsilon_2}{8}+\frac{\varepsilon_2}{4} = \frac{\varepsilon_2}{2} \tag{1.292g}$$

Finally, applying inequality 1.291a, 1.291b, and 1.292g to inequality 1.290, we get

$$\left|\sum_{n=1}^{\infty}(V_{h_n}(\mathbf{x})(k)-V_{h_n}(\mathbf{v})(k))\right| < \frac{\varepsilon_2}{4}+\frac{\varepsilon_2}{4}+\frac{\varepsilon_2}{2} = \varepsilon_2 \tag{1.293}$$

for all $k \geq k_1 > k_3 = k_2$. Given any positive numbers ε_1 and ε_2 such that inequality 1.286 holds, the latter implies inequality 1.293 for all $k \geq k_1 > k_3 = k_2$. This ends the proof of point 2 of Theorem 1.8.

Let us now start with the proof of point 3 of Theorem 1.8. For this purpose, we choose such $p \in \mathbb{N}$ that the following

$$\sum_{n=p+1}^{\infty}\sigma_1(\sigma_2\rho)^n = \frac{\sigma_1(\sigma_2\rho)^{p+1}}{1-\sigma_2\rho} < \frac{\varepsilon}{2} \tag{1.294}$$

holds, where ε is any positive real number, and σ_1, σ_2, ρ are as described in point 1 of Theorem 1.7 and in the first sentence of Theorem 1.8.

Using inequality 1.294, the convergence of the series $S(\mathbf{x})$ proven in point 1, and the result given by inequality 1.283j, we can write

$$\left\|S(\mathbf{x})-\sum_{n=1}^{p}V_{h_n}(\mathbf{x})\right\| = \left\|\sum_{n=p+1}^{\infty}V_{h_n}(\mathbf{x})\right\| \leq \sum_{n=p+1}^{\infty}\sigma_1(\sigma_2\|\mathbf{x}\|)^n \leq \sum_{n=p+1}^{\infty}\sigma_1(\sigma_2\rho)^n < \frac{\varepsilon}{2} \tag{1.295}$$

Moreover, we can write the following:

$$\begin{aligned}\left\|S(\mathbf{x})-\sum_{n=1}^{p}V_{T_q h_n}(\mathbf{x})\right\| &= \left\|S(\mathbf{x})-\sum_{n=1}^{p}V_{h_n}(\mathbf{x})+\sum_{n=1}^{p}V_{h_n}(\mathbf{x})-\sum_{n=1}^{p}V_{T_q h_n}(\mathbf{x})\right\| \\ &\leq \left\|S(\mathbf{x})-\sum_{n=1}^{p}V_{h_n}(\mathbf{x})\right\|+\left\|\sum_{n=1}^{p}V_{h_n}(\mathbf{x})-\sum_{n=1}^{p}V_{T_q h_n}(\mathbf{x})\right\|\end{aligned} \tag{1.296}$$

Furthermore, observe that each of the components of the vector $\sum_{n=1}^{p}(V_{h_n}(\mathbf{x}) - V_{T_q h_n}(\mathbf{x}))$ occurring on the extreme right-hand side of inequality 1.296 has the form

$$\left[\sum_{n=1}^{p}(V_{h_n}(\mathbf{x}) - V_{T_q h_n}(\mathbf{x}))\right]_r = \sum_{n=1}^{p}[(V_{h_n}(\mathbf{x}) - V_{T_q h_n}(\mathbf{x}))]_r, \; r = 1, \ldots, m \quad (1.297a)$$

In the next step, consider the component $[(V_{h_n}(\mathbf{x}) - V_{T_q h_n}(\mathbf{x}))]_r$ in Equation 1.297a. This component can be expressed as

$$\begin{aligned}[(V_{h_n}(\mathbf{x}) - V_{T_q h_n}(\mathbf{x}))]_r &= \sum_{i_1=-\infty}^{k} \cdots \sum_{i_n=-\infty}^{k} \sum_{s=1}^{m^n} h_{rs}^{(n)}(k-i_1, \ldots, k-i_n)(x(i_1)\cdots x(i_n))_s \\ &- \sum_{i_1=-\infty}^{k} \cdots \sum_{i_n=-\infty}^{k} \sum_{s=1}^{m^n} T_q h_{rs}^{(n)}(k-i_1, \ldots, k-i_n)(x(i_1)\cdots x(i_n))_s\end{aligned} \quad (1.297b)$$

where $(x(i_1)\cdots x(i_n))_s$ means the sth component of the vector $\mathbf{x}(i_1) \otimes \cdots \otimes \mathbf{x}(i_n)$, according to the simplified notation 1.121. Furthermore, after substituting new variables $i'_1 = k - i_1, \cdots, i'_n = k - i_n$, into Equation 1.297b and dropping the prime at each i'_i, we get from Equation 1.297b

$$\begin{aligned}[(V_{h_n}(\mathbf{x}) - V_{T_q h_n}(\mathbf{x}))]_r &= \sum_{i_1=0}^{\infty} \cdots \sum_{i_n=0}^{\infty} \sum_{s=1}^{m^n} h_{rs}^{(n)}(i_1, \ldots, i_n) \cdot (x(k-i_1)\cdots x(k-i_n))_s \\ &- \sum_{i_1=0}^{\infty} \cdots \sum_{i_n=0}^{\infty} \sum_{s=1}^{m^n} T_q h_{rs}^{(n)}(i_1, \ldots, i_n) \cdot (x(k-i_1)\cdots x(k-i_n))_s\end{aligned} \quad (1.297c)$$

Note that applying the triangle inequality and the bound on $x(k - i_1)\cdots x(k - i_n)$ given by inequality 1.281g to Equation 1.297c, we obtain

$$\begin{aligned}&\left|[(V_{h_n}(\mathbf{x}) - V_{T_q h_n}(\mathbf{x}))]_r\right| \\ &\leq \sum_{i_1=0}^{\infty,} \cdots \sum_{i_n=0}^{\infty} \sum_{s=1}^{m^n} \left|h_{rs}^{(n)}(i_1, \ldots, i_n) - T_q h_{rs}^{(n)}(i_1, \ldots, i_n)\right| \cdot \left|(x(k-i_1)\ldots x(k-i_n))_s\right| \\ &\leq (\|\mathbf{x}\|)^n \cdot \sum_{i_1=0}^{\infty} \cdots \sum_{i_n=0}^{\infty} \sum_{s=1}^{m^n} \left|h_{rs}^{(n)}(i_1, \ldots, i_n) - T_q h_{rs}^{(n)}(i_1, \ldots, i_n)\right|, \; r = 1, \ldots, m\end{aligned} \quad (1.297d)$$

The expression in inequality 1.297d dependent upon the nonlinear impulse responses $h_{rs}^{(n)}(i_1, \cdots, i_n)$ can be rewritten as

$$\sum_{i_1=0}^{\infty} \cdots \sum_{i_n=0}^{\infty} \sum_{s=1}^{m^n} \left| h_{rs}^{(n)}(i_1, \ldots, i_n) - T_q h_{rs}^{(n)}(i_1, \ldots, i_n) \right| \tag{1.298a}$$
$$= \sum_{s=1}^{m^n} \left(\sum_{i_1=0}^{\infty} \cdots \sum_{i_n=0}^{\infty} \left| h_{rs}^{(n)}(i_1, \ldots, i_n) - T_q h_{rs}^{(n)}(i_1, \ldots, i_n) \right| \right)$$

Furthermore, using the definition of the mapping T_q given by definition 1.279d in the expression in round brackets in Equation 1.298a allows us to write the bracketed term in the form

$$\sum_{i_1=0}^{\infty} \cdots \sum_{i_n=0}^{\infty} \left| h_{rs}^{(n)}(i_1, \ldots, i_n) \right| - \sum_{i_1=0}^{q} \cdots \sum_{i_n=0}^{q} \left| h_{rs}^{(n)}(i_1, \ldots, i_n) \right| \tag{1.298b}$$

Note that the expression 1.298b has been obtained by additionally applying the obvious equality $|0| = |a - a| = |a| - |a| = 0$ to the components

$$\left| h_{rs}^{(n)}(i_1, \ldots, i_n) - h_{rs}^{(n)}(i_1, \ldots, i_n) \right|, i_1, \ldots, i_n \in \{1, \ldots, q\} \tag{1.298c}$$

in the bracketed expression in Equation 1.298a.

Now observe that the value of expression 1.298b goes to zero, when q goes to infinity. On the other hand, this causes the value of expression 1.298a to go to zero, and also the bound on the extreme right-hand side of inequality 1.297d. Moreover, the above holds for all k, all values of $r = 1, \cdots, m$, and all values of $n = 1, \cdots, p$. Given any $\varepsilon > 0$, we can find such q that

$$\left\| \sum_{n=1}^{p} V_{h_n}(\mathbf{x}) - \sum_{n=1}^{p} V_{T_q h_n}(\mathbf{x}) \right\| < \frac{\varepsilon}{2} \tag{1.299}$$

holds. Using then inequalities 1.295 and 1.299 in inequality 1.296, we get, finally

$$\left\| S(\mathbf{x}) - \sum_{n=1}^{p} V_{T_q h_n}(\mathbf{x}) \right\| < \frac{\varepsilon}{2} + \frac{\varepsilon}{2} < \varepsilon \tag{1.300}$$

Note that inequality 1.300 is nothing other than inequality 1.279e, which ends the proof of point 3 of Theorem 1.8.

To prove point 4 of Theorem 1.8, observe that, given p and $\varepsilon > 0$, we are able to choose such value $\rho_1 \in (0, \rho)$ that the inequality

$$\left\| S(\mathbf{x}) - \sum_{n=1}^{p} V_{h_n}(\mathbf{x}) \right\| = \left\| \sum_{n=p+1}^{\infty} V_{h_n}(\mathbf{x}) \right\| \tag{1.301}$$

$$\leq \sum_{n=p+1}^{\infty} \sigma_1(\sigma_2\|\mathbf{x}\|)^n \leq \sum_{n=p+1}^{\infty} \sigma_1(\sigma_2\rho_1)^n = \frac{\sigma_1(\sigma_2\rho_1)^{p+1}}{1-\sigma_2\rho_1} < \frac{\varepsilon}{2}$$

will be satisfied for all $\|\mathbf{x}\| < \rho_1$. We also stress at this point the difference between inequalities 1.295 and 1.301. The first is satisfied for signals $\|\mathbf{x}\| < \rho$, with p not being fixed, in contrast to the latter, which is satisfied generally for signals smaller in the amplitude, $\rho_1 < \rho$, and with the fixed parameter p.

As before, given $\varepsilon > 0$, we are able to choose such q_0 that, for all $q \geq q_0$ and all $\mathbf{x} \in B_{\delta 1}$, inequality 1.299 is satisfied. Using this, and inequality 1.301 in inequality 1.296, we get inequality 1.300, thereby proving point 4 of Theorem 1.8.

Note that the result in point 4 is interesting in the respect that, given p and ε, we can achieve the needed accuracy represented by the truncated series by choosing an accordingly high value of q and restricting ourselves to signals of the suitably small values of the norm $\|\mathbf{x}\| < \rho_1$.

Comparison of the approximating Volterra series operator $\hat{V}$ from Section 1.9, given by Equation 1.220a, with the approximating Volterra series $\sum_{n=1}^{p} V_{T_q h_n}(x)$ of this section, defined for scalar-valued sequences, shows that both the approximating series are identical except the constant component $h^{(0)}$. In fact, identifying L and M-1 in Equation 1.220a with p and q in inequality 1.279e, respectively, we get

$$(\hat{V}x)(k) = h^{(0)} + \sum_{n=1}^{L} \sum_{i_1=0}^{M-1} \cdots \sum_{i_n=0}^{M-1} h^{(n)}(i_1, \ldots, i_n)x(k-i_1)\cdots x(k-i_n) \tag{1.302a}$$

$$= h^{(0)} + \sum_{n=1}^{p} (V_{T_q h_n} x)(k)$$

with $V_{T_q h_n} x$ given by

$$(V_{T_q h_n} x)(k) = \sum_{i_1=-\infty}^{k} \cdots \sum_{i_n=-\infty}^{k} T_q h_n^{(n)}(k-i_1, \ldots, k-i_n)x(i_1)\cdots x(i_n) \tag{1.302b}$$

which, after introducing new variables $i'_1 = k - i_1, \cdots, i'_n = k - i_n$, dropping the prime at each i'_i, and finally applying the definition of the mapping T_q, 1.279d, takes on the following form:

$$(V_{T_q h_n} x)(k) = \sum_{i_1=0}^{\infty} \cdots \sum_{i_n=0}^{\infty} T_q h^{(n)}(i_1, \ldots, i_n)x(k-i_1)\cdots x(k-i_n) \tag{1.302c}$$

$$= \sum_{i_1=0}^{q} \cdots \sum_{i_n=0}^{q} h^{(n)}(i_1, \ldots, i_n)x(k-i_1)\cdots x(k-i_n)$$

It follows also from Equations 1.302a and 1.302c that

$$\sum_{n=1}^{L}(V_{T_{(M-1)}h_n}x)(k) = \sum_{n=1}^{L}\sum_{i_1=0}^{M-1}\cdots\sum_{i_n=0}^{M-1}h^{(n)}(i_1, \ldots, i_n)\cdot x(k-i_1)\ldots x(k-i_n) \quad (1.302d)$$

We recall as well at this point that the approximating Volterra series operator $\hat{V}$ of Section 1.9 has been obtained considering systems with the property of fading memory. On the other hand, the approximating Volterra series of this section, $\sum_{n=1}^{p} V_{T_q h_n}(x)$, relates to a class of systems described by Equations 1.264 and satisfying three assumptions formulated in Equations 1.266 to 1.268.

The most important result of this section, following from Theorem 1.8, is the proof of the existence of associated or , as referred to in Reference 3, extended representations for the class of systems mentioned above. This is so because, when $(\mathbf{x}(k) - (S_e S_t \mathbf{x})(k)) \to \boldsymbol{\theta}_m$ as $k \to \infty$, where $\mathbf{x}(k) \in l^\infty(\mathbb{Z})$, it follows from point 2 of the theorem that

$$(S(\mathbf{x})(k) - S(S_e S_t \mathbf{x})(k)) \to \boldsymbol{\theta}_m \quad (1.303)$$

as $k \to \infty$, with $S_t\mathbf{x}(k) \in l^\infty(\mathbb{Z}_+)$. And we identify the series $S(\mathbf{x})(k)$ and $S(S_e S_t \mathbf{x})(k)$ in relation 1.303 with the associated (extended) representation and the strict-sense representation, respectively. Hence, the associated expansion approaches the strict-sense one for times k high enough.

1.12 OTHER APPROXIMATIONS

The purpose of this section is to show that, for systems possessing the property of approximately-finite memory or fading memory, besides the Volterra series approximations, there exist other approximations. These approximations have appeared in the literature for the first time in recent years. One example, Theorem 1.5 of section 1.9, regards causal and time-invariant operators N: $l^\infty(\mathbb{Z}_+) \to l^\infty(\mathbb{Z}_+)$, having the property of approximately-finite memory in the sense of the definition AFM1 and, additionally, the property of the functional $N(\cdot)(k) : l^\infty(\mathbb{Z}_+) \to \mathbb{R}$, as continuous.

It follows from inequality 1.238a that the approximator of the operator $(Nx)(k)$, which we call here the lattice map approximator, has the following form:

$$(\hat{N}x)(k) = L(\mathbf{d} + \mathbf{C}(\mathbf{P}_a x)(k)),\ k \in \mathbb{Z}_+ \quad (1.304a)$$

where the vector $(\boldsymbol{P}_a x)(k)$, according to Equation 1.238b, is given as

$$(\mathbf{P}_a x)(k) = \begin{bmatrix} x(k) \\ x(k-1) \\ \vdots \\ x(k-a) \end{bmatrix} \tag{1.304b}$$

and $\mathbf{C}$ is a real matrix, consisting of m rows and $(a + 1)$ columns. Moreover, $\mathbf{d}$ in Equation 1.304a is a real vector consisting of m components, and L means the lattice map, such that it generates its output value $L\mathbf{z}$ from the components of the vector $\mathbf{z} = [z_1, z_2, \ldots, z_m]^T$ by performing a finite number of the lattice operations defined by expressions 1.238c and 1.238d. On this occasion, we recall that the operations of addition and multiplication by scalar are not allowed in the lattice map.

Note that Theorem 1.5 of Section 1.9 is formulated for scalar-valued sequences. This theorem is given in Reference 4 for vector-valued sequences. For the purpose of this section, however, the formulation of Section 1.9 suffices. Here we illustrate approximations other than the Volterra series approximation for nonlinear systems with only one input and one output, so as not to complicate matters too much. In this context, the number m mentioned above at the descriptions of the matrix $\mathbf{C}$ and vector $(\mathbf{P}_a x)(k)$, occurring in Equations 1.304a and 1.304b, does not mean the number of components in the input signal vector, as in the previous section. Here, m means the number of affine components needed in the approximation for assumed accuracy. Loosely speaking, this number is, in some sense, the counterpart of the number L or p in the approximations given by Equation 1.302a. On the other hand, the number a in Equations 1.304a and 1.304b can be identified with the numbers $M - 1$ and q in Equation 1.302a, expressing in some sense the memory length of a system.

With regard to the lattice map, let us illustrate it using an example of the vector $\mathbf{z}$ mentioned above with $m = 3$ components. Then the input vector to the lattice map has the following form:

$$\mathbf{z} = \begin{bmatrix} z_1 \\ z_2 \\ z_3 \end{bmatrix}$$

but the map itself would look as shown in Figure 1.53.

According to the literature,[33] the lattice operations given by expressions 1.238c and 1.238d can be expressed as

$$y \vee z = \max(y, z) = \frac{1}{2}(y + z + |y - z|) \tag{1.305a}$$

and

$$y \wedge z = \min(y, z) = \frac{1}{2}(y + z - |y - z|) \tag{1.305b}$$

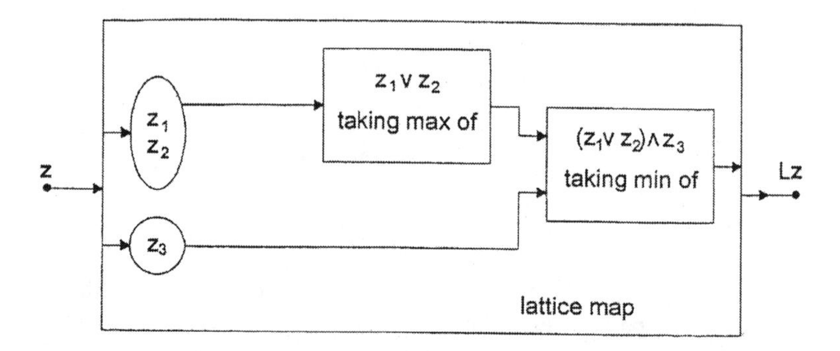

FIGURE 1.53 Illustration of lattice map with input vector with three components and two lattice operations given by expressions 1.238c and 1.238d.

Thus, either one of the lattice operations needs for its realization two adders and one subtractor or one adder and two subtractors, one multiplier, and one functional block realizing the absolute-value nonlinearity. All of these functional blocks are known as basic blocks used in realization of prescribed nonlinear characteristics (operations) in the literature on realization of nonlinear circuits and systems.[34]

Observe now that the vector $\mathbf{d} + \mathbf{C}(\mathbf{P}_a x)(k)$ can be written as

$$\mathbf{d} + \mathbf{C}(\mathbf{P}_a x)(k) = \begin{bmatrix} d_1 \\ d_2 \\ \vdots \\ d_m \end{bmatrix} + \begin{bmatrix} c_{10} & c_{11} & \dots & c_{1a} \\ c_{20} & c_{21} & \dots & c_{2a} \\ \vdots & \vdots & \vdots & \vdots \\ c_{m0} & c_{m1} & \dots & c_{ma} \end{bmatrix} \cdot \begin{bmatrix} x(k) \\ x(k-1) \\ \vdots \\ x(k-a) \end{bmatrix} \tag{1.306a}$$

$$= \begin{bmatrix} d_1 + \sum_{i=0}^{a} c_{1i} x(k-i) \\ d_2 + \sum_{i=0}^{a} c_{2i} x(k-i) \\ \vdots \\ d_m + \sum_{i=0}^{a} c_{mi} x(k-i) \end{bmatrix} = \begin{bmatrix} d_1 + \sum_{i=0}^{a} c_1(i) x(k-i) \\ d_2 + \sum_{i=0}^{a} c_2(i) x(k-i) \\ \vdots \\ d_m + \sum_{i=0}^{a} c_m(i) x(k-i) \end{bmatrix}$$

where it has been assumed that

$$c_j(i) = c_{ji}, \quad j = 1, \dots, m \text{ and } i = 0, \dots, a \tag{1.306b}$$

in the vector on the extreme right-hand side of Equation 1.306a. It follows from the above relations that each of the components of the vector $\mathbf{d} + \mathbf{C}(\mathbf{P}_a x)(k)$ represents an affine system, that is, a system described by a linear-plus-constant function. And as shown by using Equation 1.306b, each of these affine systems can be assumed to be a system described in the form of a discrete-time convolution, as for linear finite impulse response systems (see Section 1.10), plus a constant component.

Summarizing then, we can say that Theorem 1.5 of Section 1.9 proves the existence of the approximator, which is shown schematically in Figure 1.54.

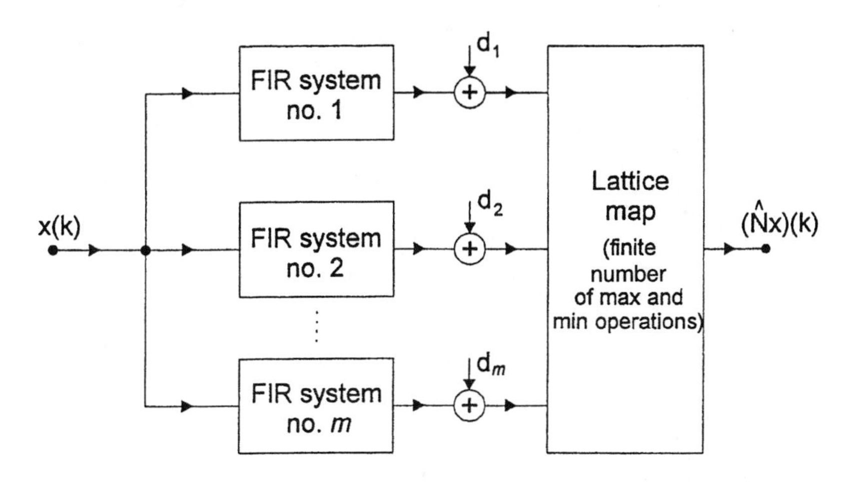

FIGURE 1.54 Structure of the lattice map approximator $(\hat{N}x)(k)$.

Finally, let us comment on the definition of the time-invariance property used in Theorem 1.5. Sandberg's time-invariance definition of the operator N: $l^\infty(\mathbb{Z}_+) \rightarrow l^\infty(\mathbb{Z}_+)$, used in Reference 4, where proof of Theorem 1.5 has been presented, is somewhat different from that given by Equation 1.159 in Section 1.7. Sandberg's definition can be formulated with the use of the delay operator U_τ, defined by Equation 1.158 in Section 1.7, in the following way:

$$(Nx)(k) = 0 \text{ for } k < k_0 \tag{1.307a}$$

and

$$(Nx)(k) = (U_{k_0} N U_{-k_0} x)(k) \text{ for } k \geq k_0 \tag{1.307b}$$

whenever $s(l) = 0$ for $l < k_0$.

Note that the defining Equation 1.307b is equivalent to the definition given by Equation 1.159. According to Equation 1.159, we can write $U_{-k_0}N = NU_{-k_0}$. Furthermore, applying the delay operator U_{k_0} to both sides of the latter equation, we arrive at Equation 1.307b. On the other hand, Equation 1.307a does not follow from Equation 1.159 at all. Consequently, some kinds of operators, such as operator $(Nx)(k) = N_0 + x(k) + x(k - 1)$, where $N_0 \neq 0$, are excluded. In the latter for instance, for $k = k_0 - 1$, we have $(Nx)(k_0 - 1) = N_0 + x(k_0 - 1) + x(k_0 - 2) = N_0 \neq 0$ $(xk) = 0$ for $k < k_o$. So the condition 1.307a is not satisfied. Note, however that the above restriction can be easily removed by taking into account the operator $(Nx)(k) - N_0$, where N_0 is the value of the operator $(Nx)(k)$ for those $k < k_0$ for which $x(k) = 0$, $k < k_0$. Then the operator $(Nx)(k) - N_0$ satisfies both the time-invariance defining Equations 1.307a and 1.307b. After finding some approximation for this operator, we add the constant N_0 to it, to get the approximation for the operator $(Nx)(k)$ itself.

Another approximation follows from the theorem presented by Sandberg in Reference 20. The theorem for nonlinear single-input single-output systems, that is, for systems operating on scalar-valued input and output sequences is presented as follows:

Theorem 1.9

Let N: $l^\infty(\mathbb{Z}_+) \to l^\infty(\mathbb{Z}_+)$ be any causal and time-invariant operator, having approximately-finite memory in the sense of the definition AFM1. Moreover, let the functional $N(\cdot)(k) : l^\infty(\mathbb{Z}_+) \to \mathbb{R}$, be continuous for each $k \in \mathbb{N} = \{1, 2, 3, \ldots\}$ on the ball B_+. Furthermore, let $\sigma : \mathbb{R} \to \mathbb{R}$ denote any continuous mapping having the property: $\sigma(x) \to 1$, when $x \to \infty$, and $\sigma(x) \to 0$, when $x \to -\infty$. For any $\varepsilon > 0$, there exist then such m and $a \in \mathbb{N}$, real numbers $\alpha_1, \ldots, \alpha_m, \beta_1, \ldots, \beta_m$, and a real matrix η consisting of m rows and $(a + 1)$ columns such that

$$\left|(Nx)(k) - \sum_{j=1}^{m} a_j \sigma(\beta_j + \eta_j(\mathbf{P}_a x)(k))\right| < \varepsilon, \quad k \in \mathbb{Z}_+ \tag{1.308}$$

holds for all $x \in B_+$. The vector η_j, $j = 1, \ldots, m$, in inequality 1.308 means the row vector made of the jth row of the matrix η.

The proof of Theorem 1.9 is omitted here, but can be found in Reference 20.

Observe now that inequality 1.308 allows us to write the approximator of the operator $(Nx)(k)$, which we call here the sigmoid function approximator, in the form

$$(\hat{N}x)(k) - \sum_{j=1}^{m} a_j \sigma(\beta_j + \eta_j(\mathbf{P}_a x)(k)), \quad k \in \mathbb{Z}_+ \tag{1.309a}$$

where the vector $(\mathbf{P}_a x)(k)$ is given by Equation 1.304b, and the mapping σ is an ordinary function satisfying the requirements of Theorem 1.9. One of the possible choices for such a function, called a sigmoid function in the literature on neural networks,[35] is illustrated in Figure 1.55. Note further that a function given by the expression

$$\sigma_1(x) = \frac{1}{2}(\mathrm{tgh}(x) + 1) = \frac{1}{2}\left(\frac{e^x - e^{-x}}{e^x + e^{-x}} + 1\right) = \frac{e^x}{e^x + e^{-x}} \tag{1.309b}$$

satisfies the requirements of the sigmoid function as referred to in Theorem 1.9. Similarly, the function

$$\sigma_2(x) = \frac{1}{\pi}\text{arc ctg}(-x) \tag{1.309c}$$

with function ctg(·) taken for angles from the first two quadrants fulfills the above requirements as well. Other expressions for functions having the shape as sketched in Figure 1.55 can also be found.

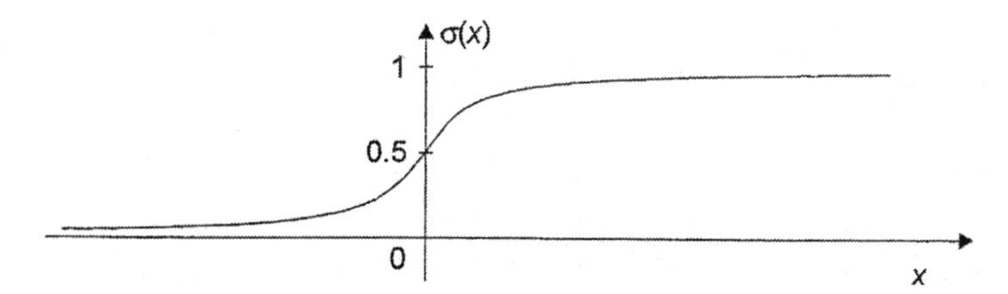

FIGURE 1.55 Sigmoid function: one of the possible choices for function $\sigma(x)$ occurring in Equation 1.309a.

Return to the sigmoid function approximator given by Equation 1.309a and observe that the expressions $B_j + \eta_j\,(\mathbf{P}_a x)(k)$, $j = 1, \ldots, M$, occurring in Equation 1.309a can be rewritten as

$$B_j + \eta_j(\mathbf{P}_a x)(k) = \beta_j + [\eta_{j0} \ldots \eta_{ja}]\begin{bmatrix} x(k) \\ x(k-1) \\ \vdots \\ x(k-a) \end{bmatrix} \tag{1.310a}$$

$$= \beta_j + \sum_{i=0}^{a} \eta_{ji} x(k-i) = \sum_{i=0}^{a} \eta_{j(i)} x(k-i), \quad j = 1, \ldots, m$$

where it has been assumed that

$$\eta_{ji} = \eta_j(i), \quad j = 1, \ldots, m, i = 0, \ldots, a \tag{1.310b}$$

in the expression on the extreme right-hand side of Equation 1.310a. Again, it follows from Equation 1.310a that each of the expressions $\boldsymbol{\beta}_j + \eta_j\,(\mathbf{P}_a x)(k)$ represents an affine system, that is, a system described by a linear-plus-constant function. Furthermore, using expression 1.310b, each of these affine systems can be represented by a discrete-time convolution, as for linear finite impulse response systems, plus a constant component.

Then, using the above interpretation of the expressions $\beta_j + \eta_j (\mathbf{P}_a x)(k)$ in the expression for the sigmoid function approximator, we can present this approximator graphically, as shown in Figure 1.56.

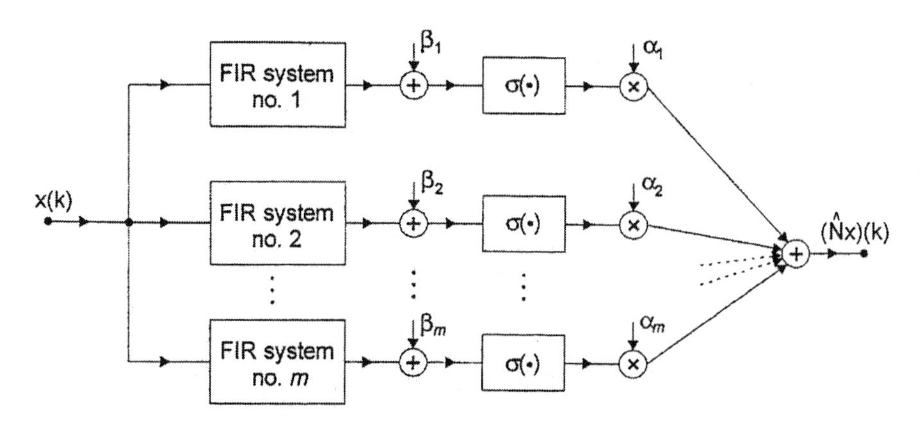

FIGURE 1.56 Structure of the sigmoid function approximator $(\hat{N}x)(k)$.

Comparison of structures of Figure 1.54 and Figure 1.56 shows high similarity between them. The only difference lies in the fact that in the structure of Figure 1.56, in place of the lattice map we have the summation of outputs coming from m sigmoid functions and multiplied by the corresponding coefficients α_j, $j = 1, \ldots, m$. Again, in the structure of Figure 1.56, the number m is in some sense, the counterpart of the number L or p in the approximations given by Equation 1.302a. On the other hand, the number a in Equations 1.309a and 1.310a can be identified with the numbers $M - 1$ and q in Equation 1.302a, expressing, in some sense, the memory length of a system.

In Reference 2, one more theorem is presented regarding the problem of approximation of nonlinear discrete-time systems with the use of so-called radial basis functions.[36,37] Using the formulation of Sandberg,[2] a simplified version for scalar-valued sequences follows.

Theorem 1.10

Let N: $l^\infty(\mathbb{Z}_+) \to l^\infty(\mathbb{Z}_+)$ denote any causal and time-invariant operator, possessing approximately-finite memory in the sense of the definition AFM1, and let the functional $N(\cdot)(k) : l^\infty(\mathbb{Z}_+) \to \mathbb{R}$ be continuous for each $k \in \mathbb{N}$ on the ball B_+. Moreover, let R mean a continuous and bounded map from $\mathbb{R}^l$ to $\mathbb{R}$ that is Lebesgue integrable[38] with

$$\int_{\mathbb{R}^l} R(\mu)d\mu \neq 0 \qquad (1.311a)$$

Then, for any $\varepsilon > 0$, there exist m, l, and $a \in \mathbb{N}$, real numbers $\alpha_1, \ldots, \alpha_m$, column vectors

$$\boldsymbol{\beta}_1 = \begin{bmatrix} \beta_{11} \\ \beta_{21} \\ \vdots \\ \beta_{l1} \end{bmatrix}, \boldsymbol{\beta}_2 = \begin{bmatrix} \beta_{12} \\ \beta_{22} \\ \vdots \\ \beta_{l2} \end{bmatrix}, \ldots, \boldsymbol{\beta}_m = \begin{bmatrix} \beta_{1m} \\ \beta_{2m} \\ \vdots \\ \beta_{lm} \end{bmatrix}$$

consisting of elements β_{ij}, $i = 1, \ldots, l$, $j = 1, \ldots, m$, real numbers, and a real matrix $\mathbf{Y}$ of order $l \times (a + 1)$ such that

$$\left| (Nx)(k) - \sum_{s=1}^{m} \alpha_s R(\mathbf{Y}(\mathbf{P}_a x)(k) + \boldsymbol{\beta}_s) \right| < \varepsilon, \quad k \in \mathbb{Z}_+ \tag{1.311b}$$

holds for all $x \in B_+$.

In expression 1.311a, the notion of Lebesgue integral occurs. For the purposes of this book, suffice it to know that this integral is some extension of the well-known ordinary Riemann integral. The explanation of the Lebesgue integral requires knowledge of the measure theory beyond the scope of this book. The interested reader is referred to Reference 38, where the general theory of integration using the notion of measure is presented. In particular in Reference 38, the Lebesgue integral is discussed in detail.

The boundedness of the map R in Theorem 1.10 means that, for bounded input sequences, it produces bounded output sequences. We also draw the reader's attention to the notation used for column vectors $\boldsymbol{\beta}_s$, $s = 1, \ldots, m$, in expression 1.311b. To this end, observe that these vectors can be considered as columns of the following matrix $\boldsymbol{\beta}$:

$$\boldsymbol{\beta} = \begin{bmatrix} \boldsymbol{\beta}_1 & \boldsymbol{\beta}_2 & \ldots & \boldsymbol{\beta}_m \end{bmatrix} = \begin{bmatrix} \beta_{11} & \beta_{12} & \ldots & \beta_{1m} \\ \beta_{21} & \beta_{22} & \ldots & \beta_{2m} \\ \vdots & \vdots & \vdots & \vdots \\ \beta_{l1} & \beta_{l2} & \ldots & \beta_{lm} \end{bmatrix} \tag{1.311c}$$

Thus, the index s at $\boldsymbol{\beta}_s$ means the corresponding column number in the matrix $\boldsymbol{\beta}$. In this context, note also that the indices $j = 1, \ldots, m$ at the row vectors η_j, used just before, meant the corresponding row numbers in the matrix η. Hence, it follows from the above that the notation of the form η_j and $\boldsymbol{\beta}_s$ is common for both the row and column vectors. To make this notation unique, it must be specified every time, whether it concerns a row or a column vector.

One of the possible choices for the function $R(\mathbf{w})$, meeting the conditions stated in Theorem 1.10, is the following function

$$R(\mathbf{w}) = e^{-\gamma(\|\mathbf{w}\|)} \tag{1.311d}$$

where $\|\cdot\|$ means any norm on $\mathbb{R}^l$ and the mapping $\gamma : \langle 0, \infty) \rightarrow \langle 0, \infty)$ is continuous. Moreover, the mapping γ satisfies the inequality: $\gamma(\|\mathbf{w}\|) \geq g \, \|\mathbf{w}\|^2$ for some positive constant g.

Because the function $R(\mathbf{w})$ depends upon the norm of the vector $\mathbf{w}$, which is non-negative, it is referred to as a radial function. Consequently, we call the approximation following from Theorem 1.10 the radial basis function approximation.

Example 1.8

Let us choose for the function $\gamma(\|\mathbf{w}\|)$ the quadratic one,

$$\gamma(\|\mathbf{w}\|) = (\|\mathbf{w}\|)^2$$

Then observe that this function satisfies the conditions required by Theorem 1.10. That is, it is continuous and fulfills the inequality: $\gamma(\|\mathbf{w}\|) \geq g \, \|\mathbf{w}\|^2$, because here

$$\gamma(\|\mathbf{w}\|) = (\|\mathbf{w}\|)^2 \geq g\|\mathbf{w}\|^2$$

holds, when the constant $| \; 0 < g \leq 1$. And finally, substitute $\gamma(\|\mathbf{w}\|) = \|\mathbf{w}\|^2$ into Equation 1.311d, which gives

$$R(\mathbf{w}) = e^{-\|\mathbf{w}\|^2}$$

We call the above function $R(\mathbf{w})$ a Gaussian function because of its shape, which resembles the shape of the function

$$\frac{1}{\sigma\sqrt{2\pi}} e^{\frac{(w-\mu)^2}{2\sigma^2}}$$

where μ is the mean value and σ^2 is the variance of a random variable w. The latter function is the probability density function of the variable w. Thus, the Gaussian function $\exp\left(-\|\mathbf{w}\|^2\right)$ with $\mathbf{w}$ identified with the vector $\mathbf{Y}(\mathbf{P}_a x)(k) + \boldsymbol{\beta}_s$ is one of the possible choices for the function $R(\cdot)$ in inequality 1.311b.

With regard to the Gaussian function $R(\mathbf{w}) = \exp\left(-\|\mathbf{w}\|^2\right)$, used for the radial basis function approximation in inequality 1.311b, we point out that its shape differs from that of the sigmoid function used in the sigmoid function approximation given by inequality 1.308. Observe from Figure 1.55 that the sigmoid function is a continuous strictly increasing function. In contrast to this, the shape of the Gaussian function considered is that of a half-bell, because $\|w\|$ is always non-negative, as shown in Figure 1.57.

The proof of Theorem 1.10 is omitted here. It can be found in Reference 20.

From inequality 1.311b, it follows that the radial basis function approximator has the following form:

$$(\hat{N}x)(k) = \sum_{s=1}^{m} \alpha_s R(\mathbf{Y}(\mathbf{P}_a x)(k) + \boldsymbol{\beta}_s), \quad k \in \mathbb{Z}_+ \tag{1.312a}$$

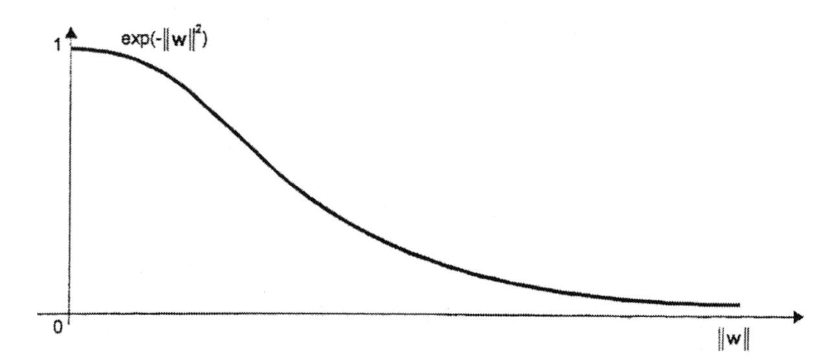

FIGURE 1.57 Sketch of the Gaussian function $\exp\left(-|w|^2\right)$.

Observe now that the vector $\mathbf{Y}(\mathbf{P}_a x)(k) + \boldsymbol{\beta}_s$ occurring in Equation 1.312a, can be, for a given k, rewritten in the explicit form as

$$\mathbf{Y}(\mathbf{P}_a x)(k) + \boldsymbol{\beta}_s = \begin{bmatrix} y_{10} & y_{11} & \cdots & y_{1a} \\ y_{20} & y_{21} & \cdots & y_{2a} \\ \vdots & \vdots & \vdots & \vdots \\ y_{l0} & y_{l1} & \cdots & y_{la} \end{bmatrix} \cdot \begin{bmatrix} x(k) \\ x(k-1) \\ \vdots \\ x(k-a) \end{bmatrix} + \begin{bmatrix} \beta_{1s} \\ \beta_{2s} \\ \vdots \\ \beta_{ls} \end{bmatrix} \tag{1.312b}$$

$$= \begin{bmatrix} \sum_{i=0}^{a} y_{1i} x(k-i) + \beta_{1s} \\ \sum_{i=0}^{a} y_{2i} x(k-i) + \beta_{2s} \\ \vdots \\ \sum_{i=0}^{a} y_{li} x(k-i) + \beta_{ls} \end{bmatrix}, \quad k \in \mathbb{Z}_+, s = 1, \ldots, m$$

Then using the following notation:

$$y_j(i) = y_{ji}, \; j = 1, \ldots, l \text{ and } i = 0, \ldots, a \tag{1.312c}$$

in Equation 1.312b, we arrive at

$$\mathbf{Y}(\mathbf{P}_a x)(k) + \boldsymbol{\beta}_s = \begin{bmatrix} \sum_{i=0}^{a} y_1(i)x(k-i) + \beta_{1s} \\ \sum_{i=0}^{a} y_2(i)x(k-i) + \beta_{2s} \\ \vdots \\ \sum_{i=0}^{a} y_l(i)x(k-i) + \beta_{ls} \end{bmatrix} \tag{1.312d}$$

Thus, it follows from Equation 1.312d that each of the components of the vector $\mathbf{Y}(\mathbf{P}_a x)(k) + \boldsymbol{\beta}_s$ represents an affine system. Moreover, each of the above affine systems possesses the description in the form of a discrete-time convolution, as for linear finite impulse response systems, plus a constant component.

As the radial function $R(\cdot)$ in the approximator given by Equation 1.312a, the Gaussian function $\exp\left(-\|\cdot\|^2\right)$ from Example 1.8, with the norm 1.130 for vectors, can be chosen. Then we get

$$\exp(-\|\mathbf{Y}(\mathbf{P}_a x)(k) + \beta_s\|^2) = \exp\left(-\left(\max_{1 \le j \le l}\left|\sum_{i=0}^{a} y_j(i)x(k-i) + \beta_{js}\right|\right)^2\right) \tag{1.312e}$$

for a given k, and for each $s = 1, \ldots, m$.

The radial basis function approximator is presented graphically in Figure 1.58.

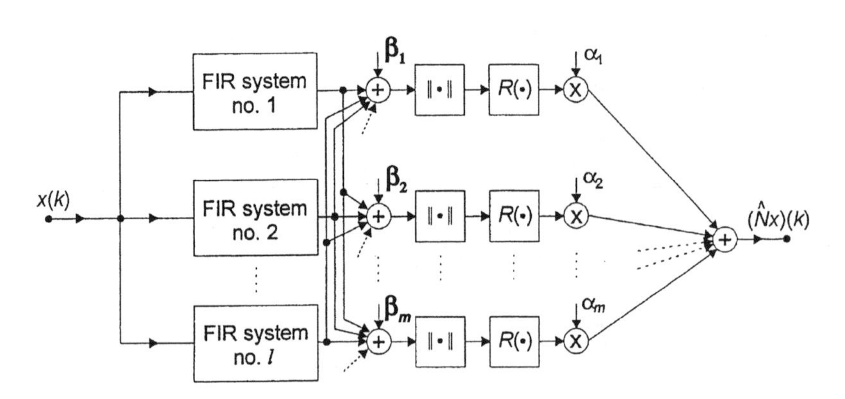

FIGURE 1.58 Structure of the radial basis function approximator $(\hat{N}x)(k)$ with the mapping R realized as given by Equation 1.311d.

Comparing the structure of Figure 1.58 with the structures of other approximators presented in Figure 1.54 and Figure 1.56, observe the differences and similarities existing between them. The main difference lies in the nonlinear part of the structure of Figure 1.58, consisting now of additions of vector-valued coefficients $\boldsymbol{\beta}_1, \ldots, \boldsymbol{\beta}_m$, calculations of the vector norm, and calculations of the value of the radial function $R(\cdot)$ for the corresponding vector norm values. Moreover, we have the number l of linear FIR subsystems in the structure of Figure 1.58, and this number is different from m, the number of coefficients α_s, $s = 1, \ldots, m$. As before, the number a here can be considered that number which expresses, in some sense, the memory length of the system approximated.

As shown in Reference 20, the theory and results regarding the lattice map, sigmoid function, and radial basis function approximations, especially Theorem 1.5 of Section 1.9, can be used to derive the Volterra series approximation. The Volterra series approximation can be viewed as a special case of the approximations mentioned above. A very important difference between these approximations and the Volterra series approximation lies in the fact that the latter uses for approximation nonlinearities of the polynomial type in contrast to the addition of some constant coefficients together with the lattice map (Figure 1.54), in contrast to the addition of some constant coefficients together with the sigmoid function and multiplication by some constant coefficients (Figure 1.56), and in contrast to the addition of some vector-valued constant coefficients together with the radial function and multiplication by some constant coefficients (Figure 1.58), used in the lattice map approximation, in the sigmoid function approximation, and in the radial basis function approximation, respectively. Another difference lies in the fact that the Volterra series approximator contains only one linear FIR subsystem (see Figures 1.36, 1.38, 1.39, and 1.49), in contrast to the structures of the other approximators containing more than one linear FIR subsystem.

2 Nonlinear Echo Cancellation

2.0 INTRODUCTION

This chapter is devoted to the problem of nonlinear echo cancellation. First, to introduce the reader to the topic, some fundamentals of echo cancellation with emphasis on adaptive cancellers are presented. We will explain how echo arises in telecommunication systems, and what are its sources. The similarities and differences between the echoes arising in voice and data transmission systems are given as well. The principle of a linear transversal filter is explained, and the principle of adaptation of its coefficients to adjust to characteristics of an echo path is discussed, too. Some basic configurations for cancelling echo are presented, especially for the digital subscriber loop occurring in the integrated services digital network (ISDN). The principle of achieving the full-duplex communication on two wires with the use of echo cancellers is presented from many points of view.

In Section 2.2, the problem of nonlinear echo cancellation in baseband is discussed in detail. First, sources of nonlinearities that occur in practice in an echo path are given. Then, the methods of nonlinear echo cancellation published in the literature are briefly described. Specifically, advantages and disadvantages of these methods are presented. Afterward, the behavior of linear and nonlinear digital cancellers working with binary input signals and in a nonlinear echo environment is analyzed in detail. A structure of the nonlinear canceller is based on the Volterra series description for binary signals developed in Section 1.10. For the purpose of analysis, the basic notions regarding stochastic processes are introduced. These are random variables, their outcomes, probabilities, random sequences, expected value of a random variable, autocorrelation function, wide-sense stationarity of a stochastic process, and so on. Finally, the nonlinear echo canceller structures based on the lattice map approximator, using the description of the sigmoid function approximator, or applying the form of the radial basis function approximator, are described in Section 2.2.

Section 2.3 is devoted to the discussion of structures of interleaved and passband nonlinear transversal filters. First, the structure for linear transversal filters for achieving higher output sampling than that at their input is explained in detail. The description of this structure is then used to extend to the case of interleaved nonlinear transversal filters. For the purpose of discussion of nonlinear echo cancellers for passband applications, some notions regarding the quadrature amplitude modulation (QAM), used in transmission in the passband, are first introduced. These are the Hilbert transform filter, analytic signal, phase splitter, and equivalent lowpass representation. Furthermore, general structures of the QAM transmitter and receiver are presented. Using the theory related to analytic signals, the structures for linear echo cancellers with single lowpass or passband complex-valued transversal filters, and their interleaved versions, are developed. Finally, these structures are extended to the nonlinear case.

2.1 ADAPTIVE CANCELLERS

The purpose of this section is to present principles of adaptive cancellers in the context of their application to echo cancellation. Hence, we start our considerations with an explanation of how the echo arises in a telecommunication system, and what the reasons are for trying to cancel its influence on the behavior of a system. To this end, consider Figure 2.1, where a simplified connection for voice transmission over a telephone network is presented. The connection shown is typical for connecting two subscribers in a telephone network.

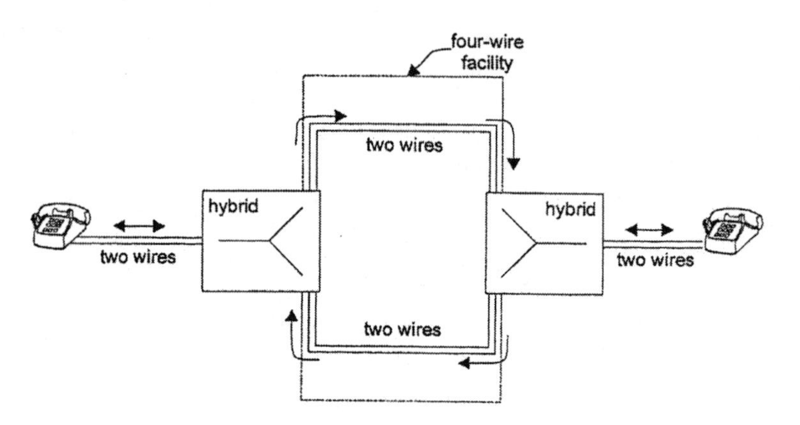

FIGURE 2.1 Voice transmission over a telephone network.

It consists of two two-wire parts on the ends and a four-wire facility in the middle. Each of the two-wire parts on the ends contains the subscriber loop and, eventually, a part of the local network. The transmission in these parts takes place in two directions, as shown in Figure 2.1. The four-wire facility consists of two separate two-wire connections, each of them only for one direction of transmission. This is shown schematically in Figure 2.1, where the upper two wires are for the transmission from the left to the right, and the lower two wires are for the opposite direction of transmission. The coupling of the middle part of the connection, with both end-parts, on the left- and right-hand side, takes place through the so-called hybrids. One of the possible realizations of the hybrid, using two separating amplifiers and one differential amplifier,[39] is shown in Figure 2.2.

The operation of the hybrid in Figure 2.2 is as follows: the signal $x'_1(t)$ coming from the four-wire facility is fed as the input signal to the first separating voltage amplifier of the hybrid (of gain equal to one), and then goes through the second separating voltage amplifier to the transformer. So it appears at the transformer as the voltage $x'_1(t)$ modified by the voltage divider consisting of the impedances R

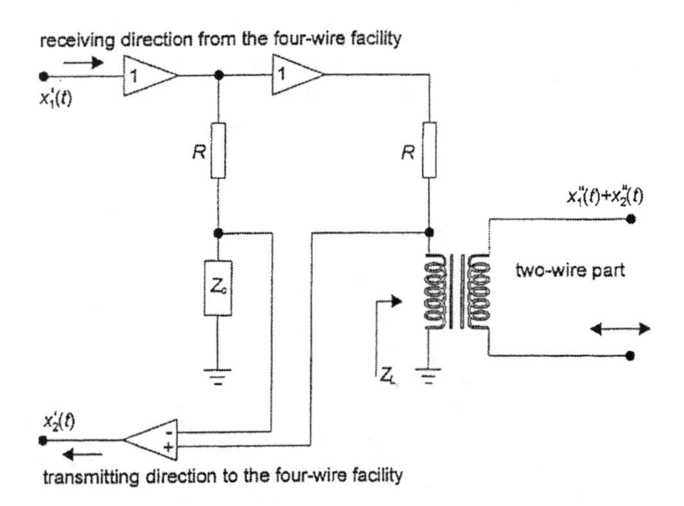

FIGURE 2.2 Realization of the hybrid, which uses two separating amplifiers and one differential amplifier.

and Z_L, where the latter means the input impedance of the two-wire transmission line shown on the extreme right in Figure 2.2. The separating behavior of the separating amplifiers in Figure 2.2 is illustrated in Figure 2.3, showing clearly "voltage transferring" in only one direction, from the left to the right. Further, the signal $x''_1(t)$ on the right-hand side of the transformer in Figure 2.2, which corresponds to the signal $x'_1(t)$, is transmitted through two wires to the customer premise on the right.

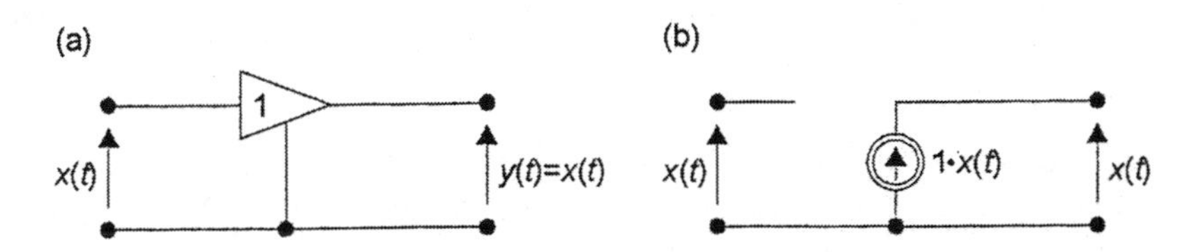

FIGURE 2.3 (a) Symbol of the voltage separating amplifier, (b) equivalent scheme with voltage-controlled voltage source of gain equal to one.

The voice signal $x''_2(t)$ from this customer to the hybrid appears on the left-hand side of the transformer only as the input signal to the "plus" input of the differential amplifier. It does not go to the "minus" input of the differential amplifier on the path through the separating amplifier. On the other hand, the signal $x'_1(t)$, modified by the voltage divider consisting of the impedances R and Z_c, appears on the "minus" input of the differential amplifier. However, when the compensating impedance Z_c equals the impedance Z_L, then the undesired leakage of the signal $x'_1(t)$ to the output of the differential amplifier is perfectly attenuated. The signal $x'_2(t)$ on the left of Figure 2.2 corresponds exclusively with the signal $x''_2(t)$ on the right. In practice,

Z_c differs more or less from Z_L depending on the realization of Z_c and properties (which are not fixed) of the two-wire transmission line on the right of Figure 2.2. Therefore, some part of the signal $x'_1(t)$ at the upper left corner of Figure 2.2 leaks to the signal $x'_2(t)$ at the lower left corner of Figure 2.2.

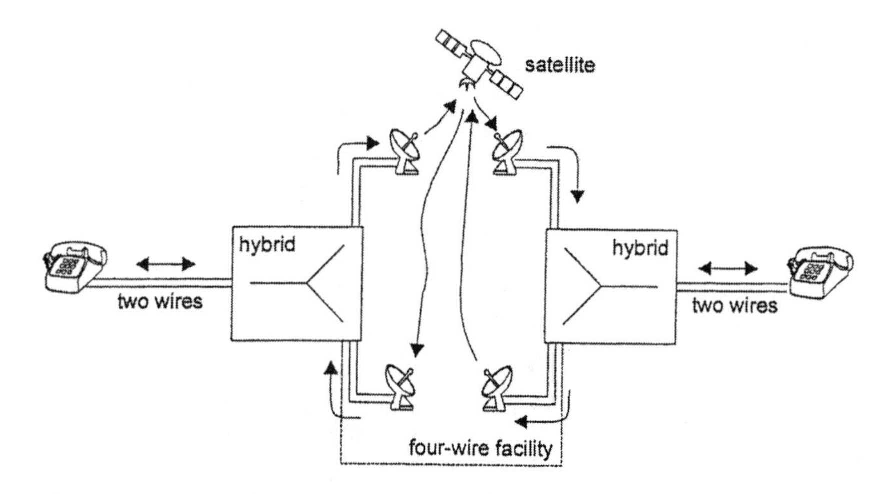

FIGURE 2.4 Long-distance call with the use of a satellite as the transmission medium.

This leakage can be expressed as the attenuation of the signal $x'_1(t)$ at the output of the differential amplifier with respect to its level at the input of the separating amplifier at the upper left corner of Figure 2.2, which can be as small as 6 to 10 dB.

The scheme of two wires/hybrid/four-wire facility/hybrid/two wires, seen in the center of Figure 2.1, is quite general. Note that the transmission through a satellite, as shown in Figure 2.4, can be also classified as belonging to this category.

With the principle of operation of the hybrid, as shown in Figure 2.2, and its description in mind, we can examine the transmission paths in Figure 2.1 and Figure 2.4 in convention[24,40] presented schematically in Figure 2.5.

Figure 2.5 shows a situation where the talker on the left-hand side speaks and the listener on the right-hand side listens. The basic desired transmission path, called the talker speech path, is shown in Figure 2.5a. Here, the speech goes through both hybrids and arrives, as needed, at the listener's site. Furthermore, Figure 2.5b shows that a part of the signal arriving at the right-hand side hybrid, because of the mismatching of the hybrid impedances Z_c and Z_L, leaks to another two wires of the four-wire facility and comes back to the talker as the talker echo. As shown in Figure 2.5c, the leakage can occur on both hybrids, so that a part of the talker speech, after making a round-trip in the four-wire facility, comes to the listener as the echo. This echo, arriving at the listener's site, is called the listener echo. Finally, observe that the talker and listener shown in Figure 2.5 can change their roles. Thus, all that has been written above about one site, in Figure 2.5, regards also another site.

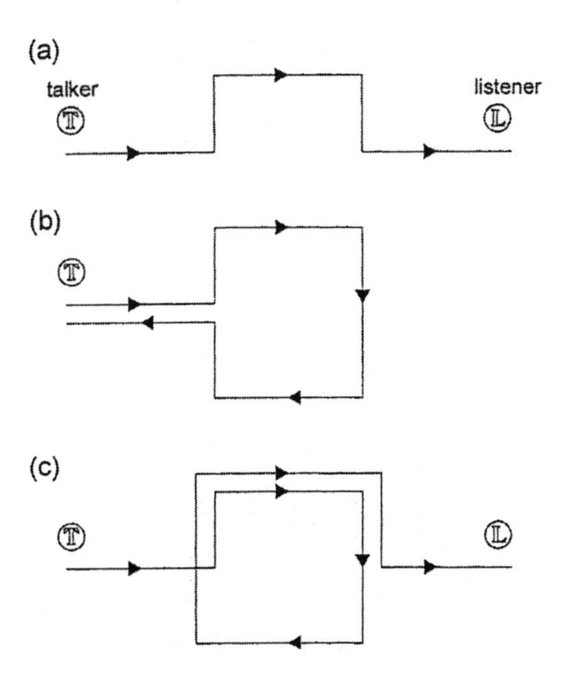

FIGURE 2.5 Basic transmission path and echo transmission paths in the telephone network: (a) talker speech path, (b) talker echo, (c) listener echo.

To prevent the arising echo, echo suppressors[40] can be used, as illustrated in Figure 2.6.

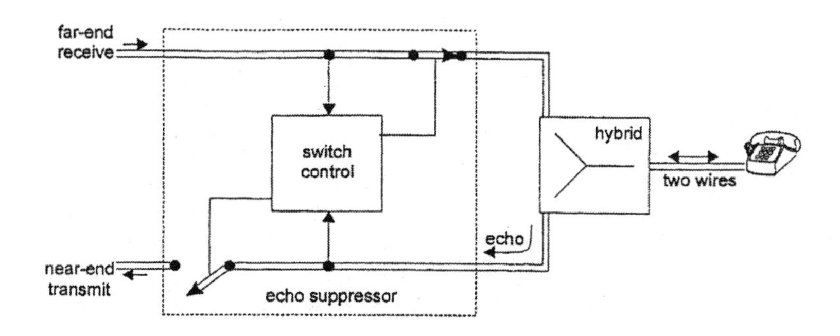

FIGURE 2.6 Principle of operation of the echo suppressor.

As shown in Figure 2.6, the echo suppressor consists of two switches and a controlling circuitry. The task of the controlling circuitry is to recognize who is speaking at a given moment, and to close one of the switches at that moment. The second switch then remains opened. Figure 2.6 shows such a situation, where the far-end is active, and this state is recognized by the controlling circuitry. The upper switch is closed, allowing the received far-end signal to go through the hybrid to the listener on the right-hand side of Figure 2.6. At the same time, the lower switch is opened, preventing the echo signal from making trips in the four-wire facility, as shown in Figs. 2.5 b and c. When the near-end becomes active, the states of the switches reverse. And we see from the above, the suppressor works correctly when persons on the opposite sides of the telephone connection do not speak simultaneously. When, for example, the listener wants suddenly to interject a point, it is likely that his or her message will not arrive to the talker on the other side of the telephone connection, because the switches will not switch for transmitting this message. In this case, a canceller will work better.

The basic difference in the operation of echo suppressors and echo cancellers is illustrated in Figure 2.7. In this figure, the situation is presented where both the far-end and near-end talkers speak simultaneously. In Figure 2.7a, no near-end transmitted signal is presented because, compared with Figure 2.6, the upper switch of the suppressor is closed, and the lower switch is opened. This is so because we assumed while before that only the far-end talker was speaking. Thus, in Figure 2.7a, both the leaked echo signal and near-end talker speech are blocked. Note that this is not the case in Figure 2.7b. In the echo canceller, an echo replica is synthesized and then subtracted from the transmitted signal. As a result, in the ideal case of building a perfect replica, there remains after performing the above operations, as shown in Figure 2.7b, only the speech of the near-end talker.

At first glance, the concept of echo cancellation as shown in Figure 2.7b can look very strange. However, it really works.[41,42,43,44] It is possible to remove the unwanted echo signal in an adaptation process such that the level of the echo remaining does not hinder the correct receipt of the wanted signal.

To explain the principle of operation of an adaptive canceller, let us now redraw the scheme of Figure 2.7b, as shown in the next figure, Figure 2.8. Here, the echo path is presented schematically by the corresponding block, called an echo path, of which the input signal is the far-end signal x, and of which the output signal is simply the echo signal e. The echo signal e adds to the near-end signal v, giving the resulting signal $e + v$. Furthermore, the echo path is modeled in some way in an adaptive echo canceller, for example, as a finite impulse response (FIR) filter. The input signal to the echo canceller is the far-end signal x, and its output signal is the estimated echo replica $\hat{e}$. The parameters of the echo path model, built in the echo canceller, are adapted in the adaptation process to make the signal $v + e - \hat{e}$ as similar to the near-end signal as possible.

Of course, in practice, cancellation of the echo is needed on both sides of a telephone connection. This is illustrated in Figs. 2.9a and b with the use of the scheme of Figure 2.8. In Figure 2.9a, the four-wire facility occurring between the ports AB and CD is marked by dashed lines and redrawn in Figure 2.9b for connections with a very long signal delay as, for example, experienced on satellite connections. The amount of 300 milliseconds is a typical value of delay in the long delay satellite channel of Figure 2.9b, for one direction of transmission.

Note that the echo signal before cancellation by the canceller on the left-hand side of Figure 2.9c makes a round trip lasting about 600 ms. And the canceller considered has to be able to synthesize such a long delay, which causes a high

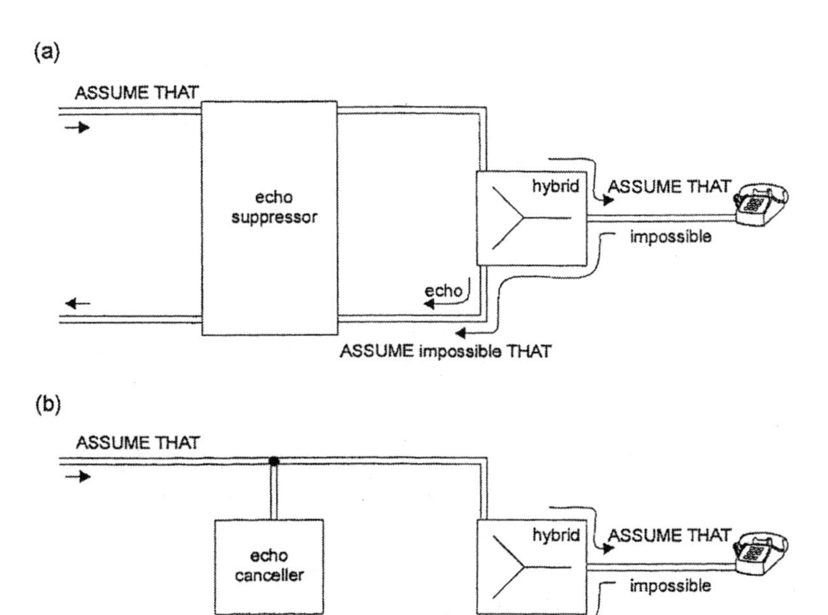

FIGURE 2.7 Illustration of the principle of operation of an echo suppressor (a), and of an echo canceller (b).

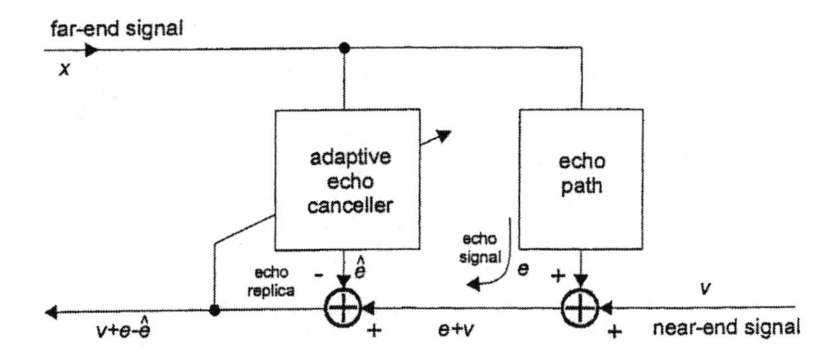

FIGURE 2.8 Principle of operation of an adaptive canceller.

realization cost, making the solution impracticable. This can be changed by moving the canceller from the left-hand side to the right-hand side in Figure 2.9c. The canceller position on the right-hand side is much better because here the echo signal has to make only a short round trip from the upper to the lower canceller port. In the canceller realization, it reflects through the need to synthesize a much shorter delay than before, which is realizable.

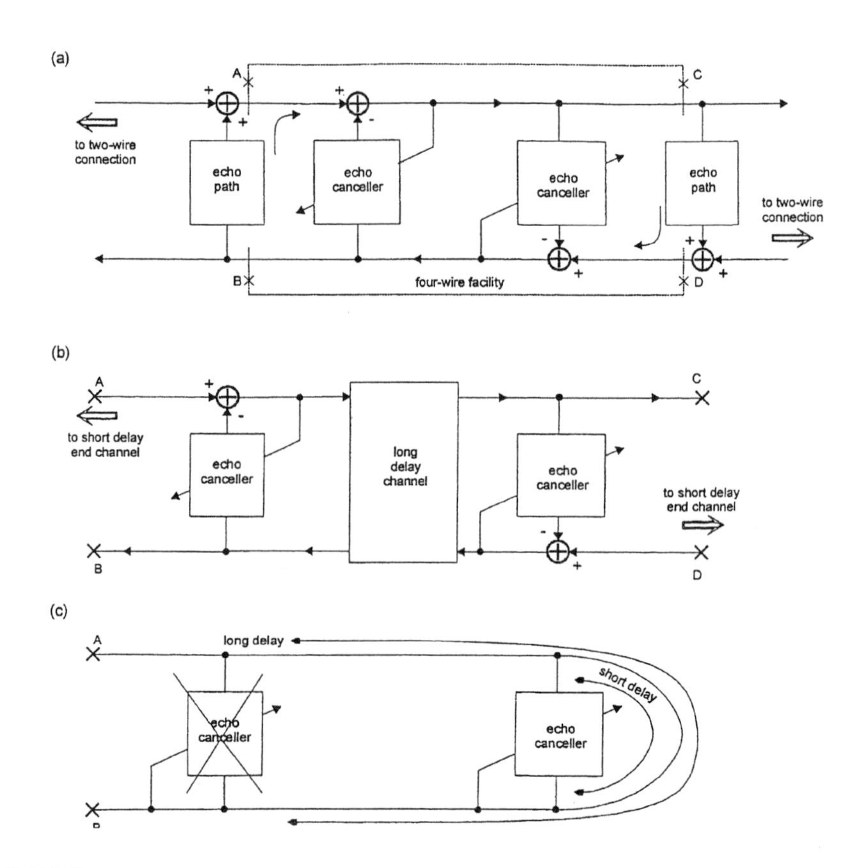

FIGURE 2.9 (a) Illustration of echo cancellation on both sides of a telephone connection, (b) long delay channel occurring between the cancellers as, for example, in connection through a satellite, (c) correct placing of the canceller with respect to the delay in echo path.

Note that placing the echo canceller on the right-hand side of Figure 2.9c corresponds to the position of the canceller on the right-hand side of Figure 2.9b. This is the exact same position. Moreover, with regard to Figure 2.9b, we say that the canceller sees a short delay end channel from the port CD to the right.

The configuration shown in Figure 2.9b, because of the properties just described, has been given the special name of a split echo canceller configuration by Messerschmitt in Reference 40. In fact, the echo canceller for both transmission directions in the middle of Figure 2.9a, between the dashed lines, is split by the long delay channel in Figure 2.9b into two distinct cancellers located near the ends of the connection.

At this point, it worth noting that the words *near end* and *far end* are used in two different meanings in the literature on echo cancellation. This point is illustrated in Figure 2.10. In this figure, the points A, B, C, and D are shifted to the left and to the right accordingly, such that the rectangle between the dashed lines encompasses also the echo paths, outside it in Figure 2.9a. Figure 2.10 considers the echo canceller configuration from the point of view of the listener at point B. For him or her, the near end and far end, understood spatially, are as shown in Figure 2.10a. They are used in such a way, for example, in References 43 and 45. However, taking into account the length of the trip made by the talker signals coming from points D and A to the listener at point B, we see that the first point is nearer than the second one. This is so because the trip for the talker signal from A to B is twice as long as that from D to B (see Figure 2.10b). The above terminology is used in Reference 40.

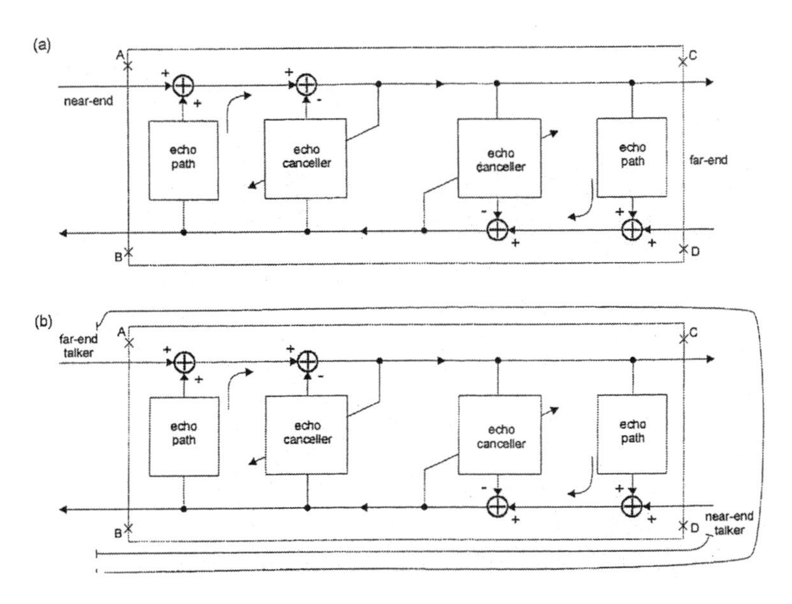

FIGURE 2.10 (a) "Near-end" and "far-end" understood spatially in the echo canceller configuration, (b) another point of view, taking into account the length of the trip made by the talker signal arriving at the listener at point B.

We will now go to data transmission, and restrict ourselves here to digital data transmission over analog channels. In this transmission, digital cancellers are used in most solutions proposed in the literature. The basic structure of the linear digital canceller is shown in Figure 2.11. This structure works with digital signals: the reference signal $x(k)$, the echo signal $e(k)$, and the near-end talker (according to the terminology assumed in Figure 2.10b) signal $v(k)$. It is linear because this is exactly the same structure as that of the linear finite impulse response (FIR) system (compare with Figure 1.48). The $x(k)$ is called the reference signal because, with reference to it, the echo replica is constructed; it is the input signal to the echo path.

If we use the expression 1.248 relating the input signal with the output signal of a FIR system through the linear convolution sum, we can write the corresponding relation for the echo replica $\hat{e}(k)$ in Figure 2.11 as

$$\hat{e}(k) = \sum_{i=0}^{M_{\hat{e}}-1} c(i)x(k-i) \tag{2.1a}$$

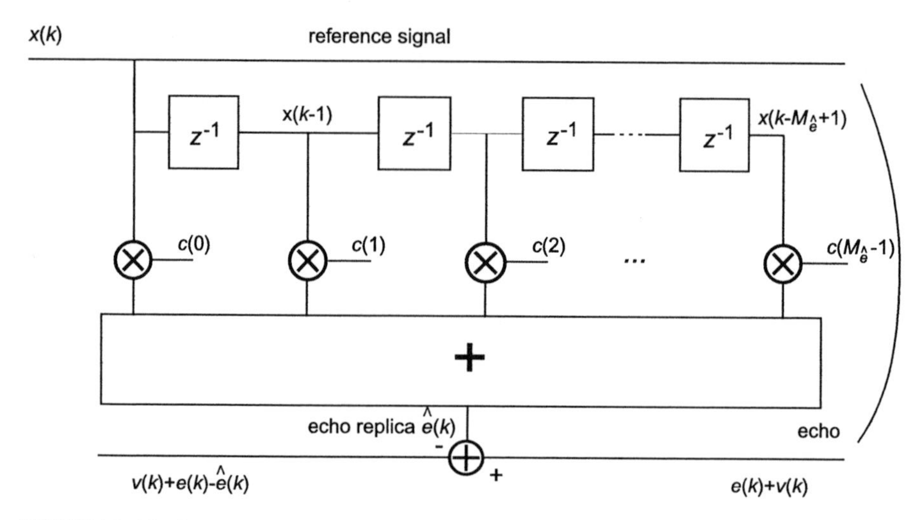

FIGURE 2.11 Structure of the linear digital canceller.

Of course, the structure of the canceller presented in Figure 2.11 is not an adaptive one; its impulse response, given by the coefficients $c(i)$, $i = 0, \ldots, M_{\hat{e}} - 1$, does not change with the elapsing time. To make this structure adaptive, we must allow the coefficients $c(i)$ to depend upon time, that is, to allow the following form: $c(k,i)$. For such coefficients determining the canceller impulse response, Equation 2.1a modifies to

$$\hat{e}(k) = \sum_{i=0}^{M_{\hat{e}}-1} c(k, i)x(k-i) \tag{2.1b}$$

It is also convenient to rewrite Equation (2.1b) in the equivalent forms

$$\hat{e}(k) = \sum_{i=0}^{M_{\hat{e}}-1} c_i(k)x(k-i) \tag{2.1c}$$

and

$$\hat{e}_k = \sum_{i=0}^{M_{\hat{e}}-1} c_i(k)x_{k-i} \tag{2.1d}$$

A rule for making the coefficients $c(k,i) = c_i(k)$, $i = 0, \ldots, M_{\hat{e}} - 1$, in Equations 2.1b to d dependent upon time can be chosen in many ways. For example, it can be the rule of the stochastic iteration algorithm,[46]

$$c(k+1, i) = c(k, i) + 2\alpha r(k)x(k-i) \tag{2.2a}$$

or

$$c_i(k+1) = c_i(k) + 2\alpha r(k)x(k-i), \quad i = 0, \ldots, M_{\hat{e}} - 1 \tag{2.2b}$$

where α is an amplification constant, and $r(k)$ means the residual signal given by

$$r(k) = v(k) + e(k) - \hat{e}(k) \tag{2.2c}$$

(see Figure 2.11).

Note that, with the time-dependent coefficients $c(k,i)$, the canceller structure of Figure 2.11 must be redrawn, as shown in Figure 2.12. Then, for correct understanding, the notational convention used in Equations 2.1c and d and 2.2b must be used.The fundamental difference in the coefficients c in the structures presented in Figures 2.11 and 2.12a lies in the fact that, in the first case, the time variable i occurs in parentheses as $c(i)$, $i = 0, \ldots, M_{\hat{e}} - 1$, and, in the second case, this time variable is "shifted" to form an index at $c_i(k)$, $i = 0, \ldots, M_{\hat{e}} - 1$. The form $c_i(k)$ is needed to express the change in the coefficient values with the changing time, as the adaptation process proceeds. When the echo path can be considered as a linear system possessing a linear impulse response of the approximately-finite length, then, after some time elapsed in the adaptation process, the values of the coefficients $c_i(k)$ correspond quite well with the values of the corresponding samples of the impulse response of the echo path. In other words, assuming the echo path response described by

$$e(k) = \sum_{i=0}^{M_e - 1} g_i x_{k-i} \tag{2.3a}$$

and the echo replica, described by Equation 2.1d with the parameter M_e satisfying inequality $M_{\hat{e}} \geq M_e$, we get

$$e(k) - \hat{e}(k) = \sum_{i=0}^{M_e - 1} (g_i - c_i(k))x_{k-i} + \sum_{i=M_{\hat{e}}}^{M_{\hat{e}} - 1} c_i(k)x_{k-i} \rightarrow 0 \tag{2.3b}$$

in the adaptation process, when $k \rightarrow \infty$. This is so because, in the adaptation process illustrated

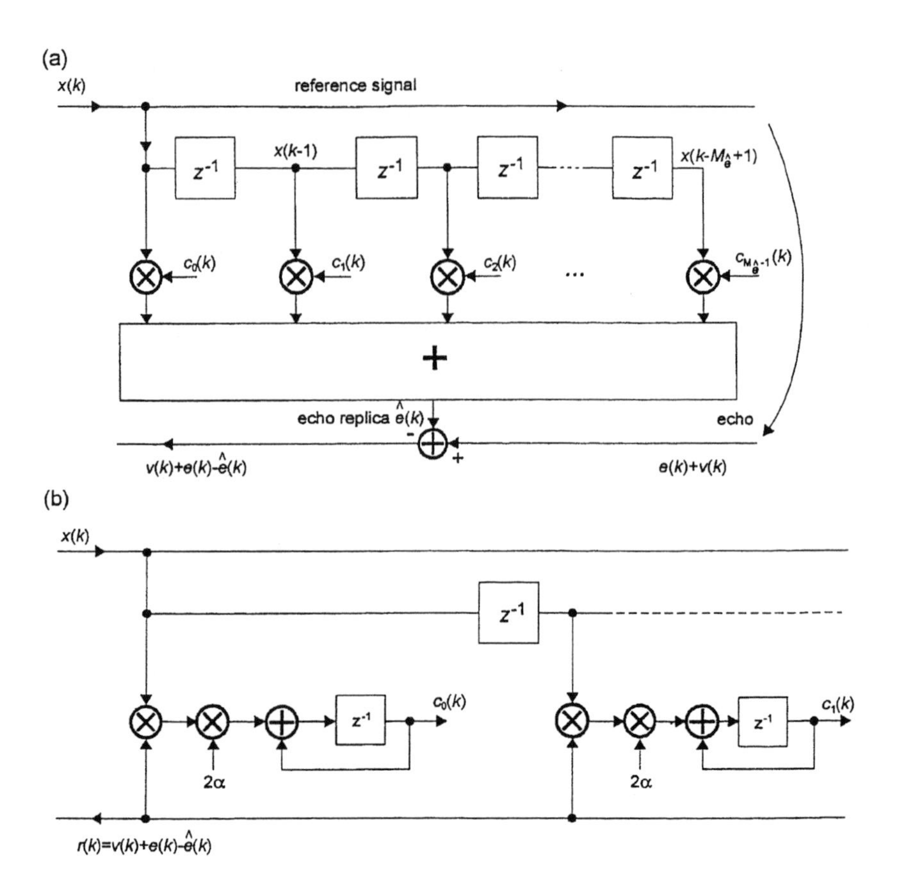

FIGURE 2.12 (a) Structure of the linear digital canceller with time-dependent coefficients, (b) calculation of the coefficients $c_i(k)$ for successing time instants according to the stochastic iteration algorithm.

in Figure 2.12, the following:

$$c_i(k) \to g_i, \quad i = 0, \ldots, M_e - 1, \quad \text{as} \quad k \to \infty \tag{2.3c}$$

and

$$c_i(k) \to 0, \quad i \geq M_e, \quad \text{as} \quad k \to \infty \tag{2.3d}$$

takes place. In Equations 2.3, g_i, $i = 0, \ldots, M_e - 1$, mean the samples of the echo path impulse response, and M_e and $M_{\hat{e}}$ are the memory lengths of the echo path and the echo canceller, respectively, in the engineer's sense explained in Section 1.9.

Of course, the description of the adaptation process given by Equations 2.3b, c, and d is highly simplified, because it does not take into account statistics of the reference signal $x(k)$ and the effect of noise. The probabilistic character of the adaptation process will be considered later.

With regard to the terminology used, we draw the reader's attention to the fact that a direct realization of the FIR filter, as shown in Figure 2.11 or, for a given k, in Figure 2.12, is called the transversal filter in the digital communication literature.

Data signals can be transmitted on a two-wire line in only one direction, in some frequency band; this mode of operation is called a half-duplex. However, the transmission in both directions can also occur simultaneously on a two-wire connection, occupying the same frequency band. This mode of operation is called the full-duplex transmission. The echo creates real problems only in the latter case because the signal transmitted in one direction can interact with the signal transmitted in the opposite direction. This is not the case in the half-duplex transmission. In fact, we have then a situation as shown in Figure 2.5a. The paths of Figs. 2.5b and 2.5c do not occur, because there is no receiver on the transmitting end.

In Reference 40, two important applications of full-duplex transmission are mentioned: digital transmission on the subscriber loop and digital transmission in voiceband. The first type of transmission is illustrated schematically in Figure 2.13. Over the subscriber loop, the digital voice and data are transmitted.

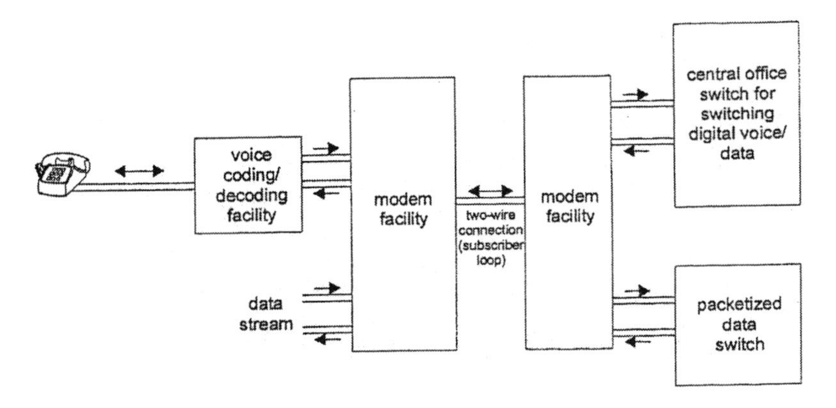

FIGURE 2.13 Digital transmission on the subscriber loop.

As shown by the block named the voice coding/decoding facility, the voice transmission requires the analog-to-digital (A/D) and digital-to-analog (D/A) conversions to be performed on the customer premises. Data signals coming from or to the modem facility in Figure 2.13 do not need any signal conversion between the above facility and the customer premises. The modems shown in Figure 2.13 perform operations of modulation and demodulation of analog impulses by the stream of digits (most often binary). This is needed because only analog signals can be transmitted on two wires of the subscriber loop.[40] The modem facility on the right-hand side of the subscriber loop in Figure 2.13 is connected to the central office switch, which enables digital voice or data stream switching. This facility also has connections for transmission of packetized data; that is, data packets used in data networks.

The digital subscriber loop as shown in Figure 2.13 and described above occurs in the integrated services digital network (ISDN), which integrates voice and data services to provide them to the customer over a common facility.[40,48] At its basic level, the ISDN provides transmission rate of 144 kbit/s for each direction. This rate is described as the basic interface and called 2B+D, which means that the interface consists of two B channels at 64 kbit/s each and one channel D at 16 kbit/s. The B channels are devoted to transmission of the digital voice and/or data, while the D channel is an additional channel for data, which can be transmitted at a lower rate (4 times lower, 64 kbit/s: 4 = 16 kbit/s, than in the B channel). Additionally, framing and control data are transmitted at the rate of 16 kbit/s, which gives altogether 144 + 16 = 160 kbit/s. With regard to Figure 2.13, that means that the transmission rates are 160 kbit/s on the subscriber loop in each direction. Moreover, the transmission on the digital subscriber loop is the so-called baseband transmission,[40] that is, without the use of carriers. This is because the frequency characteristic of the two-wire connection of the digital subscriber loop in Figure 2.13 has a shape of a lowpass filter. For example, to transmit at the rate of 160 kbits/s, it must provide bandwidth at least in the range of 0 ÷ 40 kHz.[24] On the other hand, the voiceband channel can be considered as a passband channel, having the character of a bandpass filter. It provides bandwidth in the range of 300 ÷ 3300 Hz. By the way, comparing the bandwidth 3300 − 300 = 3000 Hz with the previous value of 40 kHz for the digital subscriber loop, note that the bandwidth provided by the voiceband channel is much smaller than that provided by the digital subscriber loop. This is rather the rule.

Communication between two full-duplex voiceband data modems is shown schematically in Figure 2.14. In this figure, two modems are connected to the public telephone network via two wires representing telephone channels of bandpass frequency characteristic with the passband in the range mentioned just before; that is, between 300 ÷ 3300 Hz. They work in the full-duplex mode of operation.

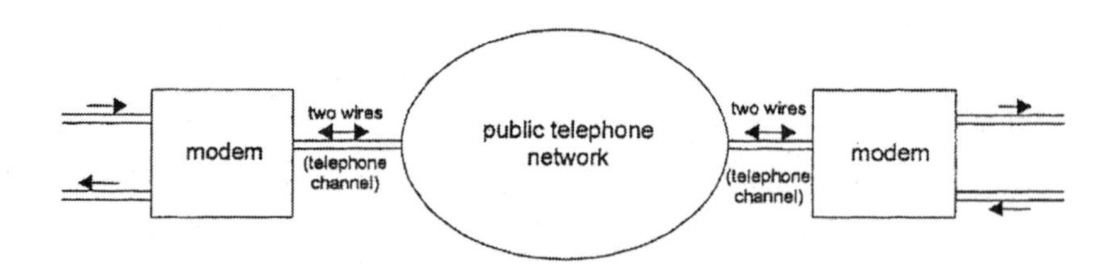

FIGURE 2.14 Two full-duplex voiceband data modems communicating over a public telephone network.

Comparing Figure 2.13 with Figure 2.14, we see that the common feature there are two modems (modem facilities) communicating in the full-duplex mode of operation via two wires, and connected to each other, directly or indirectly, through them. To reduce an inherent echo in such a connection, echo cancellers are needed on its both ends. This is shown in Figure 2.15, where hybrids are parts of modems, or modem facilities, as in Figure 2.13. Moreover, in Figure 2.15, double lines used to mark two-wire connections are replaced by single ones for simplicity, as was already done in the schematic drawings in Figs. 2.8 to 2.12.

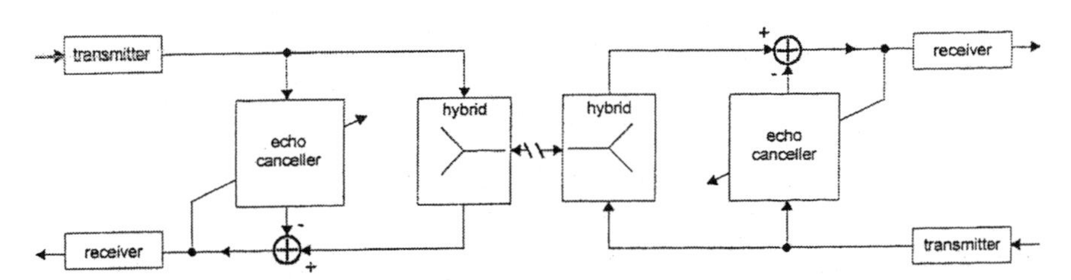

FIGURE 2.15 General scheme for echo cancellation in full-duplex data transmission.

Comparison of the structure of Figure 2.15 with the structure presented in Figure 2.9 shows that placing an echo canceller is principally the same in both figures. That is, the echo canceller is placed on the same side of the hybrid in both structures. The main difference between the configurations lies "in the middle of the connection": in Figure 2.9b, this is a four-wire facility building a long delay channel; but, in Figure 2.15, this is a short delay channel on two wires. Moreover, it follows from the discussion that the configuration of Figure 2.15 describes correctly both types of full-duplex data transmission shown in Figs. 2.13 and 2.14. The configuration of Figure 2.15 encompasses also echo cancellation in both transmission directions.

The structure in Figure 2.15 for echo cancellation in full-duplex data transmission is highly simplified. Nevertheless, it can be used in analyses aimed at finding basic characteristics of the cancellation process. This structure and the more detailed structures derived from it have been studied intensively in the literature.[48-53]

There are a transmitter and a receiver on each end of the configuration shown in Figure 2.15. The task of the hybrids is to provide a virtual four-wire connection between the above elements; two virtual wires are devoted to one direction and the next two to another direction (see Figure 2.16a). This virtual four-wire connection is, however, not ideal because of the impairments of hybrids. These impairments are responsible for the signal leakage between the upper and lower part of a hybrid, as already discussed in the description of the hybrid presented in Figure 2.2. This leakage is shown schematically in Figure 2.16b.

The signal leakage in the hybrid or, in other words, the feedthrough in the hybrid, of the transmitted signal from the transmitter to the local receiver can be as high as 6 to 10 dB, as already mentioned in the discussion of the structure in Figure 2.2. In terms of signal attenuation, this will mean 6 to 10 dB of attenuation between the transmitter and local receiver. Only 6 to 10 dB, means that the above leakage will be the main echo component to be cancelled by each of the cancellers in Figure 2.15.

Assume now, after Reference 40, that the two-wire channel attenuation from the transmitter to the receiver on the other side of the connection in Figure 2.15 is about 40 to 50 dB. This value is typical for both types of transmission: in the digital subscriber loop and in the voiceband data transmission. Furthermore, assume that

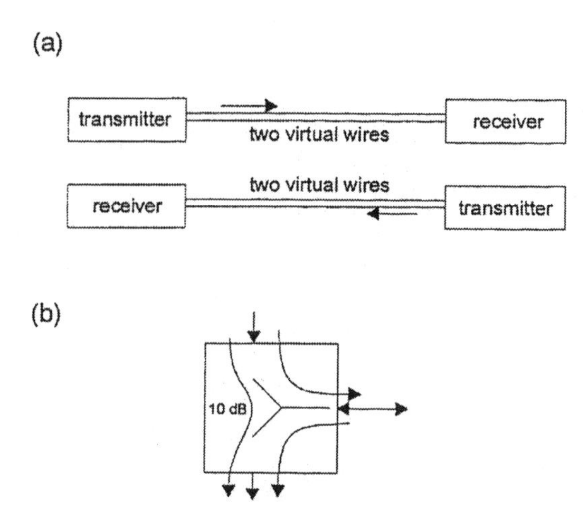

FIGURE 2.16 (a) Virtual four-wire connection, (b) signal leakage in a hybrid.

the attenuation provided by the hybrid is about 10 dB. Then, assuming that the transmitted signals on both sides of the connection are at the same level, we get the level of the local echo (local feedthrough) signal on each of the connection ends about 30 to 40 dB higher than the level of the receiver signal on these ends. For reliable data transmission, this signal should be, however, at least 20 dB under the level of the received signal. Therefore, the echo cancellers in Figure 2.15 will have to provide the attenuation of the local feedthrough signal of an order of 50 to 60 dB. The achievement of this goal can be critical[40,54,55,56] because of the occurrence of the inherent nonlinearities associated with the echo path. The next section will be devoted to explanation of the problem in more detail.

2.2 NONLINEAR ECHO CANCELLATION IN DATA TRANSMISSION

To show the sources of nonlinearities in the echo path, we must present the general scheme for echo cancellation of Figure 2.15 in more detail. In Figure 2.17, we show the more detailed structure of a full-duplex digital subscriber loop transceiver, that is, of an arrangement consisting of the transmitter and receiver. Note that such an arrangement occurs on the ends of the connection in Figure 2.16a; so this connection can be viewed as the one between two transceivers.

A similar structure as that shown in Figure 2.17 was used in Reference 24 in considerations regarding echo cancellation. This structure contains all the details needed in our explanations and analyses. Of course, it still remains a simplified structure because, for example, it does not contain a detection block in the lower, receiver path. For more details regarding the structure of the full-duplex digital subscriber loop transceiver, interested readers are referred to Reference 48.

From Figure 2.17, it is evident that the adaptive canceller works with digital signals (other solutions are also possible, for example, see Reference 54). The

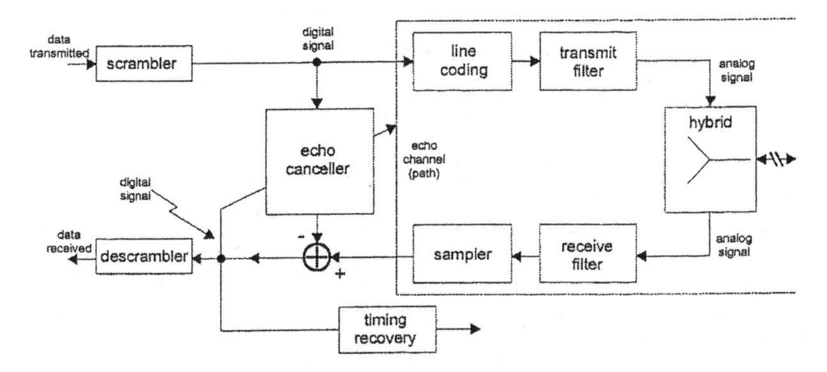

FIGURE 2.17 More detailed structure of a full-duplex digital subscriber loop transceiver.

elements of the echo path, which belong to the transceiver, lie between the dashed lines in the figure. We included here the hybrid, transmit and receive filters, the block in which line coding of the transmitted signal is performed, and the sampler in the receiving path, which converts the received analog signal into the digital one. A block where the timing recovery is performed is also shown. The task of this block is to recover a clock from the received signal, to sample it properly for getting a discrete-time sequence of data symbols in the received path. The input signal to the timing recovery block is the same signal that is used for adaptation purposes in the canceller. This is the sum of the canceller output signal and the received signal after sampling. Furthermore, on the extreme left-hand side of Figure 2.17, we have the blocks named scrambler and descrambler. The scrambler occurs in the transmitting path, and its task is to make more random the data to be transmitted. The scrambled signals, much more random than the original ones, make it easier to achieve accurate timing recovery.[24] They also enable achieving dc balance of the signal. On the other hand, the descrambler performs the inverse operation to the scrambling made by the transmitter on the opposite side of the transmission connection. After performing the descrambling operation, the original form of the received data sequence is achieved.

In the line coding block in Figure 2.17, coding of the transmitted signal, to adjust in some way to the transmission line characteristics, is performed. Generally speaking, such coding is used to control the spectral characteristic of the transmitted signal.[24] One common goal here is the introduction of a spectral zero at dc, thereby enabling transmission of a baseband signal over a channel that rejects dc components.

The task of the transmit filter in Figure 2.17 is to attenuate the higher frequency components of the signal for avoiding radio-frequency interference (RFI) and crosstalk between the different channels. Finally, the receive filter in the scheme of Figure 2.17 is to prevent aliasing effects in the subsequent sampler.

With the structure of the full-duplex digital subscriber loop described, we now can consider the sources of nonlinearities in it. In Reference 54, the sources are identified as follows:

(1) Nonlinear characteristics of data converters
(2) Nonlinearity associated with an imbalance in the positive and negative pulses transmitted
(3) Saturation in transformers of hybrids

Depending upon the concrete realization of the full-duplex subscriber loop transceiver, a greater or lesser number of analog-to-digital (A/D) and digital-to-analog (D/A) converters is used.[53, 54] In the implementations, these converters do not behave ideally. Their characteristics are, in practice, not exactly linear. The nonlinearities occurring in the converters are well documented in the literature, (see References 54, 57-60, for example). They are not discussed here. For our purposes in this book, suffice it to know that such imperfections occur, making the echo path, in effect, nonlinear

The nonlinearity mentioned in Reference 54 as the nonlinearity associated with an imbalance in the positive and negative pulses transmitted should be interpreted mathematically as occurrence of the dc component in the echo path. That is, the echo path is represented by a nonlinear operator, say, E_c. Assume that the input sequence $x(k)$ to this operator is an exactly symmetric sequence with respect to the time axis. The operator E_c then gives as the output signal something like this: $(E_c x)(k) = e_0 + e_f(k)$, where e_0 is the dc component in the echo path and $e_f(k)$ stands for the echo dc-free component. Furthermore, observe that such an operator does not obey relation 1.13 for linear systems (operators) because we have $E_c((\alpha x k)) = e_0 + e'_f(k) \neq \alpha(e_0 + e_f(k))$, where α is some real number and $e'_f(k)$ is the echo dc-free component when the input sequence to the operator E_c is equal to $\alpha x(k)$. Even when the component e_f would be linear, that is, $e'_f(k)$ would be equal to $\alpha e_f(k)$, the remaining dc component would still determine the nonlinear character of the operator E_c.

It has been reported in the literature[39, 54] that, for the levels and shapes of signals used in digital transmission, the saturating effects can occur in transformers. The transformer cannot then be considered as a linear element. It contributes in this case to the nonlinear characteristic of the echo path.

In the voiceband data transmission systems, we have similar sources of nonlinearities in the echo path, as in the case of the digital subscriber loop.

In the literature, a few interesting approaches to the problem of nonlinear echo cancellation are presented. Thomas[61] was first to systematically study the problem. He used the Volterra series method. Next, to study the problem were Coker and Simkins.[62] They, like Thomas, applied the Volterra series.

The work of Holte and Stueflotten[51] had a great impact on further studies and works on this topic. They presented the memory compensation principle, or table look-up method, that applies the following property: when the operator describing the echo path has the fading memory or, equivalently (as explained in Section 1.9), approximately-finite memory, say, of length M, then a set of possible values of the echo path responses is finite. It consists of 2^M elements for binary signals. These values can, of course, be stored in a digital memory, and afterward accessed in the echo cancellation process; the last M transmitted bits form an address to the corresponding value of the echo path response saved in memory.

Referring to Figure 2.11 that shows the structure of the linear digital canceller based on the transversal filter, we now redraw the above structure in the form incorporating the use of the memory compensation principle. The corresponding scheme is presented in Figure 2.18.

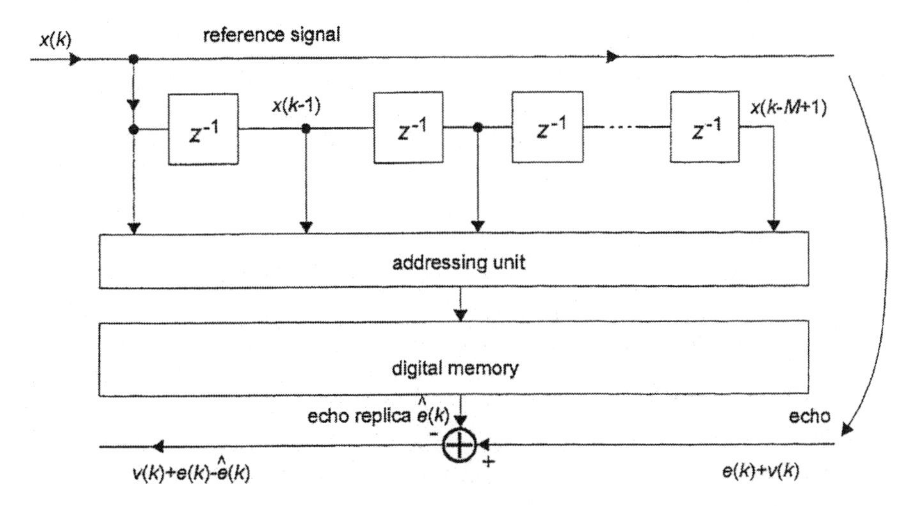

FIGURE 2.18 Structure of the digital canceller using the memory compensation principle.

It follows clearly from Figure 2.18 that the corresponding sequence of the last M bits $x(k)$, $x(k - 1)$, ..., $x(k - M + 1)$ taken from the reference sequence $x(k)$ forms the address to the digital memory. (We recall here that, according to the simplified notation assumed in Section 1.7, $x(k)$ is also equivalently used instead of $\{x(k)\}$.) At this address, the corresponding value of the echo replica for the discrete-time point k is found in the digital memory, where it was stored before in a learning process of the echo channel. Then $\hat{e}(k)$ is substracted from the incoming signal $e(k) + v(k)$ in the received path, similarly as in Figure 2.11.

Variations of the usage of the table look-up method to nonlinear echo cancellation have been presented in many publications after 1981. For example, Smith, Cowan, and Adams in Reference 55 have used the so-called transpose distributed arithmetic[63, 64] for an effective realization of the nonlinear echo canceller. More recently, Weruaga-Prieto and Figueiras-Vidal[65] have presented some results for a combined canceller, using the table look-up method for realization of its nonlinear part and FIR filter for realization of its linear part.

Other approaches to the problem of nonlinear echo cancellation are the following: the use of the canonical piecewise-linear function description[66] and the application of the neural network models[35] to model the nonlinear behavior of the echo path.

Since this book is devoted first of all to presentation of the application of the discrete Volterra series in solving nonlinear problems encountered in telecommunications, we present, in what follows, the application of this series to nonlinear echo cancellation in greater detail. However, we will also describe its relationship with the other methods mentioned. The description of the Volterra series method of echo cancellation presented in this section is based mostly on References 23, 56, and 67; and its main objective is to serve as an illustrative example, which can be used in different ways in practical applications. Also, all the analyses based on the above description, which will be presented, illustrate the calculations typically performed in studies of linear and nonlinear echo cancellers.

Assume that the nonlinear echo path in our case possesses fading memory or equivalently (as explained in Section 1.9), approximately-finite memory. Using the results of Section 1.9, we can model (approximate) its response by the discrete Volterra series. Assume additionally that an echo canceller to be constructed for the above echo path will work with binary signals. To model its behavior, we can choose a special version of the Volterra series, which was derived in Section 1.10, just for binary signals.

By the way, note that the property of possessing fading memory is crucial for the echo path modeling, independent of the method used. For example, see in the table look-up method[51] that, if the memory length M would go to infinity, then the digital memory size of the canceller, 2^M, would also go to infinity, thereby making the method quite useless.

Not to complicate our calculations too much, assume that the nonlinearities of the echo path are strong enough such that inequality $L \geq M_e$ holds, where L is a strength measure of these nonlinearities and $M = M_e$ is a memory length measure of the echo path (see Section 1.10). Then we can approximate the echo path response using the Volterra series approximator given by Equation 1.247d. Hence, the description of the echo path is, in this case,

$$e(k) = g_{00} + \sum_{i_1=0}^{M_e-1} g(i_1)a(k-i_1) + \sum_{\substack{i_1 = 0,\, i_2 = 0 \\ i_1 \neq i_2}}^{M_e-1} g(i_1, i_2)a(k-i_1)a(k-i_2) + \cdots \tag{2.4}$$

$$+ g(0, 1, \ldots, M_e - 1)a(k)a(k-1)\cdots a(k-M_e+1)$$

where somewhat different notation, as in Equation 1.247d is used. Here, g_{00} is the dc component in the echo path, which corresponds to the dc component $d_0^{(0)}$ in Equation 1.247d. Furthermore, $g(0)$, $g(1)$, …, $g(M_e - 1)$, $g(0, 1)$, $g(0, 2)$, …, $g(0, 1, \ldots, M_e - 1)$ correspond to the coefficients $d_0^{(1)}, d_1^{(1)}, \cdots, d_{M_e-1}^{(1)}, d_{01}^{(2)}, d_{02}^{(2)}, \ldots, d_{01\cdots(M_e-1)}^{(M_e)}$, in Equation 1.247d (with $M_e = M$), respectively. Moreover, the last M_e samples of the echo path input signal are denoted here otherwise than in Equation 1.247d because they are not the samples but the transmitted symbols, according to the terminology used in telecommunications. So, $a(k)$, $a(k - 1)$, …, $a(k - M_e + 1)$ correspond to $x(k)$, $x(k - 1)$, …, $x(k - M_e + 1)$ in Equation 1.247d (with $M_e = M$), respectively. These symbols take on, in our case, only two distinct values because they are binary. In other words, using the telecommunications terminology, [24] we say that the alphabet set for the above symbols possesses only two elements.

Another reason for changing the notation for the transmitted symbols, as done above, follows: We want to avoid some misinterpretations, which would follow from eventually identifying the input sequence $x(k)$ with a vector containing only M_e elements taken from it. We will construct such a vector in what follows because it will be very useful in our further considerations.

Using the echo path description given by Equation 2.4, we shall now investigate the influence of nonlinearities on echo cancellation in the digital subscriber loop, in which a linear canceller in the form of a digital adaptive transversal filter, as shown in Figure 2.12, is used. In our calculations regarding the above environment, we shall assume that the memory length of this canceller is so chosen that $M_{\hat{e}} = M_e = M$ holds. Moreover, we shall use the vector notation

$$\mathbf{a}(k) = [a(k), a(k-1), \ldots, a(k-M+1)]^{\mathrm{T}} \tag{2.5a}$$

for the transmitted data vector, consisting of M elements,

$$\mathbf{u}(k) = [1, a(k)a(k-1), a(k)(k-2), \ldots \\ a(k-M+2)a(k-M+1), \ldots, a(k)a(k-1)\ldots a(k-M+1)]^{\mathrm{T}} \tag{2.5b}$$

for the auxiliary transmitted data vector, consisting of $2^M - M$ elements,

$$\mathbf{g}_L = [g(0), g(1), \ldots, g(M-1)]^{\mathrm{T}} \tag{2.5c}$$

for the vector associated with the linear part of the echo path response, consisting of M elements,

$$\mathbf{g}_N = [g_{00}, g(0, 1), g(0, 2), \ldots, g(M-2, M-1), \ldots, g(0, 1, 2, \ldots, M-1)]^{\mathrm{T}} \tag{2.5d}$$

for the vector associated with the nonlinear part of the echo path response, consisting of $2^M - M$ elements,

$$\mathbf{c}(k) = [c_0(k), c_1(k), \ldots, c_{M-1}(k)]^{\mathrm{T}} \tag{2.5e}$$

for the vector of the adaptive transversal filter coefficients, consisting of M elements.

In Equations 2.5, the symbol "T" stands for the vector transpose. Moreover, we assumed in these equations to use M, where, in fact, $M_{\hat{e}}$ and M_e should stand, because, in the case considered, all three parameters are equal to one another. Also, for good readability, the elements in vectors are separated from each other by commas. The number of elements in each of the vectors is given as well. In this context, note that the total number of elements in both the vectors **a** and **u** is equal to 2^M, and similarly, we have 2^M elements in both the vectors $\mathbf{g}_L$ and $\mathbf{g}_N$ taken together. The above corresponds of course, with the total number of components of the sum on the right-hand side of Equation 2.4. To see that this number is also equal to 2^M, observe that the n-fold sum in Equation 1.247d or 2.4 represents a summation over all combinations of n from the M indices, 0, 1, …, $M - 1$. Hence, the total number of components in Equation 1.247d or 2.4 is given by the expression

$$\sum_{n=0}^{M} \binom{M}{n} = 2^M \tag{2.6}$$

where $\binom{M}{n}$ is a Newton symbol (for details see, for example Reference 9).

By the way, an immediate conclusion follows from comparison of Equation 2.6 with the previously mentioned size of 2^M of the digital memory in an echo canceller based on the table look-up method. Note that both numbers are identical. This means that, when all the components of the expansion 2.4 are used in the construction of a canceller based on it, then this canceller is equivalent to that using the table look-up principle.[23]

Using now the vectors defined by Equations 2.5, we can rewrite Equation 2.4 in a compact form as

$$e(k) = \mathbf{a}^{\mathrm{T}}(k) \cdot \mathbf{g}_L + \mathbf{u}^{\mathrm{T}}(k) \cdot \mathbf{g}_N \tag{2.7a}$$

Moreover, for the echo replica $\hat{e}(k)$ of our linear canceller considered, we can write

$$\hat{e}(k) = \mathbf{a}^{\mathrm{T}}(k) \cdot \mathbf{c}(k) \tag{2.7b}$$

With the use of Equations 2.7a and 2.7b, the residual signal, defined by Equation 2.2c, can be written as

$$\begin{aligned} r(k) &= e(k) - \hat{e}(k) + v(k) = e(k) - \hat{e}(k) + s(k) + n(k) \\ &= \mathbf{a}^{\mathrm{T}}(k) \cdot (\mathbf{g}_L - \mathbf{c}(k)) + \mathbf{u}^{\mathrm{T}}(k) \cdot \mathbf{g}_N + s(k) + n(k) \end{aligned} \tag{2.8}$$

where $v(k) = s(k) + n(k)$. In Equation 2.8, we assume that the echo signal $e(k)$ consists of all echo components, that is, of the signal leakage through the local hybrid, and of the components coming from the reflections along the line. On the other hand, the signal $v(k)$ consists of all the other components present in the received signal. Of these components, the most important is the data stream $s(k)$ coming from the other connection end. This signal is corrupted by noise and all the possible interactions between the transmitted and received data streams. It is denoted here $n(k)$.

Having described the nonlinear echo path and the linear echo canceller, we can now start the analysis of the adaptation process of the canceller. The objective of this analysis will be to achieve some view into the dynamic behavior of the linear canceller working in a nonlinear environment. We shall restrict ourselves here to investigation of only one adaptation algorithm; that is, of the stochastic iteration one.[46]

Consider the difference between the echo signal and echo replica. Using Equations 2.7a and 2.7b, we get

$$e(k) - \hat{e}(k) = \mathbf{a}^{\mathrm{T}}(k) \cdot (\mathbf{g}_{\mathrm{L}} - \mathbf{c}(k)) + \mathbf{u}^{\mathrm{T}}(k) \cdot \mathbf{g}_{N} \tag{2.9}$$

To go ahead, we need now one notion from the probability theory, namely, the notion of a mean value, also called, a mathematical expectation or an expected value. The definition of this notion can be found in any texbook on probability theory and its applications, but it can also be found in textbooks on digital communication (for example, in Reference 24).

Let X be a random variable. The outcome of this random variable, we then denote x, using a small letter. Proceeding further, let an event E_v be a set of possible outcomes taken from the so-called sample space Ω of all the possible outcomes. In the probability theory, an event is assigned a probability we denote here as $P(E_v)$. This probability takes on the values only from the range $\langle 0,1 \rangle$; that is, $0 \le P(E_v) \le 1$ holds.

At this point, we must make a distinction between the continuous-valued and discrete-valued random variables. For the first ones, we define the cumulative probability distribution function, $P_c(x)$ as the probability of the event $X \le x$, that is,

$$P_c(x) = P(X \le x) \tag{2.10a}$$

(By the way, the same definition of the cumulative probability distribution function as that given by definition 2.10a holds also for discrete-valued random variables. However, it will not be used here in further derivations for this type of random variables.)

The probability density function of the random variable X is then defined as

$$p(x) = \frac{d}{dx} P_c(x) = \frac{d}{dx} P(X \le x) \tag{2.10b}$$

To define the probability density function of a discrete-valued random variable X, we now follow the means presented by Lee and Messerschmitt in Reference 24. To this end, we denote the probability of an outcome $x \in \Omega$ (that is, belonging to the discrete-valued sample space Ω of the discrete-valued random variable X) by $P(X = x)$. Then, the probability density function of X can be defined as

$$p(x) = \sum_{z \in \Omega} P(X = z)\delta(x - z) \tag{2.11}$$

where the summation is over all the possible outcomes of the random variable X, and δ represents the one-dimensional discrete Dirac impulse, having the property $\delta(0) = 1$, and the zero value otherwise.

Using the definitions of the probability density function given by Equations 2.10b and 2.11, we can define the mean, or expected value, of a random variable X as

$$E(X) = \int_{-\infty}^{\infty} xp(x)dx \tag{2.12a}$$

for continuous-valued variables, and

$$E(X) = \sum_{z \in \Omega} xp(x) \tag{2.12b}$$

for discrete-valued variables. The operation $E(\cdot)$ defined in Equations 2.12a and 2.12b is called the mathematical expectation.

Returning to Equation 2.9, let us now treat the difference $e(k) - \hat{e}(k)$ for a given k as a random variable. (For simplicity of notation, we use also small letters to denote both a random variable and an outcome related with it, where this does not create any confusion.) There are good arguments for such treatment. Refer to the scheme of Figure 2.17, where the block called "scrambler" occurs in the transmitting path. The task of this block is just to make the occurrence of the transmitted symbols for given discrete time points as random as possible. Hence, the difference $e(k) - \hat{e}(k)$ for a given k can really be treated as a random variable. Moreover, it is a function of other random variables to be the transmitted data symbols. It is also a function of echo canceller coefficients, which, we can assume, change randomly. Moreover, when we have a series of random variables, as in the case of $e(k) - \hat{e}(k)$, depending upon the time, we speak then about a random (time) series and/or a random or stochastic process. In other words, a random (stochastic) discrete-time process $\{X(k)\}$ is a sequence of random variables, which are related with the corresponding discrete-time points. In the case of continuous time, we have a "continuous" time series $\{X(t)\}$, of which values are random. The outcomes of these processes, on the other hand, are denoted with the use of small letters $\{x(k)\}$ and $\{x(t)\}$, respectively. Moreover, the braces at $\{X(k)\}$, $\{X(t)\}$, $\{x(k)\}$, and $\{x(t)\}$ are omitted, when this does not lead to any confusion. The above notation is consistent with that used in Section 1.7 in the deterministic case.

Example 2.1

The notion of discrete- and continuous-time random (stochastic) processes is illustrated in Figure 2.19. This is done by plotting examples of outcomes for these processes. It is assumed in the plots that the random variable $x(k)$ for a given k assumes only two values, -1 and $+1$. That is, the sample space of this random variable is discrete and consists of two elements -1 and $+1$. The series of random variables $x(k)$ for the discrete-time points $k = \ldots, 0, 1, 2, 3, \ldots$ can appear as shown in Figures 2.19a, b, c, and so on. These sequences, $\{x(k)\}_1$, $\{x(k)\}_2$, $\{x(k)\}_3$, ..., form the space of outcomes of the random process $\{X(k)\}$. The same can be said about the signals $\{x(t)\}_1$, $\{x(t)\}_2$, $\{x(t)\}_3$ in Figures 2.19d, e, and f, three examples of outcomes of the continuous-time random process $\{X(t)\}$.

By the way, see in Figure 2.19 that the signals $\{x(k)\}_1$, $\{x(k)\}_2$, $\{x(k)\}_3$ are the signals $\{x(t)\}_1$, $\{x(t)\}_2$, $\{x(t)\}_3$ sampled at the discrete-time points $k = \ldots, 0, 1, 2, 3, \ldots$. This reflects the fact that the discrete-time process $\{X(k)\}$ was chosen here to be a sampled process $\{X(t)\}$.

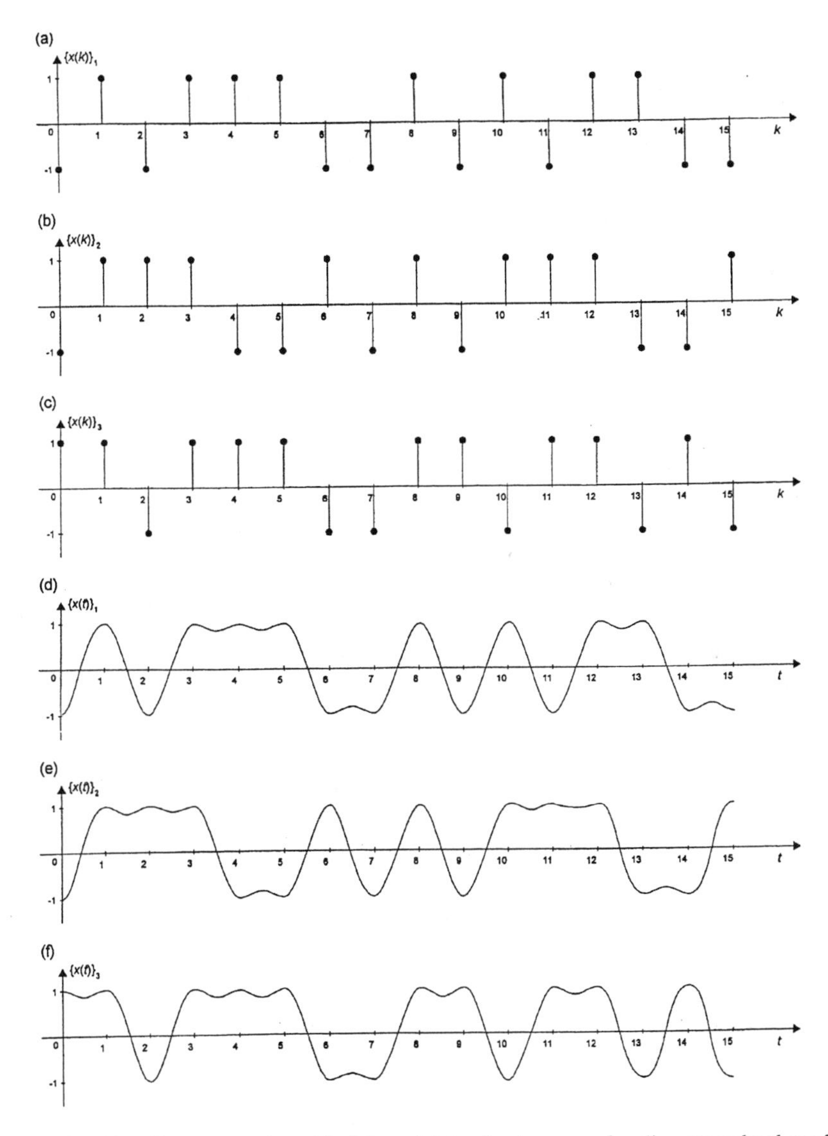

FIGURE 2.19 Three examples, (a), (b), and (c), of outcomes of a discrete-valued random process $\{X(k)\}$; and, for comparison, three examples, (d), (e), and (f), of a continuous-valued random process $\{X(t)\}$.

Example 2.2

It was assumed in the previous example that the elements $x(k)$ of the sequences $\{x(k)\}_i$, $i = 1, 2, 3\ldots$, took on only two values. Therefore, the underlying random process $\{X(k)\}$ was both the discrete-time and discrete-valued one.

However, in telecommunications, we also have situations when a discrete-time random process possesses the continuous-valued amplitude. Figure 2.20a shows an example of the outcome, denoted $\{x(t)\}_1$, of continuous-time and continuous-valued random process $\{X(t)\}$. (Note that this is $\{x(t)\}_1$ of Figure 2.19d modified slightly.)

The sequence presented in Figure 2.20b is the discrete-time continuous-valued outcome of the random process $\{X(k)\}$, the sampled-in-time random process $\{X(t)\}$ mentioned above. That is, the outcome $\{x(k)\}_1$ presented in Figure 2.20b is obtained from that of Figure 2.20a through sampling the time.

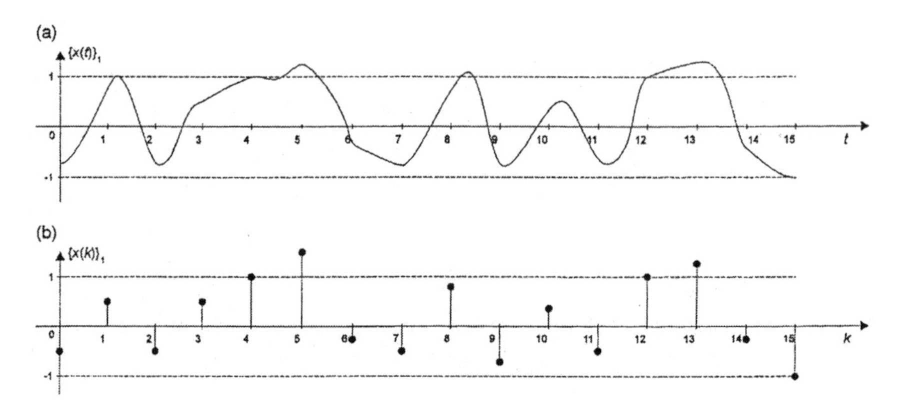

FIGURE 2.20 Example of an outcome of continuous-time continuous-valued random process $\{X(t)\}$, (a); the corresponding outcome of discrete-time continuous-valued random process $\{X(k)\}$, obtained through sampling the time in the previous one, (b).

Now take the square of the random variable $e(k) - \hat{e}(k)$ for a given k and use one of the definitions of the mean value, Equations 2.12a or 2.12b, depending on whether the new variable $(e(k) - \hat{e}(k))^2$, k fixed, is a continuous-valued or a discrete-valued random variable. Moreover, apply Equation 2.9. We get

$$mes(k) \stackrel{df}{=} E((e(k)-\hat{e}(k))^2) = E((\mathbf{a}^{\mathrm{T}}(k)\cdot(\mathbf{g}_L-\mathbf{c}(k)) + \mathbf{u}^{\mathrm{T}}(k)\cdot\mathbf{g}_N)^2) \quad (2.13)$$

where $mes(k)$ is used to denote the mean value of $(e(k) - \hat{e}(k))$ squared.

Squaring the expression in Equation 2.13, we arrive at

$$\begin{aligned} mes(k) = {} & E((\mathbf{a}^{\mathrm{T}}(k)\cdot\mathbf{g}_L-\mathbf{c}(k)))^2 \\ & + 2(\mathbf{a}^{\mathrm{T}}(k)\cdot(\mathbf{g}_L-\mathbf{c}(k))\cdot\mathbf{u}^{\mathrm{T}}(k)\mathbf{g}_N) + (\mathbf{u}^{\mathrm{T}}(k)\cdot\mathbf{g}_N)^2) \end{aligned} \quad (2.14)$$

Furthermore, without loss of generality, we can assume that the random variable $x = e(k) - \hat{e}(k)$ for a given k is continuous-valued. The fundamental theorem of expectation operation[68] says that, for any function $f(x)$ of such random variable, the following,

$$E(f(x)) = \int_{-\infty}^{\infty} f(x)p(x)dx \tag{2.15}$$

holds. From Equation 2.15, it follows immediately that

$$\begin{aligned} E(f(x)) &= E(f_1(x) + f_2(x) + \ldots + f_n(x)) \\ &= E(f_1(x)) + E(f_2(x)) + \ldots + E(f_n(x)) \end{aligned} \tag{2.16a}$$

when $f(x) = f_1(x) + f_2(x) + \ldots + f_n(x)$, and

$$E(\alpha f(x)) = \alpha E(f(x)) \tag{2.16b}$$

where α is a real number.

Explanation 2.1

Note that the random variable $Y = (e(k) - \hat{e}(k))^2$ was considered in Equation 2.13. Take into account the random variable $X = (e(k) - \hat{e}(k))$. Hence, the relationship between these random variables is given by the function $Y = f(X) = (X)^2$.

The fundamental theorem of expectation operation given by Equation 2.15 says that the expected values calculated in Equations 2.13 and 2.14 can be taken with regard to the random variable Y as well as to the random variable X. This is so because the equality

$$\int_{-\infty}^{\infty} y p_Y(y)dy = \int_{-\infty}^{\infty} f(x)p_X(x)dx$$

holds, where p_Y and p_X are the probability density functions of the variables Y and X, respectively.

Returning now to Equation 2.14 and using relations 2.16a and 2.16b in it, we get

$$\begin{aligned} mes(k) &= E((\mathbf{a}^{\mathrm{T}}(k) \cdot (\mathbf{g}_L - \mathbf{c}(k)))^2) \\ &+ 2E(\mathbf{a}^{\mathrm{T}}(k) \cdot (\mathbf{g}_L - \mathbf{c}(k)) \cdot \mathbf{u}^{\mathrm{T}}(k)\mathbf{g}N) + E((\mathbf{u}^{\mathrm{T}}(k) \cdot g_N)^2) \end{aligned} \tag{2.17}$$

Note that $\mathbf{a}^{\mathrm{T}}(k) \cdot (\mathbf{g}_L - \mathbf{c}(k))$ can also be written as $(\mathbf{g}_L - \mathbf{c}(k))^{\mathrm{T}} \cdot \mathbf{a}(k)$. Hence, the first component on the right-hand side of Equation 2.17 has the following equivalent form:

$$E((\mathbf{g}_L - \mathbf{c}(k))^{\mathrm{T}} \cdot \mathbf{a}(k) \cdot \mathbf{a}^{\mathrm{T}}(k) \cdot (\mathbf{g}_L - \mathbf{c}(k))) \tag{2.18}$$

To proceed further, consider the expression $\mathbf{a}(k) \cdot \mathbf{a}^{\mathrm{T}}(k)$ in expression 2.18. This expression forms a matrix

$$\boldsymbol{a}(k)\cdot\mathbf{a}^{\mathrm{T}}(k) = \begin{bmatrix} a(k) \\ a(k-1) \\ \vdots \\ a(k-M+1) \end{bmatrix} [a(k), a(k-1), \ldots, a(k-M+1)]$$

$$= \begin{bmatrix} a(k)a(k), a(k)a(k-1), & \ldots, & a(k)a(k-M+1) \\ a(k-1)a(k), a(k-1)a(k-1), & \ldots, & a(k-1)a(k-M+1) \\ \vdots & \vdots & \vdots \\ a(k-M+1)a(k), a(k-M+1)a(k-1), & \ldots, & a(k-M+1)a(k-M+1) \end{bmatrix} \tag{2.19}$$

As already mentioned, we treat the elements of the vector $\mathbf{a}(k)$ as random variables in our consideration. Hence, the matrix given by Equation 2.19 is, in fact, a matrix consisting of elements that are products of random variables. For such objects, our definitions 2.12a and 2.12b, and relation 2.15, regarding the expectation operation, need some extension. First, we define what we mean under the notion of the expected value of a matrix. This simply means taking the expected values of its elements. Applying the above definition to the matrix given by Equation 2.19, we get

$$E(\mathbf{a}(k)\cdot\mathbf{a}^{\mathrm{T}}(k))$$

$$= \begin{bmatrix} E(a(k)a(k)), E(a(k)a(k-1)), & \ldots, & E(a(k)a(k-M+1)) \\ E(a(k-1)a(k)), E(a(k-1)a(k-1)), & \ldots, & E(a(k-1)a(k-M+1)) \\ \vdots & \vdots & \vdots \\ E(a(k-M+1)a(k)), E(a(k-M+1)a(k-1)) & \ldots, & E(a(k-M+1)a(k-M+1)) \end{bmatrix} \tag{2.20}$$

where the expectation operation E is taken with respect to the "M-dimensional" random vector-valued variable, represented by the vector $\mathbf{a}(k)$ consisting of M "one-dimensional" random scalar-valued variables. More precisely, when $f(\mathbf{x})$, where $\mathbf{x} = [x_1, x_2, \ldots, x_n]^{\mathrm{T}}$, is a scalar-valued function of n random variables $x_1, x_2, \ldots, x_n$, then the expected value of $f(\mathbf{x})$, is given by

$$E(f(\mathbf{x})) = \underbrace{\int_{-\infty}^{\infty}\int_{-\infty}^{\infty}\cdots\int_{-\infty}^{\infty}}_{n\text{ times}} f(\mathbf{x})p(x_1, x_2, \ldots, x_n)dx_1dx_2\ldots dx_n = \int_{\Omega} f(\mathbf{x})p(\mathbf{x})d\Omega \tag{2.21a}$$

for continuous-valued variables, and

$$E(f(\mathbf{x})) = \sum_{x_1}\sum_{x_2}\cdots\sum_{x_n} f(\mathbf{x})p(x_1, x_2, \ldots, x_n) = \sum_{x\in\Omega} f(\mathbf{x})p(\mathbf{x}) \quad (2.21b)$$

for discrete-valued variables. Relation 2.21a is an extension of the relation 2.12a and expectation fundamental theorem given by Equation 2.15. On the other hand, relation 2.21b is a similar extension of the relation 2.12b and the expectation fundamental theorem

$$E(f(\mathbf{x})) = \sum_{x\in\Omega} f(x)p(x) \quad (2.22)$$

for one-dimensional discrete-valued random variables, being an equivalent of Equation 2.15 for one-dimensional continuous-valued random variables. Moreover, the integral $\int_{\Omega}\ldots d\Omega$ in Equation 2.21a represents a shorthand notation for the n-dimensional integral, as shown in the above equation. This integral is over the n-dimensional sample space Ω of the random variables $X_1, X_2, \ldots, X_n$.

Similarly, $\sum_{x\in\Omega}$ stands for the n summations over the n-dimensional sample space Ω of the discrete-valued random variables $X_1, X_2, \ldots, X_n$. The joint probability density functions of n random variables $p(x_1, x_2, \ldots, x_n) = p(\mathbf{x})$ in Equations 2.21a and 2.21b are the corresponding generalizations of the previously given expressions 2.10b and 2.11. That is,

$$p(x_1, x_2, \ldots, x_n) = \frac{\partial^n P_c(x_1, x_2 \ldots, x_n)}{\partial x_1 \partial x_2 \cdots \partial x_n} \quad (2.23a)$$

with the joint cumulative probability distribution function $P_c(x_1, x_2, \ldots, x_n)$ given by

$$P_c(x_1, x_2, \ldots, x_n) = P(X_1 \le x_1, X_2 \le x_2, \ldots, X_n \le x_n) \quad (2.23b)$$

for continuous-valued random variables, and

$$p(x_1, x_2, \ldots, x_n) \quad (2.23c)$$

$$= \sum_{z_1\in\Omega}\sum_{z_2\in\Omega}\cdots\sum_{z_n\in\Omega} P(X_1 = z_1, X_2 = z_2, \ldots, X_n = z_n)\delta_M(x_1 - z_1, x_2 - z_2, \ldots, x_n - z_n)$$

for discrete-valued random variables, with $\delta_M(x_1 - z_1, x_2 - z_2, \ldots, x_n - z_n)$ which could be called a multi-dimensional discrete Dirac impulse, having the property: $\delta_M(0, 0, \ldots, 0) = 1$, and the zero value otherwise.

Equation 2.23b is the generalization of expression 2.10a. Moreover, the function $P(X_1 \le x_1, X_2 \le x_2, \ldots, X_n \le x_n)$ in Equation 2.23b and the function $P(X_1 = z_1, X_2 = z_2,$

$\ldots, X_n = z_n)$ in Equation 2.23c are the generalizations of the one-variable functions $P(X \le x)$ in definition 2.10a and $P(X = z)$ in definition 2.11), respectively. The function $P(X_1 \le x_1, X_2 \le x_2, \ldots, X_n \le x_n)$ expresses the probability of the event ($X_1 \le x_1$ and $X_2 \le x_2$ and … and $X_n \le x_n$). Similarly, $P(X_1 = z_1, X_2 = z_2, \ldots, X_n = z_n)$ expresses the probability of the event ($X_1 = z_1$ and $X_2 = z_2$ and … and $X_n = z_n$) .

Let us now return to the matrix given by Equation 2.20. Assume that the transmitted symbols $a(k)$, $a(k - 1)$, …, $a(k - M + 1)$ in this matrix are random and binary, and assume that they take on the values $+1$ or -1. We then have to work in Equation 2.20 with the discrete-valued random variables and must use the expression 2.21b for calculation of the mean value, with the function $f(x)$ being quadratic.

We say that the random variables $X_1, X_2, \ldots, X_n$ are independent, or statistically independent, when the probability functions $P(X_1 \le x_1, X_2 \le x_2, \ldots, X_n \le x_n)$ and $P(X_1 = z_1, X_2 = z_2, \ldots, X_n = z_n)$ can be expressed as

$$P(X_1 \le x_1, X_2 \le x_2, \ldots, X_n \le x_n) = P_1(X_1 \le x_1)P_2(X_2 \le x_2)\cdots P_n(X_n \le x_n) \quad (2.24a)$$

and

$$P(X_1 = z_1, X_2 = z_2, \ldots, X_n = z_n) = P_1(X_1 = z_1)P_2(X_2 = z_2)\cdots P_n(X_n = z_n) \quad (2.24b)$$

respectively, where the probability $P_1(X_1 \le x_1)$ is the probability of the event $X_1 \le x_1$, and so on. Moreover, the relations 2.24 are equivalent to the corresponding relations between the probability density functions,

$$p(x_1, x_2, \ldots, x_n) = p_1(x_1)p_2(x_2)\cdots p_n(x_n) \quad (2.25)$$

where $p_1(x_1)$ is the probability density function of the random variable X_1, and so on. For further calculations, assume that the transmitted symbols $a(k)$, $a(k - 1)$, …, $a(k - M + 1)$ satisfy the property of statistical independence stated above. That is, the probability density function defined by Equation 2.23c can be expressed, in our case, as

$$p(a(k), a(k-1), \ldots, a(k-M+1)) = p_1(a(k))p_2(a(k-1))\cdots p_M(a(k-M+1)) \quad (2.26)$$

according to Equation 2.25.

The use of relation 2.26, as we shall see, will simplify our calculations of the expected values in the matrix 2.20. To this end, consider, for instance, the first element of the matrix 2.20, $E(a(k)a(k))$. Using Equation 2.21b, it can be written as

$$E(a^2(k)) = \sum_{a(k)} \sum_{a(k-1)} \cdots \sum_{a(k-M+1)} a^2(k)p(a(k), a(k-1), \ldots, a(k-M+1)) \quad (2.27a)$$

and applying Equation 2.26 as

$$E(a^2(k)) \tag{2.27b}$$

$$= \left(\sum_{a(k)} a^2(k)p_1(a(k))\right)\left(\sum_{a(k-1)} p_2(a(k-1))\right)\ldots\left(\sum_{a(k-M+1)} p_M(a(k-M+1))\right)$$

where $f(\mathbf{x})$ of Equation 2.21b is the quadratic function $a^2(k)$, and $a(k)$, $a(k-1)$, ..., $a(k-M+1)$ also stand for the corresponding random variables, according to the notational convention assumed.

Take now the sum of $p(x)$ given by Equation 2.11 over all the possible outcomes of the random variable X, that is, for all $x \in \Omega$, where Ω is the discrete-valued sample space of the random variable X. We then get

$$\sum_{x \in \Omega} p(x) = \sum_{x \in \Omega}\sum_{z \in \Omega} P(X=z)\delta(x-z) = \sum_{x \in \Omega} P(X=x) = 1 \tag{2.28}$$

because the event that the random variable X takes on one of the values $x \in \Omega$ is a certain event.

Note that the relation 2.28 can be applied to each of the expressions $\sum_{a(k-1)} p_2(a(k-1))$, ..., $\sum_{a(k-M+1)} p_M(a(k-M+1))$ occurring in Equation 2.27b. In fact, each of these expressions is equal to one. Consequently, Equation 2.27b simplifies to

$$E(a^2(k)) = \sum_{a(k)} a^2(k)p_1(a(k)) \tag{2.29}$$

Explanation 2.2

Assume that our random variables $a(k)$, $a(k-1)$, ..., $a(k-M+1)$ (here the notational convention with the use of small letters for random variables is used) take on only two values, -1 or $+1$, with the same probability, $1/2$. The sample spaces Ω for these random variables consist of only two elements, -1 and $+1$. Moreover, relation 2.28, for instance, for $a(k-1)$, looks then as

$$\sum_{a(k-1) \in \{-1,1\}} p_2(a(k-1)) = \sum_{a(k-1) \in \{-1,1\}} \sum_{z \in \{-1,1\}} P(A(k-1)=z)\delta(a(k-1)-z)$$

$$= \sum_{a(k-1) \in \{-1,1\}} P(A(k-1)=a(k-1)) = P(A(k-1)=-1) + P(A(k-1)=1) = \frac{1}{2} + \frac{1}{2} = 1$$

Observe also that in the above equation, two different notations, $A(k-1)$, for the random variable, and $a(k-1)$ for its outcome, are used to avoid any calculation errors that could arise by using exclusively small letters.

Knowing that the sample space of the random variable $a(k)$ consists of two elements, -1 and $+1$, we can rewrite Equation 2.29 as

$$E(a^2(k)) = (-1)^2 \cdot \frac{1}{2} + (1)^2 \cdot \frac{1}{2} = 1 \tag{2.30}$$

So the element $E(a^2(k))$ in the matrix 2.20 is simply equal to 1.

Consider now the next element in the matrix 2.20, $E(a(k)a(k-1))$. Using Equation 2.21b, it can be written as

$$E(a(k)a(k-1)) = \sum_{a(k)} \sum_{a(k-1)} \cdots \sum_{a(k-M+1)} a(k)a(k-1) \cdot p(a(k), a(k-1), \ldots, a(k-M+1)) \tag{2.31a}$$

and furthermore, applying Equation 2.26 as

$$E(a(k)a(k-1)) = \left(\sum_{a(k)} \sum_{a(k-1)} a(k)a(k-1)p_1(a(k))p_2(a(k-1))\right) \cdot \left(\sum_{a(k-2)} p_3(a(k-3))\right) \cdots \left(\sum_{a(k-M+1)} p_M(a(k-M+1))\right) \tag{2.31b}$$

By the appliction of relation 2.28 to the sums $\sum_{a(k-2)} p_3(a(k-3)), \ldots, \sum_{a(k-M+1)} p_M(a(k-M+1))$ in Equation 2.31b, the latter simplifies to

$$E(a(k)a(k-1)) = \sum_{a(k)} \sum_{a(k-1)} a(k)a(k-1)p_1(a(k))p_2(a(k-1)) = \left(\sum_{a(k)} a(k)p_1(a(k))\right)\left(\sum_{a(k-1)} a(k-1)p_2(a(k-1))\right) \tag{2.31c}$$

Finally, substituting the corresponding values for $a(k)$, $p_1(a(k))$, $a(k-1)$, and $p_2(a(k-1))$ in Equation 2.31c, we get

$$E(a(k)a(k-1)) = \left((-1)\frac{1}{2} + (+1)\frac{1}{2}\right)\left((-1)\frac{1}{2} + (+1)\frac{1}{2}\right) = 0 \cdot 0 = 0 \tag{2.31d}$$

So that the element $E(a(k)a(k-1))$ in the matrix 2.20 equals 0.

Using the same methodology as that used in calculations of elements $E(a^2(k)$ and $E(a(k)a(k-1))$ to calculate the remaining elements of the matrix 2.20, we arrive at such a result that the elements on the diagonal of the matrix $E(\mathbf{a}(k)\mathbf{a}^T(k))$ are ones and the remaining elements are zeros. That is, the matrix $E(\mathbf{a}(k)\mathbf{a}^T(k))$ has in this case, the following form:

$$E(\mathbf{a}(k)\cdot\mathbf{a}^{\mathrm{T}}(k)) = \begin{bmatrix} 1 & 0 & 0 & \dots & 0 \\ 0 & 1 & 0 & \dots & 0 \\ 0 & 0 & 1 & \dots & 0 \\ \vdots & \vdots & \vdots & \vdots & \vdots \\ 0 & 0 & 0 & \dots & 1 \end{bmatrix} \tag{2.32}$$

Consider now two sets of discrete-valued random variables $\{X_1, X_2, \dots, X_i\}$ and $\{X_{i+1}, X_{i+2}, \dots, X_n\}$, which are statistically independent, that is,

$$p(\mathbf{x}) = p(x_1, x_2, \dots, x_i, x_{i+1}, x_{i+2}, \dots, x_n) = p_1(x_1, x_2, \dots, x_i)\cdot p_2(x_{i+1}, x_{i+2}, \dots, x_n) \tag{2.33}$$

take into account a function $f(\mathbf{x})$ dependent upon the random variables $X_1, \dots, X_i, X_{i+1}, \dots, X_n$, and calculate the expected value of this function. Using Equations 2.21b and 2.33 gives

$$E(f(\mathbf{x})) = \sum_{X_1}\dots\sum_{X_i}\sum_{X_{i+1}}\dots\sum_{X_n} f(\mathbf{x})p_1(x_1, \dots, x_i)\cdot p_2(x_{i+1}, \dots, x_n) \tag{2.34a}$$

Furthermore, assuming in Equation 2.34a that the function $f(\mathbf{x})$ can be expressed as $f(\mathbf{x}) = f_1(x_1, \dots, x_i)\cdot f_2(x_{i+1}, \dots, x_n)$, we can rewrite expression 2.34a as

$$\begin{aligned} E(f(\mathbf{x})) &= E(f_1(x_1, \dots, x_i) f_2(x_{i+1}, \dots, x_n)) \\ &= \left(\sum_{X_1}\dots\sum_{X_i} f_1(x_1, \dots, x_i)p_1(x_1, \dots, x_i)\right)\cdot\left(\sum_{X_{i+1}}\dots\sum_{X_n} f_2(x_{i+1}, \dots, x_n)p_2(x_{i+1}, \dots, x_n)\right) \\ &= E(f_1(x_1, \dots, x_i))\cdot E(f_2(x_{i+1}, \dots, x_n)) \end{aligned} \tag{2.34b}$$

Example 2.3

Consider the following expression:

$$\mathbf{x}^{\mathrm{T}}\cdot\mathbf{A}\cdot\mathbf{x} = [x_1, x_2, x_3]\begin{bmatrix} a_{11} & a_{12} & a_{13} \\ a_{21} & a_{22} & a_{23} \\ a_{31} & a_{32} & a_{33} \end{bmatrix}\begin{bmatrix} x_1 \\ x_2 \\ x_3 \end{bmatrix}$$

Note that the above expression is a scalar-valued one, although its components are vectors and a matrix. Assume that these components represent two sets of random variables. That is, the vector

$$\mathbf{x} = \begin{bmatrix} x_1 \\ x_2 \\ x_3 \end{bmatrix}$$

represents a vector-valued random variable

$$\mathbf{X} = \begin{bmatrix} X_1 \\ X_2 \\ X_3 \end{bmatrix}$$

Similarly, the matrix

$$\mathbf{A} = \begin{bmatrix} a_{11} & a_{12} & a_{13} \\ a_{21} & a_{22} & a_{23} \\ a_{31} & a_{32} & a_{33} \end{bmatrix}$$

represents a matrix-valued random variable

$$\begin{bmatrix} A_{11} & A_{12} & A_{13} \\ A_{21} & A_{22} & A_{23} \\ A_{31} & A_{32} & A_{33} \end{bmatrix}$$

In what follows, we use the notational convention for random variables with the use of small letters; that is, we treat the components of $\mathbf{x}^{\mathrm{T}}\cdot\mathbf{A}\cdot\mathbf{x}$ as the random variables. Further, we want to calculate the expected value of this expression. So we write

$$E(\mathbf{x}^{\mathrm{T}}\cdot\mathbf{A}\cdot\mathbf{x})$$
$$=E((x_1a_{11}+x_2a_{21}+x_3a_{31})x_1+(x_1a_{12}+x_2a_{22}+x_3a_{32})x_2+(x_1a_{13}+x_2a_{23}+x_3a_{33})x_3)$$
$$=E(x_1x_1a_{11}+x_2x_1a_{21}+x_3x_1a_{31}+x_1x_2a_{12}+x_2x_2a_{22}+x_3x_2a_{32}+x_1x_3a_{13}+x_2x_3a_{23}+x_3x_3a_{33})$$

To proceed further, we need an extension of the property 2.16a to the multivariable case and discrete-valued random variables. Using 2.21b, we get such a result, because, assuming $f(\mathbf{x}) = f_1(\mathbf{x}) + \cdots + f_m(\mathbf{x})$, in this expression, we have

$$E(f(\mathbf{x})) = E(f_1(\mathbf{x})+\cdots+f_m(\mathbf{x})) = \sum_{x_1}\cdots\sum_{x_n}(f_1(\mathbf{x})+\cdots+f_m(\mathbf{x}))p(\mathbf{x}) \qquad \text{(I)}$$
$$= \sum_{x_1}\cdots\sum_{x_{n-1}}\left(\sum_{x_n}f_1(\mathbf{x})p(\mathbf{x})+\cdots+\sum_{x_n}f_m(\mathbf{x})p(\mathbf{x})\right) = \cdots$$
$$= \sum_{x_1}\cdots\sum_{x_n}f_1(\mathbf{x})p(\mathbf{x})+\cdots+\sum_{x_1}\cdots\sum_{x_n}f_m(\mathbf{x})p(\mathbf{x}) = E(f_1(\mathbf{x}))+\cdots+E(f_m(\mathbf{x}))$$

Using the above property in the previous expression, we now get

$$E(\mathbf{x}^{\mathrm{T}}\cdot\mathbf{A}\cdot\mathbf{x}) = E(x_1x_1a_{11}) + E(x_2x_1a_{21}) + E(x_3x_1a_{31}) + E(x_1x_2a_{12}) + E(x_2x_2a_{22}) + E(x_3x_2a_{32}) + E(x_1x_3a_{13}) + E(x_2x_3a_{23}) + E(x_3x_3a_{33})$$

Furthermore, we assume that the two sets of random variables, corresponding to the matrix **A** and to the vector **x**, are statistically independent. This assumption, according to relation 2.34b, allows us to rewrite the expression presented above in the following form:

$$E(\mathbf{x}^{\mathrm{T}}\cdot\mathbf{A}\cdot\mathbf{x}) = E(x_1x_1)E(a_{11}) + E(x_2x_1)E(a_{21}) + E(x_3x_1)E(a_{31}) + E(x_1x_2)E(a_{12}) + E(x_2x_2)E(a_{22}) + E(x_3x_2)E(a_{32}) + E(x_1x_3)E(a_{13}) + E(x_2x_3)E(a_{23}) + E(x_3x_3)E(a_{33})$$

Note that each of the expected values $E(a_{11})$, $E(a_{21})$, …, $E(a_{33})$, in the above expression is a real number. Using the property expressed by Equation 2.16b, we can write $E(\mathbf{x}^{\mathrm{T}}\cdot\mathbf{A}\cdot\mathbf{x})$ in an equivalent form:

$$E(\mathbf{x}^{\mathrm{T}}\cdot\mathbf{A}\cdot\mathbf{x}) = E(x_1E(a_{11})x_1) + E(x_2E(a_{21})x_1) + E(x_3E(a_{31})x_1) + \cdots + E(x_3E(a_{33})x_3)$$

Applying the result regarding the expected value of a sum of functions derived in this example, we can put the latter expression into the following form:

$$E(\mathbf{x}^{\mathrm{T}}\cdot\mathbf{A}\cdot\mathbf{x}) = E((x_1E(a_{11}) + x_2E(a_{21}) + x_3E(a_{31}))x_1) + \cdots + E((x_1E(a_{13}) + x_2E(a_{23}) + x_3E(a_{33}))x_3)$$

$$= E\left(\mathbf{x}^{\mathrm{T}}\cdot\begin{bmatrix}E(a_{11})\\E(a_{21})\\E(a_{31})\end{bmatrix}x_1 + \mathbf{x}^{\mathrm{T}}\cdot\begin{bmatrix}E(a_{12})\\E(a_{22})\\E(a_{32})\end{bmatrix}x_2 + \mathbf{x}^{\mathrm{T}}\cdot\begin{bmatrix}E(a_{13})\\E(a_{23})\\E(a_{33})\end{bmatrix}x_3\right)$$

$$= E\left(\mathbf{x}^{\mathrm{T}}\cdot\begin{bmatrix}E(a_{11}) & E(a_{12}) & E(a_{13})\\E(a_{21}) & E(a_{22}) & E(a_{23})\\E(a_{31}) & E(a_{32}) & E(a_{33})\end{bmatrix}\cdot\mathbf{x}\right) = E(\mathbf{x}^{\mathrm{T}}\cdot E(\mathbf{A})\cdot\mathbf{x})$$

The conclusion from this example is that, when the sets of random variables represented by the vector **x** and the matrix **A** are statistically independent, then we can write

$$E(\mathbf{x}^{\mathrm{T}}\cdot\mathbf{A}\cdot\mathbf{x}) = E(\mathbf{x}^{\mathrm{T}}\cdot E(\mathbf{A})\cdot\mathbf{x})$$

Note that the result of Example 2.3 is not restricted to vectors **x** having three elements and matrices **A** of order 3×3. This result can be generalized to any number n of the elements of the vector **x** and order $n\times n$ of matrix **A**, accordingly.

Using in expression 2.18 the result given by Equation 2.32 and the property derived in Example 2.3, generalized to $n = M$, we get

$$E((\mathbf{g}_L - \mathbf{c}(k))^{\mathrm{T}} \cdot \mathbf{a}(k) \cdot \mathbf{a}^{\mathrm{T}}(k) \cdot (\mathbf{g}_L - \mathbf{c}(k))) \tag{2.35}$$

$$= E((\mathbf{g}_L - \mathbf{c}(k))^{\mathrm{T}} \cdot E(\mathbf{a}(k) \cdot \mathbf{a}^{\mathrm{T}}(k)) \cdot (\mathbf{g}_L - \mathbf{c}(k)))$$

$$= E\left((\mathbf{g}_L - \mathbf{c}(k))^{\mathrm{T}} \cdot \begin{bmatrix} 1 & 0 & 0 & \dots & 0 \\ 0 & 1 & 0 & \dots & 0 \\ \vdots & \vdots & \vdots & \vdots & \vdots \\ 0 & 0 & 0 & \dots & 1 \end{bmatrix} \cdot (\mathbf{g}_L - \mathbf{c}(k)) \right)$$

$$= E((\mathbf{g}_L - \mathbf{c}(k))^{\mathrm{T}} \cdot (\mathbf{g}_L - \mathbf{c}(k)))$$

In derivation 2.35, we assumed that the transmitted symbols $a(k)$, …, $a(k - M + 1)$ were statistically independent of the adjusted transversal filter coefficients $c_0(k)$, …, $c_{M-1}(k)$. Further, both the sets were treated as the sets of random variables. Moreover, as seen in the example below, when a set of random variables represented by a vector **y** is statistically independent of a set of random variables represented by a matrix **A**, then the set represented by the vector

$$\mathbf{x} = \boldsymbol{\alpha}\mathbf{y} + \boldsymbol{\beta} = \begin{bmatrix} \alpha_1 & 0 & \cdots & 0 \\ 0 & \alpha_2 & \cdots & 0 \\ \vdots & \vdots & \vdots & \vdots \\ 0 & 0 & \cdots & \alpha_n \end{bmatrix} \cdot \begin{bmatrix} y_1 \\ y_2 \\ \vdots \\ y_n \end{bmatrix} + \begin{bmatrix} \beta_1 \\ \beta_2 \\ \vdots \\ \beta_n \end{bmatrix} \tag{2.36a}$$

where α_1, …, α_n and β_1, …, β_n are some real numbers, is statistically independent of the set represented by the matrix **A** as well. The transformation $\mathbf{x} = \boldsymbol{\alpha} y + \boldsymbol{\beta}$ given by Equation 2.36a, applied in Equation 2.35, had the following form:

$$\mathbf{x} = \mathbf{g}_L - \mathbf{c}(k) = \begin{bmatrix} -1 & 0 & \cdots & 0 \\ 0 & -1 & \cdots & 0 \\ \vdots & \vdots & \vdots & \vdots \\ 0 & 0 & \cdots & -1 \end{bmatrix} \cdot \begin{bmatrix} c_0(k) \\ c_1(k) \\ \vdots \\ c_{M-1}(k) \end{bmatrix} + \begin{bmatrix} g(0) \\ g(1) \\ \vdots \\ g(M-1) \end{bmatrix} = -\mathbf{c}(k) + \mathbf{g}_L \tag{2.36b}$$

Example 2.4

Consider a set of random variables y_1, …, y_n, and assume that these variables are statistically independent of random variables a_{11}, a_{12}, …, a_{nn}, being elements of a matrix **A** of order $n \times n$. Furthermore, assume that the random variables, represented by the vector **y**, are transformed using the relation 2.36a into a set, of which elements form a vector **x**. We show that the resulting set is also statistically independent of the set $a_{11}, a_{12}, \ldots, a_{nn}$. To this end, we start with the relation

$$p_{ya}(y_1, \ldots, y_n, a_{11}, a_{12}, \ldots, a_{nn}) = p_y(y_1, \ldots, y_n) \cdot p_a(a_{11}, a_{12}, \ldots, a_{nn}) \qquad \text{(I)}$$

describing the statistical independence of the random variable sets, see Equation 2.33. The objective is to find the probability density function $p_{xa}(x_1, \ldots, x_n, a_{11}, a_{12}, \ldots, a_{nn})$.

In the expression above, $p_y(\cdot)$ and $p_a(\cdot)$ are the probability density functions of the random variable sets $\{y_1, \ldots, y_n\}$ and $\{a_{11}, a_{12}, \ldots, a_{nn}\}$, respectively.

To proceed further, we need now to recall one result from the literature regarding the transformation of random variables (see, for example Reference 69). Applying this result in our case, we say that we have the set of random variables $y_1, \ldots, y_n$, $a_{11}, a_{12}, \ldots, a_{nn}$ with the joint probability density function $p_{ya}(y_1, \ldots, y_n, a_{11}, a_{12}, \ldots, a_{nn})$; further, we transform this set of random variables into another set $x_1, \ldots, x_n$, $a_{11}, a_{12}, \ldots, a_{nn}$ related to the previous one by the functions

$$\begin{aligned}
x_1 &= f_1(y_1, \ldots, y_n, a_{11}, a_{12}, \ldots, a_{nn}) \\
&\vdots \\
x_n &= f_n(y_1, \ldots, y_n, a_{11}, a_{12}, \ldots, a_{nn}) \\
\\
a_{11} &= f_{11}(y_1, \ldots, y_n, a_{11}, a_{12}, \ldots, a_{nn}) \\
a_{12} &= f_{12}(y_1, \ldots, y_n, a_{11}, a_{12}, \ldots, a_{nn}) \\
&\vdots \\
a_{nn} &= f_{nn}(y_1, \ldots, y_n, a_{11}, a_{12}, \ldots, a_{nn})
\end{aligned}$$

Moreover, we assume that the above functions are single-valued and invertible, and have continuous partial derivatives as well. So there exist the inverse functions

$$\begin{aligned}
y_1 &= f_1^{-1}(x_1, \ldots, x_n, a_{11}, a_{12}, \ldots, a_{nn}) \\
&\vdots \\
y_n &= f_n^{-1}(x_1, \ldots, x_n, a_{11}, a_{12}, \ldots, a_{nn}) \\
\\
a_{11} &= f_{11}^{-1}(x_1, \ldots, x_n, a_{11}, a_{12}, \ldots, a_{nn}) \\
a_{12} &= f_{12}^{-1}(x_1, \ldots, x_n, a_{11}, a_{12}, \ldots, a_{nn}) \\
&\vdots \\
a_{nn} &= f_{nn}^{-1}(x_1, \ldots, x_n, a_{11}, a_{12}, \ldots, a_{nn})
\end{aligned}$$

We assume also that the inverse functions are single-valued with continuous partial derivatives. Then, the joint probability density function p_{xa} for the set $x_1, \ldots, x_n, a_{11}, a_{12}, \ldots, a_{nn}$ is given by the following expression:

$$p_{xa}(x_1, \ldots, x_n, a_{11}, a_{12}, \ldots, a_{nn}) \tag{II}$$
$$= p_{ya}(y_1 = f_1^{-1}(\cdot), \ldots, y_n = f_n^{-1}(\cdot), a_{11} = f_{11}^{-1}(\cdot), a_{12} = f_{12}^{-1}(\cdot), \ldots, a_{nn} = f_{nn}^{-1}(\cdot))\|J\|$$

where $|J|$ is the absolute value of the determinant of the Jacobian matrix of the transformation considered,

$$\mathbf{J} = \begin{bmatrix} \frac{\partial f_1^{-1}(\cdot)}{\partial x_1} & \frac{\partial f_2^{-1}(\cdot)}{\partial x_1} & \cdots & \frac{\partial f_{nn}^{-1}(\cdot)}{\partial x_1} \\ \frac{\partial f_1^{-1}(\cdot)}{\partial x_2} & \frac{\partial f_2^{-1}(\cdot)}{\partial x_2} & \cdots & \frac{\partial f_{nn}^{-1}(\cdot)}{\partial x_2} \\ \vdots & \vdots & \vdots & \vdots \\ \frac{\partial f_1^{-1}(\cdot)}{\partial a_{nn}} & \frac{\partial f_2^{-1}(\cdot)}{\partial a_{nn}} & \cdots & \frac{\partial f_{nn}^{-1}(\cdot)}{\partial a_{nn}} \end{bmatrix}$$

Now, returning to Equation 2.36b and completing the vector $\mathbf{x}$ of random variables $x_1, \ldots, x_n$ with the remaining random variables $a_{11}, a_{12}, \ldots, a_{nn}$, we get

$$\begin{bmatrix} x_1 \\ x_2 \\ \vdots \\ x_n \\ a_{11} \\ a_{12} \\ \vdots \\ a_{nn} \end{bmatrix} = \begin{bmatrix} -1 & 0 & \cdots & \cdots & \cdots & \cdots & & 0 \\ 0 & -1 & \cdots & \cdots & \cdots & \cdots & & \\ \vdots & \vdots & & & & & & \vdots \\ \vdots & \vdots & & -1 & & & & \vdots \\ \vdots & \vdots & & & 1 & & & \vdots \\ \vdots & \vdots & & & & 1 & & \vdots \\ & & & & & & \vdots & \\ 0 & & \cdots & \cdots & \cdots & \cdots & & 1 \end{bmatrix} \begin{bmatrix} y_1 \\ y_2 \\ \vdots \\ y_n \\ a_{11} \\ a_{12} \\ \vdots \\ a_{nn} \end{bmatrix} + \begin{bmatrix} \beta_1 \\ \beta_2 \\ \vdots \\ \beta_n \\ 0 \\ 0 \\ \vdots \\ 0 \end{bmatrix} \quad \text{with}$$

$$\begin{bmatrix} y_1 \\ y_2 \\ \vdots \\ y_n \end{bmatrix} = \begin{bmatrix} c_0(k) \\ c_1(k) \\ \vdots \\ c_{M-1}(k) \end{bmatrix}, \begin{bmatrix} \beta_1 \\ \beta_2 \\ \vdots \\ \beta_n \end{bmatrix} = \begin{bmatrix} g(0) \\ g(1) \\ \vdots \\ g(M-1) \end{bmatrix} \quad \text{and } n = M$$

It follows from the above matrix equation that

$$\begin{aligned} x_1 &= f_1(\cdot) = -y_1 + \beta_1 \\ x_2 &= f_2(\cdot) = -y_2 + \beta_2 \\ &\vdots \\ x_n &= f_n(\cdot) = -y_n + \beta_n \end{aligned}$$

$$\begin{aligned} a_{11} &= f_{11}(\cdot) = a_{11} \\ a_{12} &= f_{12}(\cdot) = a_{12} \\ &\vdots \\ a_{nn} &= f_{nn}(\cdot) = a_{nn} \end{aligned}$$

Hence, the inverse functions are given by

$$\begin{aligned} y_1 &= f_1^{-1}(\cdot) = -x_1 + \beta_1 \\ y_2 &= f_2^{-1}(\cdot) = -x_2 + \beta_2 \\ &\vdots \qquad \vdots \qquad \vdots \\ y_n &= f_n^{-1}(\cdot) = -x_n + \beta_n \end{aligned}$$

$$\begin{aligned} a_{11} &= f_{11}^{-1}(\cdot) = \quad a_{11} \\ a_{12} &= f_{12}^{-1}(\cdot) = \quad a_{12} \\ &\vdots \qquad \vdots \qquad \vdots \\ a_{nn} &= f_{nn}^{-1}(\cdot) = \quad a_{nn} \end{aligned}$$

Furthermore, applying the latter result for calculation of the Jacobian matrix of the transformation, we arrive at

$$\mathbf{J} = \begin{bmatrix} -1 & 0 & \cdots & \cdots & \cdots & \cdots & 0 \\ 0 & -1 & \cdots & \cdots & \cdots & \cdots & \\ & & & & & & \vdots \\ \vdots & \vdots & -1 & & & & \vdots \\ \vdots & \vdots & & 1 & & & \vdots \\ \vdots & \vdots & & & 1 & & \vdots \\ & & & & & \ddots & \\ 0 & & \cdots & \cdots & \cdots & \cdots & 1 \end{bmatrix} \begin{matrix} 1 \\ 2 \\ \vdots \\ n \\ \\ \\ \\ nn+n \end{matrix}$$

The determinant of this Jacobian matrix is given by

$$|\mathbf{J}| = (-1)^n(1)^{nn} = (-1)^n$$

Using the above result in relation (II) of this example, we get

$$\begin{aligned} & p_{xa}(x_1, \ldots, x_n, a_{11}, a_{12}, \ldots, a_{nn}) \\ & = \left|(-1)^n\right| \cdot p_{ya}(y_1 = f_1^{-1}(.), \ldots, y_n = f_1^{-1}(.), a_{11}, a_{12}, \ldots, a_{nn}) \end{aligned}$$

Substituting then p_{ya} given by (I) into the above expression, we arrive at

$$\begin{aligned} & p_{xa}(x_1, \ldots, x_n, a_{11}, a_{12}, \ldots, a_{nn}) \\ & = \left|(-1)^n\right| \cdot p_{ya}(y_1 = f_1^{-1}(.), \ldots, y_n = f_n^{-1}(.)) p_a(a_{11}, a_{12}, \ldots, a_{nn}) \end{aligned}$$

On the other hand, using the same procedure and expressions exploited earlier to the transformation of random variables $y_1, \ldots, y_n$ into $x_1, \ldots, x_n$, it can be easily shown that

$$p_x(x_1, \ldots, x_n) = \left|(-1)^n\right| \cdot p_y(y_1 = f_1^{-1}(\cdot), \ldots, y_n = f_n^{-1}(\cdot))$$

holds. And this result, applied in the previous equation, leads finally to

$$p_{xa}(x_1, \ldots, x_n, a_{11}, a_{12}, \ldots, a_{nn}) = p_x(x_1, \ldots, x_n) p_a(a_{11}, a_{12}, \ldots, a_{nn})$$

which expresses the statistical independence between the sets of random variables $\{x_1, \ldots, x_n\}$ and $\{a_{11}, a_{12}, \ldots, a_{nn}\}$.

Consider now the second component in Equation 2.17. The expected value of $\mathbf{a}^{\mathrm{T}}(k) \cdot (\mathbf{g}_L - \mathbf{c}(k))\mathbf{u}^{\mathrm{T}}(k)\mathbf{g}_N$ in this component can be written as

$$\begin{aligned} &E(\mathbf{a}^{\mathrm{T}}(k) \cdot (\mathbf{g}_L - \mathbf{c}(k))\mathbf{u}^{\mathrm{T}}(k)\mathbf{g}_N) \\ &= E((a(k)(g(0) - c_0(k)) + \cdots + a(k-M+1)(g(M-1) - c_{M-1}(k))) \\ &\quad \cdot (g_{00} + a(k)a(k-1)g(0,1) + \cdots + a(k)a(k-1)\cdots \\ &\quad \cdot a(k-M+1) \cdot g(0,1,2,\ldots,M-1))) \end{aligned} \tag{2.37a}$$

Multiplying the expressions in Equation 2.37a and then applying the property (I) proved in Example 2.3, we get

$$\begin{aligned} &E(\mathbf{a}^{\mathrm{T}}(k) \cdot (\mathbf{g}_L - \mathbf{c}(k))\mathbf{u}^{\mathrm{T}}(k)\mathbf{g}_N) \\ &= E(a(k)(g(0) - c_0(k))g_{00}) + E(a(k)(g(0) - c_0(k))a(k)a(k-1)g(0,1)) + \ldots \\ &+ E(a(k)(g(0) - c_0(k))a(k)a(k-1)\cdots a(k-M+1)g(0,1,2,\ldots,M-1)) + \ldots \\ &\quad + E(a(k-M+1)(g(M-1) - c_{M-1}(k))a(k)a(k-1)\cdots \\ &\quad \cdot a(k-M+1)g(0,1,2,\ldots,M-1)) \end{aligned} \tag{2.37b}$$

Furthermore, using the assumed statistical independence of the sets of random variables $\{a(k), a(k-1), \ldots, a(k-M+1)\}$ and $\{c_0(k), c_1(k), \ldots, c_{M-1}(k)\}$, and the methodology used in Example 2.4, one can show that the sets of random variables $\{a(k), a(k-1), \ldots, a(k-M+1)\}$ and $\{g(0) - c_0(k), g(1) - c_1(k), \ldots, g(M-1) - c_{M-1}(k)\}$ are statistically independent as well. Applying this fact in Equation 2.37b leads to

$$\begin{aligned} &E(\mathbf{a}^{\mathrm{T}}(k) \cdot (\mathbf{g}_L - \mathbf{c}(k))\mathbf{u}^{\mathrm{T}}(k)\mathbf{g}_N) \\ &= g_{00}E(a(k))E(g(0) - c_0(k)) + g(0,1)E(a^2(k)a(k-1))E(g(0) - c_0(k)) + \cdots \\ &+ g(0,1,2,\ldots,M-1)E(a^2(k)a(k-1)\cdots a(k-M+1)) \cdot E(g(0) - c_0(k)) + \cdots \\ &+ g(0,1,2,\ldots,M-1)E(a(k)a(k-1)\cdots a^2(k-M+1))E(g(M-1) - c_{M-1}(k)) \end{aligned} \tag{2.37c}$$

Observe now that applying the statistical independence property of the transmitted symbols $a(k), \ldots, a(k-M+1$ expressed by Equation 2.26, we can write the expected values $E(a^2(k) \cdot a(k-1)), \ldots, E(a(k)a(k-1) \cdots a^2(k-M+1))$, occurring in Equation 2.37c, in the form

$$\begin{aligned} E(a^2(k)a(k-1)) &= E(a^2(k))E(a(k-1)) \\ &\vdots \qquad\qquad\qquad \vdots \end{aligned} \tag{2.38a}$$

$$E(a^2(k)a(k-1)\cdots a(k-M+1)) = E(a^2(k))E(a(k-1))\cdots E(a(k-M+1)) \tag{2.38b}$$

$$E(a(k-1)a(k)a(k-1)) = E(a(k))E(a^2(k-1)) \tag{2.38c}$$

$$\vdots$$

$$E(a(k-1)a(k)a(k-1)\cdots a(k-M+1)) = E(a(k))E(a^2(k-1))\cdots E(a(k-M+1)) \tag{2.38d}$$

$$\vdots$$

$$E(a(k)a(k-1)\cdots a^2(k-M+1)) = E(a(k))E(a(k-1))\cdots E(a^2(k-M+1)) \tag{2.38e}$$

$$\vdots$$

Moreover, note that the remaining expected values in Equation 2.37c, involving single transmitted symbols, equal zeros. This is because

$$E(a(k)) = E(a(k-1)) = \cdots = E(a(k-M+1)) = (-1)\cdot\frac{1}{2} + (1)\cdot\frac{1}{2} = 0 \tag{2.39}$$

The result (2.39) can be also used in all the expressions 2.38 because they consist of at least one multiplier of the form mentioned in Equation 2.39 times some other components. Using Equation 2.39 in the expressions 2.38 leads to the conclusion that all the latter expressions equal zeros. Furthermore, using this result and that given by Equation 2.39 in Equation 2.37c allows us to say that

$$E(\mathbf{a}^{\mathrm{T}}(k)\cdot(\mathbf{g}_L - \mathbf{c}(k))\mathbf{u}^{\mathrm{T}}(k)\mathbf{g}_N) = 0 \tag{2.40}$$

Thereby, the second component in Equation 2.17 is equal to zero.

It remains now to consider the third component of Equation 2.17. To this end, note that the product $(\mathbf{u}_k^{\mathrm{T}}\mathbf{g}_N)^2$ can be expressed as

$$\begin{aligned}
&(\mathbf{u}_k^{\mathrm{T}}\mathbf{g}_N)^2 \\
&= (g_{00} + g(0,1)a(k)a(k-1) + \cdots + g(0,1,\ldots,M-1)a(k)a(k-1)\cdots a(k-M+1)) \\
&\cdot (g_{00} + g(0,1)a(k)a(k-1) + \cdots + g(0,1,\ldots,M-1)a(k)a(k-1)\cdots a(k-M+1)) \\
&= g_{00}^2 + g^2(0,1)a(k)(a(k-1))^2 + g^2(0,2)(a(k)a(k-2))^2 + \cdots \\
&\quad g^2(M-2,M-1)(a(k-M+2)a(k-M+1))^2 + \ldots \\
&+ g^2(0,1,2,\ldots,M-1)(a(k)a(k-1)\cdots a(k-M+1))^2 + 2g_{00}g(0,1)a(k)a(k-1) + \cdots \\
&+ 2g_{00}g(0,1,\ldots,M-1)a(k)a(k-1)\cdots a(k-M+1) \\
&+ 2g(0,1)g(0,2)a^2(k)a(k-1)a(k-2) + \cdots \\
&+ 2g(0,1,\ldots,M-1)g(0,1,\ldots,M-2)a^2(k)a^2(k-1)\cdots a(k-M+1) + \cdots \\
&+ 2g(0,1,\ldots,M-1)g(1,2,\ldots,M-1)a(k)a^2(k-1)\cdots a^2(k-M+1)
\end{aligned} \tag{2.41a}$$

Hence, under the assumptions regarding the properties of the transmitted symbols, and proceeding similarly as in considering the first and second components in Equation 2.17, we get from Equation 2.41a

$$E((\mathbf{u}_k^{\mathrm{T}}\mathbf{g}_N)^2)$$
$$= g_{00}^2 + g^2(0, 1) + g^2(0, 2) + \cdots + g^2(M-2, M-1) + \cdots + g^2(0, 1, 2, \ldots, M-1) \tag{2.41b}$$

By introducing a function R_N, given by

$$R_N = g_{00}^2 + g^2(0, 1) + g^2(0, 2) + \cdots + g^2(M-2, M-1) + \cdots + g^2(0, 1, 2, \ldots, M-1) \tag{2.42a}$$

we can also write Equation 2.41b in the following form:

$$E((\mathbf{u}_k^{\mathrm{T}}\mathbf{g}_N)^2) = R_N \tag{2.42b}$$

Finally, returning to Equation 2.17 and introducing in it the results of Equations 2.35, 2.40, and 2.42b, we arrive at

$$mes(k) = E((\mathbf{g}_L - \mathbf{c}(k))^{\mathrm{T}} \cdot (\mathbf{g}_L - \mathbf{c}(k))) + R_N \tag{2.43}$$

To proceed further, let us now recall the rule for changing the transversal filter coefficients given by Equation 2.2b. We rewrite this rule in the vector-form and with the transmitted symbols $a(k - i)$ instead of $x(k - i)$, $i = 0, \ldots, M - 1$, as

$$\mathbf{c}(k+1) = \mathbf{c}(k) + 2\alpha r(k) \cdot \mathbf{a}(k) \tag{2.44a}$$

Performing the substitution of k' for $k + 1$ in Equation 2.44a, and renaming then k' as k, and substituting $r(k)$ given by the expression (2.8), we obtain

$$\mathbf{c}(k) = \mathbf{c}(k-1) + 2\alpha(\mathbf{a}^{\mathrm{T}}(k-1) \cdot (\mathbf{g}_L - \mathbf{c}(k-1)) + \mathbf{u}^{\mathrm{T}}(k-1) \cdot \mathbf{g}_N + v(k-1)) \cdot \mathbf{a}(k-1) \tag{2.44b}$$

Substituting Equation 2.44b into Equation 2.43, we arrive at

$$\begin{aligned} mes(k) = {} & E(((\mathbf{g}_L - \mathbf{c}(k-1))^{\mathrm{T}} - 2\alpha\mathbf{a}^{\mathrm{T}}(k-1)\mathbf{a}^{\mathrm{T}}(k-1)(\mathbf{g}_L - \mathbf{c}(k-1)) \\ & - 2\alpha\mathbf{a}^{\mathrm{T}}(k-1)\mathbf{u}^{\mathrm{T}}(k-1) \cdot \mathbf{g}_N - 2\alpha\mathbf{a}^{\mathrm{T}}(k-1)v(k-1)) \\ & \cdot ((\mathbf{g}_L - \mathbf{c}(k-1)) - 2\alpha\mathbf{a}(k-1)\mathbf{a}^{\mathrm{T}}(k-1)(\mathbf{g}_L - \mathbf{c}(k-1)) \\ & - 2\alpha\mathbf{a}(k-1)\mathbf{u}^{\mathrm{T}}(k-1) \cdot \mathbf{g}_N - 2\alpha\mathbf{a}(k-1)v(k-1))) + R_N \end{aligned} \tag{2.45a}$$

Note that the fact that the transpose of a sum of vectors is equal to the sum of the vectors transposed was used in Equation 2.45a. Moreover, the fact that a number multiplied by a vector (for example $\mathbf{a}^{T}(k-1)(\mathbf{g}_L - \mathbf{c}(k-1))\mathbf{a}^{T}(k-1)$, being a number multiplied by the vector $\mathbf{a}^{T}(k-1)$) can be also represented in the reverse order, that is, as this vector times the number applied.

Performing all the multiplications in Equation 2.45a, applying then the assumptions regarding the statistical independence of $\mathbf{a}$ and $\mathbf{c}$, and furthermore exploiting the fact of the binary form of the transmitted symbols, assuming values: -1 and $+1$ (which, among others, gives $\mathbf{a}^{T}(k-1)\,\mathbf{a}(k-1) = M$), we obtain from Equation 2.45a the following result:

$$\begin{aligned} mes(k) &= E((\mathbf{g}_L - \mathbf{c}(k-1))^{T} \cdot (\mathbf{g}_L - \mathbf{c}(k-1))) \cdot (1 - 4\alpha + 4\alpha^2 M) + 4\alpha^2 M R_N \\ &\quad + 8\alpha^2 M g_{00} E(v(k-1)) + 4\alpha^2 M E((v(k-1))^2) + R_N \end{aligned} \tag{2.45b}$$

Finally, using Equation 2.43 and the observation that $4\alpha^2 MR_N + R_N = (1 - 4\alpha + 4\alpha^2 M)R_N + 4\alpha R_N$ in Equation 2.45b leads to

$$mes(k) = (1 - 4\alpha + 4\alpha^2 M)mes(k-1) + 4\alpha R_N + 8\alpha^2 M g_{00} \cdot m_v + 4\alpha^2 M m_{2v} \tag{2.45c}$$

where m_v and m_{2v} are the mean values of the random variables $V(k-1)$ and $(V(k-1))^2$, respectively, that is,

$$m_v = E(V(k-1)) \tag{2.45d}$$

and

$$m_{2v} = E(V(k-1)^2) \tag{2.45e}$$

With regard to m_v and m_{2v}, using another terminology,[69] it is worth noting that they are the first and the second moment, respectively, of the random variable $V(k-1)$. These moments are independent of the discrete time k, when the random process $\{v(k-1), v(k-2), \ldots\}$, of which outcomes we denote using small letters, $\{v(k-1), v(k-2), \ldots\}$, is a discrete-time wide-sense stationary (WSS) process.

Before defining the WSS random process, we would like to draw the reader's attention to the notational convention we will also use in what follows. That is, we shall use small letters for denoting a random process, when this will not cause any confusion. Note that the same terminological convention was already used in the case of random variables.

Now returning to the definition of the WSS random process, we start with a more obvious definition of stationarity, that is with the strict-sense stationarity property. With regard to this property, we say[24, 69] that a discrete-time random process $\{X(k), X(k-1), X(k-2), \ldots\}$ is a strict-sense stationary process if the probability density function of each of the random variables $X(k)$, $X(k - 1)$, $X(k - 2)$, ... is independent of time, and the joint probability density function of any set of the above random variables depends only upon the time differences related with these variables. That is, in the case of considering, for example, $X(k)$ and $X(k - 2)$, the latter function depends only upon the difference $\Delta k = k - (k - 2) = 2$, but not upon k and/or $k = -2$. Further, we say that a discrete-time random process is a wide-sense stationary (WSS) one when the first condition is satisfied, and the second is satisfied at least for any set consisting of two random variables.

Note that the wide-sense stationarity implies that the mean value (expected value) of the random process

$$m_x = E(X(k)) = E(X(k-1)) = E(x(k-2)) = \dots \tag{2.46a}$$

does not depend upon time, that is, it is a constant equal to the value of the first moment. Furthermore, the so-called autocorrelation function (defined here for real-valued processes),

$$R_{xx}(k, k-1) = E(X(k)X(k-i)) = R_{xx}(i) \tag{2.46b}$$

is then a function of only one argument, being the time difference $i = k - (k - i)$.

To use the result (2.46b) in Equation 2.45e, observe first that, when $i = 0$, the autocorrelation function is identical with the second moment of a random variable. Therefore, $m_{2x} = R_{xx}(0) = \text{const}$. That is, the fact that $m_{2x} = \text{const}$ follows from the wide-sense stationarity property of the random process $\{X(k), X(k - 1), X(k - 2), \dots\}$.

Example 2.5

Consider the difference equation

$$x(k) = ax(k-1) + b$$

where a and b are real numbers. Substituting a new variable $k' = k - 1$ into this equation leads to

$$x(k'+1) = a(k') + b$$

Now dropping the prime at k in the above relation, we get

$$x(k+1) = ax(k) + b$$

Moreover, assume the starting point in the latter equation at $k = 0$. To solve this equation, we can use, for example, the Z-transform method.[10] What we need is the following relation[10]:

$$Z\{x(k+1)\} = z(Z\{x(k)\} - x(0))$$

where $Z\{\cdot\}$ means the one-sided Z transform (see Section 1.3). Applying this relation in the previous equation, we get

$$z(X(z) - x(0)) = Z\{ax(k) + b\} = aX(z) + \frac{bz}{z-1}$$

because $Z\{b\}$, or more precisely, $Z\{\varepsilon(k)b\}$, where $\varepsilon(k)$ is the standard step function (see Figure 1.5 in Section 1.1) equals $bz/(z-1)$. Moreover, the Z transform satisfies the linearity property.

Solving the latter equation for $X(z)$, we arrive at

$$X(z) = \frac{zx(0)}{z-a} + \frac{bz}{(z-a)(z-1)} = \frac{zx(0)}{z-a} + \frac{b}{1-a}\frac{z}{z-1} - \frac{b}{1-a}\frac{z}{z-a}$$

Exploiting the Z transform pairs

$$1 \cdot \varepsilon(k) \leftrightarrow \frac{z}{z-1}$$

and

$$a^k \leftrightarrow \frac{z}{z-a}$$

in this equation, we get finally

$$x(k) = a^k\left(x(0) - \frac{b}{1-a}\right) + \frac{b}{1-a}, \quad k = 0, 1, 2, \ldots$$

The result of Example 2.5 will now help us to solve the difference equation 2.45c. This is the difference equation of the first order of the same form as that which we solved in Example 2.5, with $x(k)$ and $x(k-1)$ now identified with $mes(k)$ and $mes(k-1)$, respectively, and with a now equal to $(1 - 4\alpha + 4\alpha^2 M)$ *and* b equal to $4\alpha R_N + 8\alpha^2 M g_{00} m_v + 4\alpha^2 M m_{2v}$, accordingly. Using the final result of Example 2.5, we get the following solution of the difference Equation 2.45c:

$$mes(k) = (1 - 4\alpha + 4\alpha^2 M)^k\left(mes(0) - \frac{4\alpha R_N + 8\alpha^2 M g_{00} m_v + 4\alpha^2 M m_{2v}}{4\alpha(1-\alpha M)}\right) + \frac{4\alpha R_N + 8\alpha^2 M g_{00} m_v + 4\alpha^2 M m_{2v}}{4\alpha(1-\alpha M)} \quad (2.47)$$

where $mes(0)$ is the initial value of $mes(k)$ at the starting point $k = 0$.

Let us draw a few conclusions from the solution given by Equation 2.47:

1. Note that the solution for the linear case, that is for a linear echo path, can be obtained from Equation 2.47 by assuming in it $R_N = 0$ and $g_{00} = 0$. We get then the following result:

$$mes_L(k) = (1 - 4\alpha + 4\alpha^2 M)^k\left(mes_L(0) - \frac{\alpha M m_{2v}}{1-\alpha M}\right) + \frac{\alpha M m_{2v}}{1-\alpha M} \quad (2.48a)$$

2. Observe from Equation 2.47 that $mes(k)$ converges, that is, $mes(k)$ decreases when $k \to \infty$, when the condition

$$|1 - 4\alpha + 4\alpha^2 M| < 1 \quad (2.48b)$$

holds. Furthermore, note from Equation 2.48a that the same condition for convergence is needed in the case of a linear echo path. Independently of the case of linear or nonlinear environment, the condition for $mes_L(k)$ or $mes(k)$ to decrease is identical.

Solving inequality 2.48b for α assumed to be a positive real number, we can express the convergence condition 2.48b equivalently as

$$0 < a < \frac{1}{M} \tag{2.48c}$$

3. According to the theory of stochastic processes (see, for example, Chapter 3 of Reference 24), $mes(k)$ or $mes_L(k)$ can be interpreted as the power of the stochastic process $e(k) - \hat{e}(k)$, with $e(k)$ as a nonlinear or linear echo response, accordingly. We call this power for $k \to \infty$ the residual echo power. It follows from Equations 2.47 and 2.48a, respectively, that the residual echo powers in the nonlinear and linear echo environments are given by

$$mes(\infty) = \frac{R_N + 2\alpha M g_{00} m_v + \alpha M m_{2v}}{1 - \alpha M} \tag{2.48d}$$

and

$$mes_L(\infty) = \frac{\alpha M m_{2v}}{1 - \alpha M} \tag{2.48e}$$

accordingly. Comparing Equation 2.48d with Equation 2.48e, we observe that $mes(\infty) > mes_L(\infty)$, that is, the residual echo power in the case of nonlinear echo environment is greater than that in the case of linear echo environment.

It is also worthwhile to compare the residual echo power with the power of the received signal (together with the additive corrupting noise). Because the latter power is nothing other than m_{2v}, we get from Equations 2.48d and 2.48e

$$rre = 10\log\left(\frac{mes(\infty)}{m_{2v}}\right) = 10\log\left(\frac{R_N + 2\alpha M g_{00} m_v}{(1 - \alpha M) m_{2v}} + \frac{\alpha M}{1 - \alpha M}\right) \tag{2.48f}$$

and

$$rre_L = 10\log\left(\frac{\alpha M}{1 - \alpha M}\right) \tag{2.48g}$$

respectively. In Equation 2.48f, *rre* means the relative residual echo power in the nonlinear echo environment, and rre_L in Equation 2.48g the similar measure for the linear echo environment. Moreover, log in the above two equations means the logarithm with a base equal to 10.

4. The rate of convergence of an adaptation algorithm can be defined, as in Reference 46, by the coefficient v_{20}, which is a number of iterations, required to reduce the residual echo by 20 dB. To find this coefficient, we must solve the following equation:

$$10\log\left(\frac{mes(v_{20})}{mes(0)}\right) = -20\text{dB} \tag{2.48h}$$

On assuming

$$mes(k) \gg \frac{R_N + 2\alpha M g_{00} m_v + aM m_{2v}}{1 - \alpha M} \tag{2.48i}$$

in Equation 2.47 or

$$mes_L(k) \gg \frac{\alpha M m_{2v}}{1 - \alpha M} \tag{2.48j}$$

in Equation 2.48a, and applying then Equations 2.47 and 2.48a in Equation 2.48h to solve for v_{20}, we get

$$v_{20} \cong \frac{-2}{\log(1 - 4\alpha + 4\alpha_2 M)} \tag{2.48k}$$

in both the cases. Because v_{20} must be an integer, we take as this number of iterations the nearest integer greater than or equal to the real number provided by expression 2.48k.

The residual echo power expressed by $mes(\infty)$ or *rre*, and the rate of convergence, v_{20}, are two very important measures of performance of any echo canceller adaptation algorithm. Between these measures, there is a fundamental tradeoff, for the class of algorithms considered, in the sense that, when the speed of convergence is made better, this has, in consequence, the worsening of the rest echo after adaptation. In other words, when the value of v_{20} decreases this immediately causes the increase of the value of $mes(\infty)$, as illustrated in Figure 2.21.

In Figure 2.21, the functions v_{20} and $mes_L(\infty)$ for the linear echo environment are sketched versus the amplification coefficient α. From the curve for v_{20}, it follows that v_{20} decreases when the amplification coefficient α increases from the values near zero to the value equal to $\frac{1}{2}\frac{1}{M}$. At the same time, as the curve for $mes_L(\infty)$ shows, the value

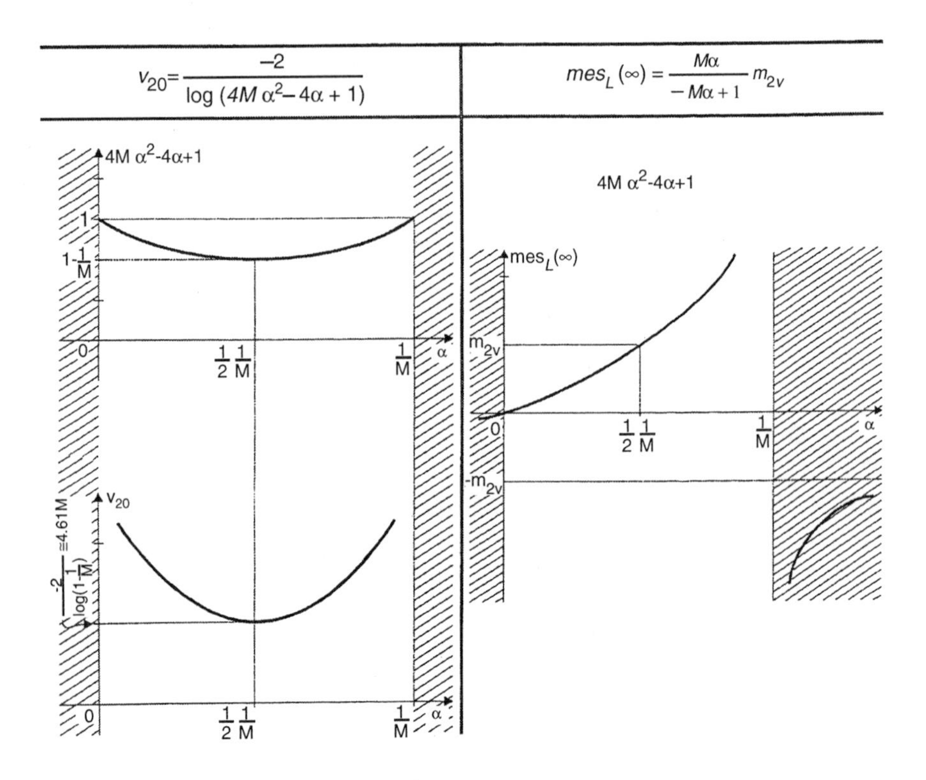

FIGURE 2.21 Sketch of the functions v_{20} and $mes_L(\infty)$ versus the amplification coefficient α.

of the rest echo power increases. So really, we have to work with the tradeoff between the parameters v_{20} and $mes_L(\infty)$.

In the range of values of $\alpha\left(0, \frac{1}{2}\frac{1}{M}\right)$, the functions $v_{20}(\alpha)$ and $mes_L(\infty)(\alpha)$ change their values in opposite directions. In the next range of $\alpha\left(\frac{1}{2}\frac{1}{M}, \frac{1}{M}\right)$, this is not the case. Here the values of both functions increase when α increases. However, the latter range is of no practical importance because $mes_L(\infty) \geq m_{2v}$ holds in it (see Figure 2.21).

The smallest value of the parameter v_{20} is approximately equal to 4, 61M (see Figure 2.21). That is, the minimal value of v_{20} is proportional to the memory length of the echo path. This result is, however, of no practical importance because the values of α must be so chosen to be very small to ensure a proper attenuation of $mes_L(\infty)$ with respect to m_{2v}, at least at the level of -20 dB. Furthermore, it can be shown that the values of α are then of order $2^{-14} \div 2^{-18}$. Moreover, for such very small values of α, good approximations for $mes_L(\infty)$ given by expression 2.48e and v_{20} given by expression 2.48k are

$$\frac{mes_L(\infty)}{m_{2v}} \cong \alpha M \tag{2.48l}$$

and

$$v_{20} \cong \frac{1.15}{\alpha} \tag{2.48m}$$

respectively.

The tradeoff between $mes_L(\infty)$ and v_{20} is evident when looking at Equations 2.48l and 2.48m.

As already mentioned, when the condition 2.48i is satisfied, then the expression for the parameter v_{20} is the same in both the cases of the linear and nonlinear echo path environments. This expression is given by expression 2.48k and illustrated on the left-hand side of Figure 2.21. The difference between the linear and nonlinear echo paths occurs with regard to the second parameter, $mes_L(\infty)$. For the nonlinear echo path, this parameter as a function of α, according to Equation 2.48d, is sketched in Figure 2.22. Comparison of the curves $mes_L(\infty)(\alpha)$ of Figure 2.21 and $mes(\infty)(\alpha)$ of Figure 2.22 shows that the latter starts with the value R_N at $\alpha = 0$, and increases more rapidly than the curve $mes_L(\infty)(\alpha)$ to achieve the value $2R_N + (m_{2v} + 2g_{00}m_v)$ for $\alpha = \frac{1}{2}\frac{1}{M}$.

Consider now the influence of the echo path nonlinearities on the residual echo power in more detail, and take for this purpose the relative measure *rre*, but with the logarithm dropped for simplicity. From Equation 2.48f, we have then

$$rre' = \frac{R'_N}{m_{2v}}\frac{1}{1-\alpha M} + \frac{g_{00}^2}{m_{2v}}\frac{1}{1-\alpha M} + \frac{2g_{00}m_v}{m_{2v}}\frac{\alpha M}{1-\alpha M} + \frac{\alpha M}{1-\alpha M} \tag{2.49a}$$

where

$$R'_N = R_N - g_{00}^2 \tag{2.49b}$$

and

$$rre = 10\log(rre') \tag{2.49c}$$

The component g^2_{00} is excluded in Equation 2.49b from R_N given by Equation 2.42a because it is responsible for the transmitted pulse asymmetry. In what follows, the influence of the pulse asymmetry is considered separately. Moreover, the relation

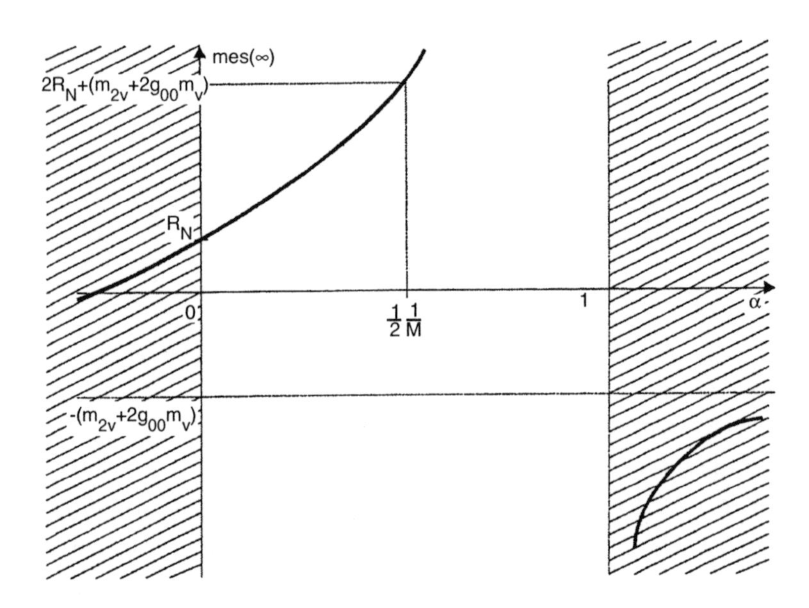

FIGURE 2.22 Sketch of the function $mes(\infty)(\alpha)$ for the nonlinear echo path.

between the newly introduced parameter rre' and the parameter rre is given by Equation 2.49c.

Proceeding further, we can use Equation 2.49a to determine when the particular component coming from the echo path nonlinearities (excluding the pulse asymmetry), from the transmitted pulse asymmetry, and from the combined transmitted and received pulse asymmetry can be neglected, compared to the component $\alpha M/(1 - \alpha M)$ as the relative residual echo power in the linear echo environment (see Equation 2.48g). Hence, we can write the following inequalities.

1. To neglect the component related with the echo path nonlinearities (excluding the pulse asymmetry), the inequality

$$\frac{R'_N}{m_{2v}} \frac{1}{1 - \alpha M} \ll \frac{\alpha M}{1 - \alpha M} \tag{2.50a}$$

must be satisfied. Furthermore, one obtains

$$\frac{R'_N}{m_{2v}} \ll \alpha M \tag{2.50b}$$

from inequality 2.50a.

2. The inequality

$$\frac{g_{00}^2}{m_{2v}}\frac{1}{1-\alpha M} \ll \frac{\alpha M}{1-\alpha M} \tag{2.51a}$$

must be satisfied to be able to neglect the component related with the transmitted pulse asymmetry. Simplifying inequality 2.51a gives

$$\frac{g_{00}^2}{m_{2v}} \ll \alpha M \tag{2.51b}$$

3. Because g_{00} as well as m_v can be negative, we take in our consideration the absolute value of the product $g_{00} \cdot m_v$, and compare it with the power of the signal $v(k)$, m_{2v}. Returning to Equation 2.49a, we need

$$\frac{2|g_{00}m_v|}{m_{2v}}\frac{\alpha M}{1-\alpha M} \ll \frac{\alpha M}{1-\alpha M} \tag{2.52a}$$

to be satisfied for neglect of the component related with the combined transmitted and received pulse asymmetry. Furthermore, we get

$$\frac{|g_{00}m_v|}{m_{2v}} \ll \frac{1}{2} \tag{2.52b}$$

from inequality 2.52a.

The importance of inequalities 2.50b, 2.51b, and 2.52b lies in the fact they determine quantitatively the conditions under which the echo path nonlinearities and the signal asymmetry do not disturb the behavior of the linear echo canceller. When these inequalities are not satisfied, then to obtain the required echo attenuation, one needs to implement a nonlinear canceller.

Note that, when only the inequality 2.51b is violated, one has to implement only one additional tap more compared to the linear echo canceller, to cancel the dc component g_{00}. Therefore the echo replica $\hat{e}(k)$ in this case will assume the following form:

$$\hat{e}(k) = \mathbf{a}^T(k) \cdot \mathbf{c}(k) + 1 \cdot c_{oo}(k) \tag{2.53a}$$

which is the modified expression 2.7b, where the dc component $c_{00}(k)$ for compensation of g_{00} is added to it. This component represents, of course, the additional tap just mentioned. Furthermore, subtracting $\hat{e}(k)$ given by Equation 2.53a from $e(k)$, expressed by Equation 2.7a, and using Equation 2.8, we get as an expression for the residual signal

$$r(k) = \mathbf{a}^T(k)\cdot(\mathbf{g}_L - \mathbf{c}(k)) + \mathbf{u}_1^T(k)\cdot(\mathbf{g}_{N1} - c_{00}(k)) + \mathbf{u}_2^T(k)\cdot\mathbf{g}_{N2} + \nu(k) \quad (2.53b)$$

with $\mathbf{u}_1(k) = 1$ and $\mathbf{u}_2(k) = [a(k)a(k-1), a(k)a(k-2), \ldots, a(k-M+2)a(k-M+1), \ldots, a(k)a(k-1)\cdots a(k-M+1)]^T$, and with $\mathbf{g}_{N1} = g_{00}$ and $\mathbf{g}_{N2} = [g(0, 1), g(0, 2), \ldots, g(M-2, M-1), \ldots, g(0, 1, 2, \ldots, M-1)]^T$. More generally, we assume here, and in what follows, that the vector $\mathbf{u}_1(k)$ represents that part of the vector $\mathbf{u}(k)$ that is involved, as shown in the expression 2.53b, in the compensation of some nonlinear part of the nonlinear echo. In Equation 2.53b, this is just the dc component $\mathbf{u}_1^T(k)\cdot\mathbf{g}_{N1} = 1\cdot g_{00} = g_{00}$, which has to be compensated by the adapted dc component $\mathbf{u}_1^T(k)\cdot c_{00}(k) = 1\cdot c_{00}(k) = c_{00}(k)$ generated by the nonlinear canceller. Moreover, the vector $\mathbf{u}_2(k)$ represents the remaining part of $\mathbf{u}(k)$, that is not involved in nonlinear compensation. So we can write

$$\mathbf{u}(k) = [\mathbf{u}_1^T(k), \mathbf{u}_2^T(k)]^T \quad (2.53c)$$

to be consistent with the definition of $\mathbf{u}(k)$ given by expression 2.5b. Similarly, to be consistent with the definition of $\mathbf{g}_N$ expressed by 2.5d, we write

$$\mathbf{g}_N = [\mathbf{g}_{N1}^T, \mathbf{g}_{N2}^T]^T \quad (2.53d)$$

That is, as in the case of $\mathbf{u}_1(k)$ and $\mathbf{u}_2(k)$, $\mathbf{g}_{N1}$ and $\mathbf{g}_{N2}$ are the corresponding parts of the vector $\mathbf{g}_N$, which are and are not involved in nonlinear compensation, respectively. Furthermore, note that, with the notation just introduced, the expression 2.9 for the difference between the echo signal and echo replica can be rewritten as

$$e(k) - \hat{e}(k) = \mathbf{a}^T(k)\cdot(\mathbf{g}_L - \mathbf{c}(k)) + \mathbf{u}_1^T(k)\cdot(\mathbf{g}_{N1} - \mathbf{c}_N(k)) + \mathbf{u}_2^T(k)\cdot\mathbf{g}_{N2} \quad (2.53e)$$

where the vector $\mathbf{c}_N(k)$ can be viewed as an complement of the vector $\mathbf{c}(k)$ for an adaptive linear transversal filter (see Equation 2.5e) to describe an adaptive nonlinear transversal filter such that the linear part of the latter filter is described by the vector of coefficients $\mathbf{c}(k)$ and its nonlinear part is described by the vector of coefficients $\mathbf{c}_N(k)$. The structure of the vector $\mathbf{c}_N(k)$ is then

$$\mathbf{c}_N(k) = [c_{00}(k), c_{01}(k), c_{02}(k), \ldots, c_{M-2,M-1}(k), \ldots, c_{0,1,2,\ldots,M-1}(k)]^T \quad (2.53f)$$

reflecting the structure of the vector $\mathbf{g}_N$ (see Equation 2.5d). That is, the element $c_{00}(k)$ in expression 2.53f corresponds to g_{00} in expression 2.5d, $c_{01}(k)$ in expression 2.53f to $g(0, 1)$ in expression 2.5d, and so on. Eventually, as in, for example, expression 2.53b, this vector does not consist of all the possible $2^M - M$ elements. It can be shorter than the vector $\mathbf{g}_N$.

The nonlinear transversal filter just introduced, with only one additional tap compared to the linear filter, for cancelling the dc component, is illustrated in Figure 2.23.

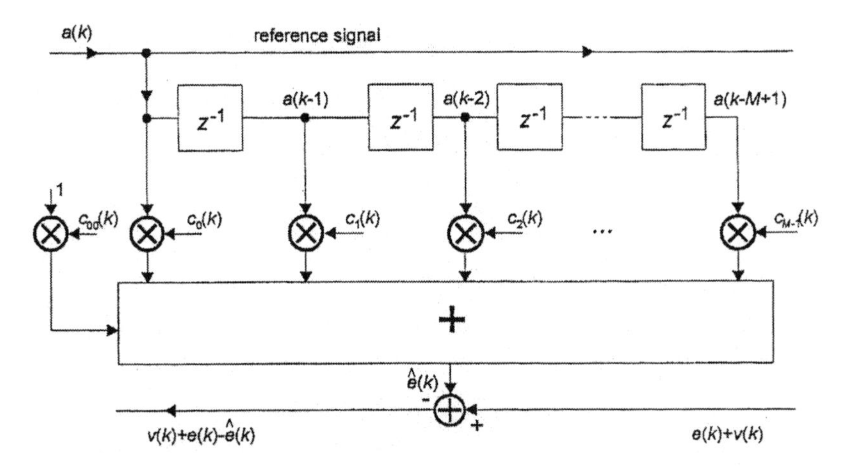

FIGURE 2.23 Adaptive nonlinear transversal filter with nonlinear part consisting of only one tap $c_{00}(k)$ for cancelling adaptatively dc component.

Consider once again the difference between the echo signal and echo replica given by Equations 2.9 or 2.53e, and take the expected value of it. That is, take into account the following:

$$E(e(k) - \hat{e}(k)) = \begin{cases} E(\mathbf{a}^T(k)(\mathbf{g}_L - \mathbf{c}(k)) + \mathbf{u}^T(k)\mathbf{g}_N) \\ \qquad \text{for a linear canceller} \\ E(\mathbf{a}^T(k)(\mathbf{g}_L - \mathbf{c}(k)) + \\ +\mathbf{u}_1^T(k)(\mathbf{g}_{N1} - \mathbf{c}_N(k)) + \mathbf{u}_2^T(k)\mathbf{g}_{N2} \\ \qquad \text{for a nonlinear canceller} \end{cases} \tag{2.54a}$$

Using the assumptions regarding the random vectors $\mathbf{a}(k)$, $\mathbf{u}(k)$, $\mathbf{c}(k)$, and extending the statistical independence property also to the random vector $\mathbf{c}_N(k)$, we get from Equation 2.54a

$$E(e(k) - \hat{e}(k)) = \begin{cases} g_{00} \quad \text{for a linear canceller} \\ E(g_{00} - c_{00}(k)) \quad \text{for a nonlinear} \\ \text{canceller possessing a tap for cancellation} \\ \text{of the dc component} \end{cases} \tag{2.54b}$$

And, for the time after adaptation, that is, substituting $k = \infty$ in Equation 2.54b, we obtain

$$E(e(\infty) - \hat{e}(\infty)) = \begin{cases} g_{00} & \text{for a linear canceller} \\ E(g_{00} - c_{00}(\infty)) & \text{for a nonlinear} \\ \text{canceller possessing a tap for cancellation} \\ \text{of the dc component} \end{cases} \tag{2.54c}$$

Moreover, let us consider

$$\begin{aligned} mes_N(k) &= E((e(k) - \hat{e}_N(k))^2) = E((\mathbf{a}^T(k)\mathbf{g}_L - \mathbf{c}(k)) + \mathbf{u}_1^T(k) \\ &\quad \cdot (\mathbf{g}_{N1} - \mathbf{c}_N(k)) + \mathbf{u}_2^T(k)\mathbf{g}_{N2})^2) \end{aligned} \tag{2.55a}$$

for the nonlinear canceller, similarly as before, in considering the corresponding expression for the linear canceller. In Equation 2.55a, $\hat{e}_N(k)$ means the echo replica generated by the nonlinear canceller. So, the function $mes_N(k)$ is nothing other than the function $mes(k)$ for this canceller.

In what follows, we provide the main steps of the derivation leading to a similar result as that given by Equation 2.43 for the linear canceller, without going into details. First, after squaring the expression in Equation 2.55a and applying rule (I) of Example 2.3, we get

$$\begin{aligned} mes_N(k) &= E((\mathbf{a}^T(k)(\mathbf{g}_L - \mathbf{c}(k)))^2) + E((\mathbf{u}_1^T(k)(\mathbf{g}_{N1} - \mathbf{c}_N(k)))^2) \\ &+ E((\mathbf{u}_2^T(k)\mathbf{g}_{N2})^2) + E(2\mathbf{a}^T(k)(\mathbf{g}_L - \mathbf{c}(k)) \cdot \mathbf{u}_1^T(k)(\mathbf{g}_{N1} - \mathbf{c}_N(k))) \\ &+ E(2\mathbf{a}^T(k)(\mathbf{g}_L - \mathbf{c}(k)) \cdot \mathbf{u}_2^T(k)\mathbf{g}_{N2}) + E(2\mathbf{u}_1^T(k)(\mathbf{g}_{N1} - \mathbf{c}_N(k)) \cdot \mathbf{u}_2^T(k)\mathbf{g}_{N2}) \end{aligned} \tag{2.55b}$$

Because of the properties assumed regarding the random vectors $\mathbf{a}(k)$, $\mathbf{u}_1(k)$, $\mathbf{u}_2(k)$, $\mathbf{c}(k)$, and $\mathbf{c}_N(k)$, the fourth, fifth, and sixth components on the right-hand side of Equation 2.55b are equal to zero. This simplifies Equation 2.55b to

$$mes_N(k) = E((\mathbf{a}^T(k)(\mathbf{g}_L - \mathbf{c}(k)))^2) + E((\mathbf{u}_1^T(k)(\mathbf{g}_{N1} - \mathbf{c}_N(k)))^2) + E((\mathbf{u}_2^T(k)\mathbf{g}_{N2})^2) \tag{2.55c}$$

On the other hand, keeping in mind that the component $E(2\mathbf{a}^T(k)(\mathbf{g}_L - \mathbf{c}(k)) \cdot \mathbf{u}_1^T(k) \cdot (\mathbf{g}_{N1} - \mathbf{c}_N(k)))$ is equal to zero, and introducing new vectors

$$\mathbf{a}_e(k) = [\mathbf{a}^T(k), \mathbf{u}_1^T(k)]^T \tag{2.55d}$$

$$\mathbf{c}_e(k) = [\mathbf{c}^T(k), \mathbf{c}_N^T(k)]^T \tag{2.55e}$$

$$\mathbf{g}_e = [\mathbf{g}_L^T, \mathbf{g}_{N1}^T]^T \tag{2.55f}$$

into Equations 2.55b or 2.55c, we can rewrite each of these expressions in the following form:

$$mes_N(k) = E((\mathbf{a}_e^T(k)(\mathbf{g}_e - \mathbf{c}_e(k)))^2) + E((\mathbf{u}_2^T(k)\mathbf{g}_{N2})^2) \tag{2.55g}$$

Note that the vectors $\mathbf{a}_e(k)$, $\mathbf{c}_e(k)$, and $\mathbf{g}_e$ given by expressions 2.55d, 2.55e, and 2.55f are the extended vectors $\mathbf{a}(k)$, $\mathbf{c}(k)$, and $\mathbf{g}_L$, respectively. These extended vectors are used to describe the behavior of the nonlinear canceller in a way similar to that done in the case of the linear canceller.

Further, proceeding with the expression in Equation 2.55g in the same way as in the case of the linear canceller, we simplify Equation 2.55g to the following form:

$$mes_N(k) = E((\mathbf{g}_e - \mathbf{c}_e(k))^T \cdot (\mathbf{g}_e - \mathbf{c}_e(k))) + R_{N2} \tag{2.55h}$$

which is the same as that in Equation 2.43 for the linear canceller. R_{N2} in Equation 2.55h is that part of R_N given by Equation 2.42a, that is not intended to be cancelled by the nonlinear canceller, while the remaining part of R_N is that which the nonlinear canceller tries to cancel. So, the following relation

$$R_N = R_{N1} + R_{N2} \tag{2.55i}$$

holds. For example, for the nonlinear canceller presented in Figure 1.23 with only one additional tap for cancelling the dc component, we shall write

$$R_{N1} = g_{00}^2 \tag{2.56a}$$

and

$$\begin{aligned} R_{N2} &= R_N - g_{00}^2 = g^2(0, 1) + g^2(0, 2) + \ldots \\ &+ g^2(M-2, M-1) + \ldots + g^2(0, 1, 2, \ldots, M-1) \end{aligned} \tag{2.56b}$$

Moreover, in this case, the R'_N defined by expression 2.49b will be equal to R_{N2}, that is, $R'_N = R_{N2}$ will hold.

In the adaptation process of the coefficients of a nonlinear transversal filter, we use the same rule as that expressed by Equation 2.44a, however, modified now to the form

$$\mathbf{c}_e(k+1) = \mathbf{c}_e(k) + 2\alpha r(k) \cdot \mathbf{a}_e(k) \tag{2.57a}$$

with the residual signal $r(k)$ modified, as in Equation 2.53b, but put into the more general form, valid not only for one nonlinear tap. Using the extended vectors $\mathbf{a}_e(k)$, $\mathbf{c}_e(k)$, and $\mathbf{g}_e$, we get for $r(k)$,

$$r(k) = \mathbf{a}_e^T(k) \cdot (\mathbf{g}_e - \mathbf{c}_e(k)) + \mathbf{u}_2^T(k) \cdot \mathbf{g}_{N2} + v(k) \tag{2.57b}$$

Furthermore, applying Equations 2.57a and 2.57b in Equation 2.55h and proceeding as in the case of the linear canceller (see Equations 2.44b and 2.45a), we get finally

$$\begin{aligned} mes_N(k) &= E((\mathbf{g}_e - \mathbf{c}_e(k-1))^T \cdot (\mathbf{g}_e - \mathbf{c}_e(k-1))) \\ &\cdot (1 - 4\alpha + 4\alpha^2 M_T) + 4\alpha^2 M_T R_{N2} \\ &+ 8\delta_{g_{00}} \alpha^2 M_T g_{00} E(v(k-1)) + 4\alpha^2 M_T E((v(k-1))^2 + R_{N2} \end{aligned} \tag{2.58a}$$

which can be also put into the following form:

$$\begin{aligned} mes_N(k) &= (1 - 4\alpha + 4\alpha^2 M_T) mes_N(k-1) + 4\alpha R_{N2} \\ &+ 8\delta_{g_{00}} \alpha^2 M_T g_{00} m_v + 4\alpha^2 M_T m_{2v} \end{aligned} \tag{2.58b}$$

In Equations 2.58a and 2.58b, M_T means the number of taps of the nonlinear canceller. For example, for the nonlinear transversal filter of Figure 2.23, this number is equal to $M+1$, that is, M linear taps plus one nonlinear, for cancelling the dc component. Furthermore, $\delta_{g_{00}}$ is an indicator of whether the element g_{00} is an element of the vector $\mathbf{g}_e$ or not. When the element g_{00} is incorporated into the vector $\mathbf{g}_e$, then $\delta_{g_{00}} = 0$; otherwise, $\delta_{g_{00}} = 1$. Terms m_v and m_{2v} in Equation 2.58b are defined by expressions 2.45d and 2.45e, respectively.

Equation 2.58b is the difference equation of the first order, so its solution has the same form as the solution of Equation 2.45c. We get for Equation 2.58b

$$\begin{aligned} mes_N(k) &= (1 - 4\alpha + 4\alpha^2 M_T)^k \\ &\left(mes_N(0) - \frac{4\alpha R_{N2} + 8\delta_{g_{00}} \alpha^2 M_T g_{00} m_v + 4\alpha^2 M_T m_{2v}}{4\alpha(1 - \alpha M_T)} \right) \\ &+ \frac{4\alpha R_{N2} + 8\delta_{g_{00}} \alpha^2 M_T g_{00} m_v + 4\alpha^2 M_T m_{2v}}{4\alpha(1 - \alpha M_T)} \end{aligned} \tag{2.59}$$

Performing an analysis of the solution given by Equation 2.59, similar to the case of the linear canceller, we conclude:

1. The convergence condition 2.48c is now slightly modified to

$$0 < \alpha < \frac{1}{M_T} \tag{2.60a}$$

   That is, in place of M, we have now M_T.
2. The residual echo power is given by

$$mes_N(\infty) = \frac{R_{N2}}{1-\alpha M_T} + \frac{2\delta_{g_{00}}\alpha M_T g_{00} m_v}{1-\alpha M_T} + \frac{\alpha M_T M_{2v}}{1-\alpha M_T} \tag{2.60b}$$

Moreover, note that applying the indicator $\delta_{g_{00}}$ in Equation 2.54c, we can rewrite this equation in a more general form as

$$E((e-\hat{e})(\infty)) = \begin{cases} g_{00} & \text{for a linear canceller} \\ g_{00} - \bar{\delta}_{00} \cdot E(c_{00}(\infty)) & \text{for a nonliner canceller} \end{cases} \tag{2.61}$$

where $\bar{\delta}_{g_{00}}$ means the inverse of $\delta_{g_{00}}$ (that is, $\bar{\delta}_{g_{00}} = 1$ when $\delta_{g_{00}} = 0$, and $\bar{\delta}_{g_{00}} = 0$ when $\delta_{g_{00}} = 1$).

After derivation of the corresponding relations for the nonlinear canceller as well, we now can make some comparisons, and from them draw conclusions. First, observe from Equation 2.54c or 2.61 that application of the nonlinear tap for cancelling the dc component in the nonlinear echo enables reduction of the mean value of the residual echo signal. Furthermore, vanishing $E((e - \hat{e})(\infty))$ is possible. This takes place when the stochastic processes represented by the scalar-valued set $\{v(k)\}$ and vector-valued set $\{\mathbf{a}(k)\}$ are jointly wide-sense stationary, which means that the vector $\mathbf{p} = E(v(k)\cdot\mathbf{a}(k))$ is independent of the discrete-time variable k. Moreover, the latter vector must be the zero vector, that is, $\mathbf{p} = \mathbf{0}$. Then, as shown in reference 40, the coefficients of the adaptive transversal filter go, in the mean value sense, into the corresponding coefficient values describing the echo path response. Among others, $E(c_{00}(\infty)) \rightarrow g_{00}$, causing $E((e - \hat{e})(\infty)) \rightarrow 0$.

With the result of Reference 40 in mind, we take this opportunity to comment on the relations 2.3b, 2.3c, and 2.3d. In describing the adaptation process more precisely, understand the results presented by the above relations in the mean value sense and be aware of the fact that the relations hold only when the vector $\mathbf{p}$, just defined, is the zero vector. When this is not the case, $E((e - \hat{e})(k))$ does not go exactly to zero as $k \rightarrow \infty$. Similarly $c_i(k)$, $i = 0, \ldots, M_e - 1$, do not go exactly to g_i as well as $c_i(k)$, $i \geq M_e$, do not go to zero values as $k \rightarrow \infty$. For more details regarding the behavior of the adaptation process from the probabilistic point of view, see Reference 40.

Looking at Equation 2.60b, we see that application of the tap $c_{00}(k)$ also causes reduction of the residual echo power, that is, of the variance of $(e - \hat{e})(k)$. More precisely, when $c_{00}(k)$ is applied, then the second component on the right-hand side of Equation 2.60b vanishes, and the first component in this equation is smaller because R_{N2} does not comprise then g^2_{00}. However, if after applying the tap $c_{00}(k)$, the first component in Equation 2.60b is still larger than the third one, it is necessary to identify the largest components of R_{N2} among the remaining $g^2(0, 1)$, $g^2(0, 2)$, …, $g^2(M - 2, M - 1)$, …, $g^2(0, 1, 2, \ldots, M - 1)$, and reconstruct the nonlinear canceller by adding the corresponding nonlinear taps into it for cancellation of the identified components. The objective of such a procedure is to achieve such a R_{N2}, say R'_{N2}, which satisfies the inequality

$$R'_{N2} \ll \alpha M_T m_{2v} \tag{2.62a}$$

Compare this inequality with inequality 2.50b.

Finally, note from Equation 2.60b that, by obtaining inequality 2.62a, we arrive with the residual echo power at the irreducible limit given by

$$\frac{\alpha M_T m_{2v}}{1 - \alpha M_T} \tag{2.62b}$$

Proceeding similarly as in Equations 2.48h to k with the solution given by Equation 2.59, we get the expression:

$$v_{20N} \cong \frac{-2}{\log(1 - 4\alpha + 4\alpha^2 M_T)} \tag{2.63}$$

for the rate of convergence of the coefficient adaptation of the nonlinear transversal filter.

Comparing the expressions 2.62b with 2.48e and 2.63 with 2.48k, we see that the corresponding expressions are identical in both the cases of the nonlinear canceller and of the linear canceller working in the linear echo environment, except that, in the first case, we have M_T in the expressions in place of M. It follows from this that the curves presented in Figure 2.21 are also valid for the nonlinear canceller, with the parameter M replaced by M_T. Hence, because $M_T = M +$ number of additional nonlinear taps, we conclude immediately from Figure 2.21 that the addition of nonlinear taps worsens slightly the rate of convergence of the adaptation process. Moreover, it also increases the value of the irreducible limit for the residual echo power because

$$\frac{M\alpha m_{2v}}{1 - M\alpha} < \frac{M_T \alpha m_{2v}}{1 - M_T \alpha}$$

holds for $0 < \alpha < \frac{1}{M_T} < \frac{1}{M}$.

Example 2.6

Let a nonlinear echo path have the memory length of $M_e = 3$. We choose three linear taps for cancellation of this echo. That is, we have $M_{\hat{e}} = M_e = M = 3$ in our case. Furthermore, assume that the transmitted pulse asymmetry is so large in our case that inequality 2.51b is violated for a chosen α, $0 < \alpha < \frac{1}{3}$. Hence, we decide to use a nonlinear canceller with a tap $c_{00}(k)$ for cancellation of the dc component. Further analysis shows, however, that the value of $R_{N2} = g^2(0, 1) + g^2(0, 2) + g^2(1, 2) + g^2(0, 1, 2)$, is still not acceptable.

By the way, according to the result of Example 1.5, the R_{N2} given above comprises all possible components when M equals 3. For instance, R_{N2} here cannot comprise such a component as $g^2(0, 1, 2, 3)$.

Assume $g^2(0, 1)$ and $g^2(0, 2)$ are the largest values in the expression for R_{N2} given above. Eliminating them from R_{N2} gives R'_{N2}, which satisfies inequality 2.62a for $M_T = 3 + 1 + 2 = 6$, and for a chosen α, $0 < \alpha < \frac{1}{6}$.

Finally, we check the correctness of our choice for α. For instance, to fulfill the requirement of the difference of 20 dB between the power levels of the residual echo signal and received signal postulated at the end of Section 2.1, we should satisfy the following inequality:

$$\frac{mes_N(\infty)}{m_{2v}} \cong \frac{\alpha M_T}{1 - \alpha M_T} \leq \frac{1}{100}(-20\text{dB})$$

(see expression 2.62b).

Solving the above inequality for $M_T = 6$ gives $\alpha < 1.684 \cdot 10^{-3}$. Thus, when the chosen value of α was greater than $1.684 \cdot 10^{-3}$, the whole procedure for R_{N2} (without the component g^2_{00}) must be repeated. Note that checking inequality 2.51b would be superfluous in such a case.

A nonlinear transversal filter of this example with three nonlinear taps for cancelling g_{00}, $g(0, 1)$, and $g(0, 2)$ is shown in Figure 2.24.

The method of nonlinear echo cancellation with the use of the nonlinear transversal filter seems to be the simplest one. However, as mentioned at the beginning of this section, there are other approaches, which are more or less equivalent to the method using the nonlinear transversal filter based on the discrete-time Volterra series. These approaches represent some alternatives for the latter, as sometimes stressed in the literature.[66] Moreover, they seem to be better suited to the work in environments incorporating strong nonlinearities. Some of these methods have been listed as, for example, the method using the memory compensation principle, [51] and the approach applying the canonical piecewise-linear function description of a nonlinear echo path, etc.

The results of Section 1.12 regarding the approximations of the response of nonlinear systems possessing the property of fading memory can be also used in modeling the nonlinear echo, which possesses the property mentioned above. Three structures of nonlinear echo cancellers, based on the approximations of Section 1.12, are presented in Figures 2.25, 2.26, and 2.27. To be consistent with the notation introduced in Section 1.12 for the corresponding approximators, the same notation regarding $d_1, \ldots, d_m$ in Figure 2.25, $\beta_1, \ldots, \beta_m$, $\sigma(\cdot)$, $\alpha_1, \ldots, \alpha_m$ in Figure 2.26, and $\boldsymbol{\beta}_1, \ldots, \boldsymbol{\beta}_m$, the norm $\|\cdot\|$, $\alpha_1, \ldots, \alpha_m$ in Figure 2.27 is used. That is, this notation is consistent with that in Figure 1.54, in Figure 1.56, and in Figure 1.58, respectively.

To make the nonlinear echo canceller structures of Figures 2.25, 2.26, and 2.27 adaptive, one allows the coefficients of the linear transversal filters in these structures to be subject to the change in an adaptation process. Also, one allows the other parameters — $d_1, \ldots, d_m$ in the structure of Figure 2.25, $\beta_1, \ldots, \beta_m$, $\alpha_1, \ldots, \alpha_m$ in the structure of Figure 2.26, and the elements of vectors $\boldsymbol{\beta}_1, \ldots, \boldsymbol{\beta}_m$, the amplification

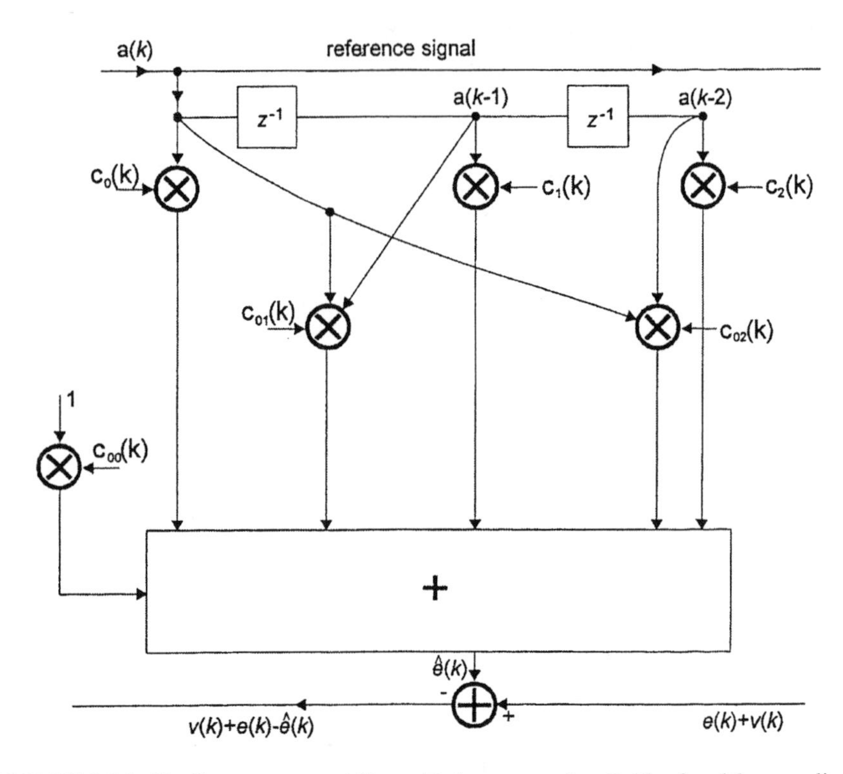

FIGURE 2.24 Nonlinear transversal filter with the memory length $M = 3$ and three nonlinear taps $c_{00}(k)$, $c_{01}(k)$, and $c_{02}(k)$.

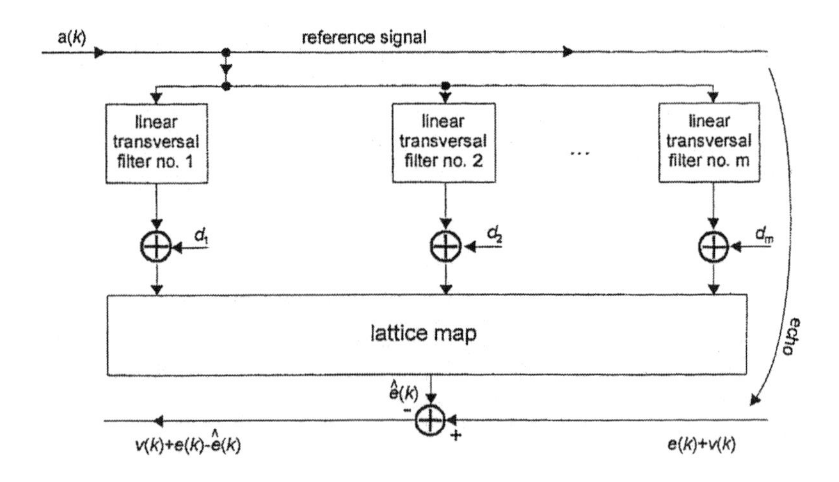

FIGURE 2.25 Nonlinear echo canceller structure based on the lattice map approximator.

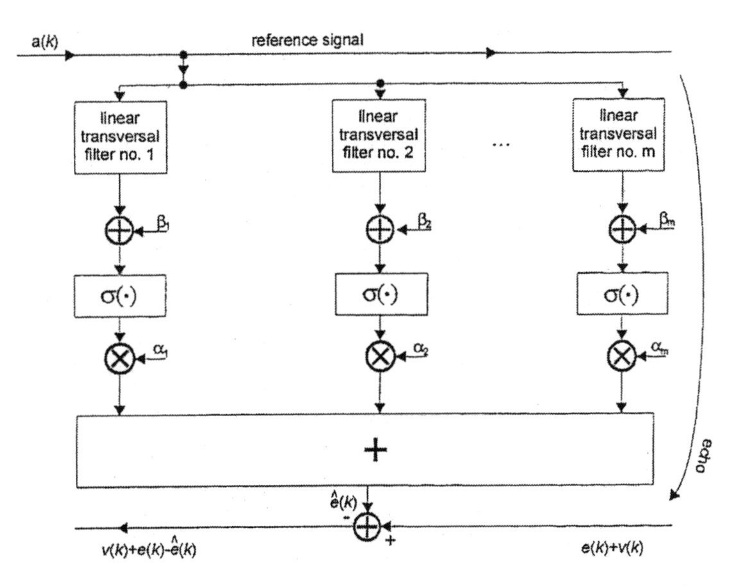

FIGURE 2.26 Nonlinear echo canceller structure based on the sigmoid function approximator.

coefficients $\alpha_1, \ldots, \alpha_m$ in the structure of Figure 2.27 to be subject to the change. Of course, one can restrict oneself to adapting the coefficients of the linear transversal filters only, letting the other parameters be unchangeable, or one can find another adaptation strategy. Moreover, the algorithm for adaptation of the coefficients and parameters just mentioned can be the stochastic iteration algorithm used in the analysis of the nonlinear transversal filter.

Note that the function *mes*(*k*), as defined in Equation 2.13, has the following form:

$$mes(k) = E((\mathbf{a}^T(k)\mathbf{g}_L + \mathbf{u}^T(k)\mathbf{g}_N - L(\mathbf{d} + \mathbf{C}\mathbf{a}(k)))^2) \tag{2.64a}$$

for the nonlinear echo canceller based on the lattice map approximator in Figure 2.25,

$$mes(k) = E\left(\left(\mathbf{a}^T(k)\mathbf{g}_L + \mathbf{u}^T(k)\mathbf{g}_N - \sum_{j=1}^{m}\alpha_j\sigma(\beta_j + \boldsymbol{\eta}_j\mathbf{a}(k))\right)^2\right) \tag{2.64b}$$

for the nonlinear echo canceller based on the sigmoid function approximator in Figure 2.26, and

$$mes(k) = E\left(\left(\mathbf{a}^T(k)\mathbf{g}_L + \mathbf{u}^T(k)\mathbf{g}_N - \sum_{s=1}^{m}\alpha_s R(\mathbf{Y}\mathbf{a}(k) + \boldsymbol{\beta}_s)\right)^2\right) \tag{2.64c}$$

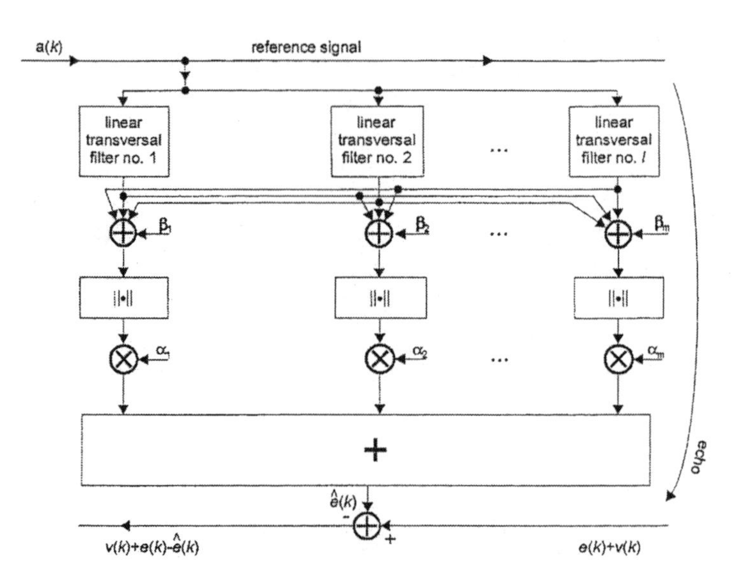

FIGURE 2.27 Nonlinear echo canceller structure based on the radial basis function approximator.

for the nonlinear echo canceller based on the radial basis function approximator in Figure 2.27. In Equation 2.64a, the lattice map L, the matrix $\mathbf{C}$, and the vector $\mathbf{d}$ are as defined in Equation 1.304a. Similarly, the function $\sigma(\cdot)$, the row vectors $\boldsymbol{\eta}_j$, $j = 1, \ldots, m$, and the coefficients α_j and β_j, $j = 1, \ldots, m$ in Equation 2.64b are as defined in Equation 1.309a. Moreover, the radial function $R(\cdot)$, the matrix $\mathbf{Y}$, the vectors $\boldsymbol{\beta}_s$, $s = 1, \ldots, m$, and the coefficients α_s, $s = 1, \ldots, m$ in Equation 2.64c are as defined in Equation 1.312a. The vector $(\mathbf{P}_a x)(k)$ occurring in Equations 1.304a, 1.309a, and 1.312a is equal to the vector of the transmitted data symbols $a(k)$ in the expressions 2.64a, 2.64b, and 2.64c. This is so because we assumed here the parameter a, being the index of $\mathbf{P}_a$ given by Equation 1.304b, equal to $M - 1$.

The expressions for the function $mes(k)$ for the linear echo canceller, Equation 2.43, or for the nonlinear echo canceller based on the nonlinear transversal filter, Equation 2.55h, are the quadratic functions of the elements of the vectors $\mathbf{c}(k)$ or $\mathbf{c}_e(k)$, respectively. These functions possess only one minimum, and the adaptation of the transversal filter coefficients collected in the vector $\mathbf{c}(k)$ or $\mathbf{c}_e(k)$ leads to achieving this minimum.

In contrast to this, a problem exists in rearranging the expressions 2.64a, 2.64b, and 2.64c in such a way that leads to getting similar expressions as 2.43 and 2.55h. Generally, this is impossible. Moreover, the nonlinear echo cancellers of the type presented in Figures 2.25, 2.26, and 2.27 can possess the functions $mes(k)$ with more than one minimum. Such a problem was reported in Reference 66 for the nonlinear echo cancellers based on the canonical piecewise-linear function description. However, the convergence of the adaptation algorithm used to a local minimum was also shown in Reference 66.

2.3 INTERLEAVED AND PASSBAND NONLINEAR TRANSVERSAL FILTERS

Timing recovery in the digital transmission imposes that the sampling rate at the echo canceller output is higher than the sampling rate at its input. The echo cancellers in form of the so-called interleaved linear transversal filters[24,40] are a suitable solution to work in the environment described above. In this section, we consider, the impact of the constraint regarding the difference between the echo canceller input and output sampling rates when a nonlinear transversal filter as the echo canceller must be applied to cancel the nonlinear echo.

Also addressed in this section are nonlinear transversal filters working in the passband. In fact, the nonlinear transversal filter discussed in the previous section was applied as a nonlinear canceller in the baseband. This application is typical for transmission in the digital subscriber loop. On the other hand, in the voiceband data transmission the carrier frequency is used, so we have to work with transmission in the passband.

Returning to the first topic, consider once again Equation 2.1a describing the echo replica generated by a canceller in the form of a linear transversal filter, and rewrite this equation in the following form:

$$\hat{e}(k) = \sum_{i=0}^{M-1} c(i)a(k-i) \tag{2.65a}$$

where, in place of $x(k - i)$ stand now the transmitted data symbols $a(k - i)$, and $M_{\hat{e}}$ is simply denoted by M, that is $M_{\hat{e}} = M$. Furthermore, assume that the sampling rate at the canceller output is R times larger than at its input. In other words, when T means the interval between the transmitter data symbols, of which reciprocal is equal to the sampling rate at the canceller input, then the sampling interval at the canceller output equals T/R. Showing explicitly the sampling interval in Equation 2.65a, we get

$$\hat{e}\left(k\frac{T}{R}\right) = \sum_{i=0}^{(M-1)R} c\left(i\frac{T}{R}\right)a\left((k-i)\frac{T}{R}\right) \tag{2.65b}$$

with the memory length of the filter impulse response equal to $M' = (M - 1)R + 1$ instead of M as it would be in the case of the sampling rate $1/T$. This point is illustrated in more detail in Figure 2.28.

Observe now that the components in Equation 2.65b can be rearranged in such a way that Equation 2.65b can be rewritten in the form

$$\hat{e}\left(k\frac{T}{R}\right) = \left(c\left(0\frac{T}{R}\right)a\left((k-0)\frac{T}{R}\right) + c\left(R\frac{T}{R}\right)a\left((k-R)\frac{T}{R}\right) + \cdots \right. \tag{2.65c}$$
$$\left. + c\left((M-1)R\frac{T}{R}\right)a\left((k-(M-1)R)\frac{T}{R}\right)\right)$$
$$+ \left(c\left(1\frac{T}{R}\right)a\left((k-1)\frac{T}{R}\right) + c\left((R+1)\frac{T}{R}\right)a\left((k-R-1)\frac{T}{R}\right) + \cdots \right.$$
$$\left. + c\left(((M-1)R+1)\frac{T}{R}\right)a\left((k-((M-1)R+1))\frac{T}{R}\right)\right) + \cdots$$
$$+ \left(c((R-1)\frac{T}{R}\right)a\left((k-R+1)\frac{T}{R}\right) + c\left((2R-1)\frac{T}{R}\right)$$
$$\left. \cdot a\left((k-2R+1)\frac{T}{R}\right) + \cdots + c\left((MR-1)\frac{T}{R}\right)a\left((k-MR+1)\frac{T}{R}\right)\right)$$
$$= \sum_{l=0}^{R-1}\sum_{i=0}^{M-1} c\left((l+iR)\frac{T}{R}\right)a\left((k-(l+iR))\frac{T}{R}\right)$$

when, additionally, the components $\sum_{(M-1)R+1}^{MR-1} c\left(i\frac{T}{R}\right)a\left((k-i)\frac{T}{R}\right)$ are added. These additional components are, of course, equal to zero when the memory length of the filter impulse response is exactly equal to $(M-1)R+1$ at the sampling rate R/T.

Performing the multiplications in the parentheses of $c(\cdot\cdot)$ and $a(\cdot\cdot)$ in Equation 2.65c, and introducing a new discrete time variable k' satisfying the relation $k\frac{T}{R} = k'T + n\frac{T}{R}$, we obtain

$$\hat{e}\left(k'T + n\frac{T}{R}\right) = \sum_{l=0}^{R-1}\sum_{i=0}^{M-1} c\left(\left(i+\frac{l}{R}\right)T\right)a\left(\left(k'-i+\frac{n-l}{R}\right)T\right), \quad 0 \le n \le R-1 \tag{2.65d}$$

It follows from Equation 2.65d that there are R phases, $\frac{0}{R}, \frac{1}{R}, \ldots, \frac{R-1}{R}$, between the neighboring time points $k'T$ and $(k'+1)T$, when the higher sampling rate R/T is applied. Let us write down explicitly the values of the echo replica $\hat{e}$, sampled at the rate R/T, that follow from Equation 2.65d. We then have

$$\left(\hat{e}(k'T) = \sum_{l=0}^{R-1}\sum_{i=0}^{M-1} c\left(\left(i+\frac{l}{R}\right)T\right)a\left(\left(k'-i+\frac{0-l}{R}\right)T\right), \text{ for } n=0\right)$$

$$\cdots$$

$$\left(\hat{e}\left(\left(k'+\frac{1}{R}\right)T\right) = \sum_{l=0}^{R-1}\sum_{i=0}^{M-1} c\left(\left(i+\frac{l}{R}\right)T\right)a\left(\left(k'-i+\frac{1-l}{R}\right)T\right), \text{ for } n=1\right)$$

$$\cdots$$

$$\left(\hat{e}\left(\left(k'+\frac{R-1}{R}\right)T\right) = \sum_{l=0}^{R-1}\sum_{i=0}^{M-1} c\left(\left(i+\frac{l}{R}\right)T\right)a\left(\left(k'-i+\frac{R-1-l}{R}\right)T\right), \text{ for } n=R-1\right) \tag{2.65e}$$

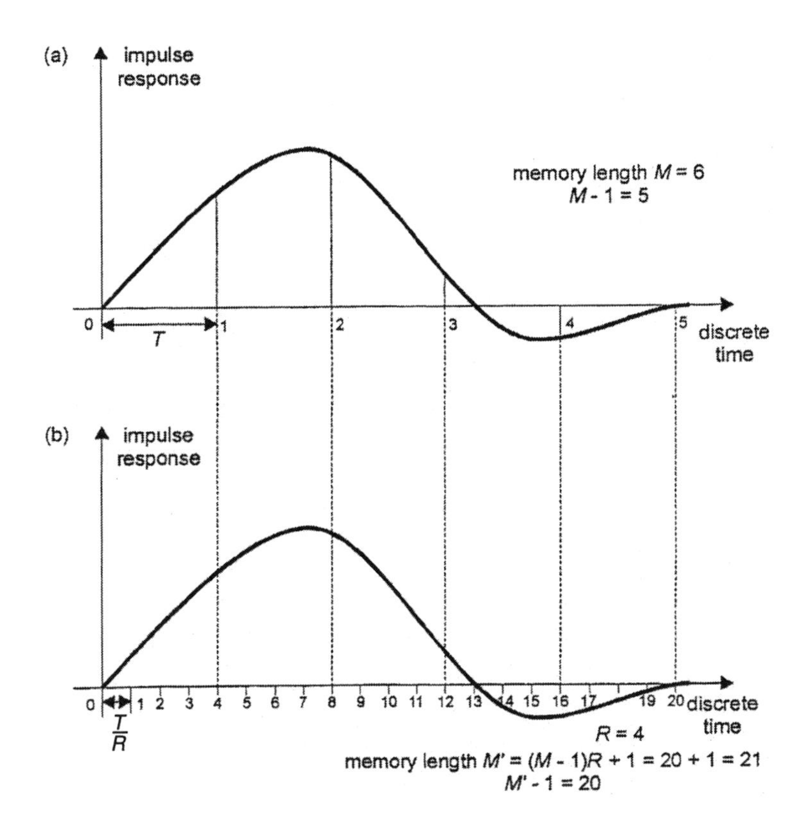

FIGURE 2.28 Determining the impulse response memory length of a linear filter at sampling rate $1/T$ (a), and at sampling rate R/T (b).

The form of $\hat{e}$ given by Equation 2.65e for the phase $0/R$ with $R = 4$ and $M = 6$ is illustrated in Figure 2.29. Note also that, for each of the remaining phases $\frac{1}{R}, \ldots, \frac{R-1}{R}$ we arrive at a similar structure as that shown in Figure 2.29 for the phase $0/R$ with $R = 4$.

Before going further, we make a remark regarding the values of samples of the impulse response of a transversal filter. For this purpose, consider again Equation 2.65a, which is, in fact, a sampled version of the convolution-integral used in description of linear continuous-time systems. In another form, showing the sampling period, this equation can be rewritten as

$$\hat{e}(kT) \cong \sum_{i=0}^{M-1} \tilde{c}(iT)a((k-i)T)\Delta(iT) \tag{2.66a}$$

where $\tilde{c}(iT)$ stands for the filter impulse response sample at the time point iT.

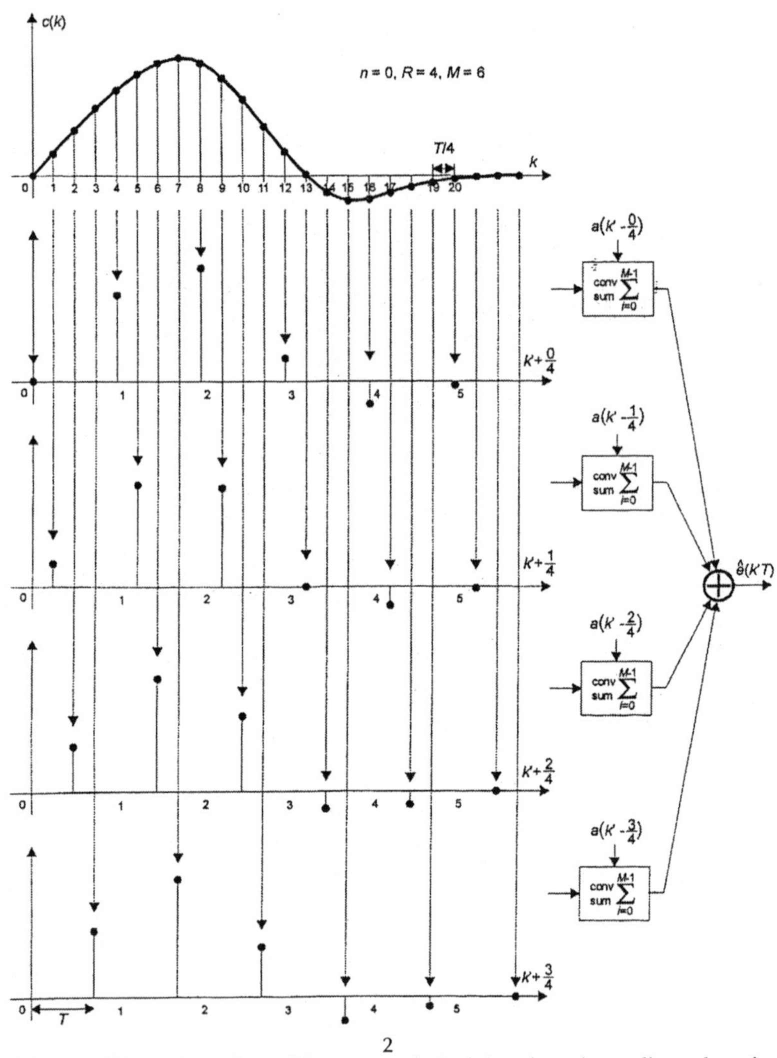

FIGURE 2.29 Illustration of possible means of obtaining the echo replica value given by Equation 2.65e with the use of R convolution summations ("conv sum" blocks in the figure) applying the samples achieved at the sampling rate $1/T$.

The symbol of "approximately equal to" points out the fact that the expression given by Equation 2.66a approximates the value of the corresponding convolution-integral. And, in this context, note that the above interpretation corresponds with the application considered because the echo path represents in fact an analog system. That is, its description is through a convolution-integral, which is replicated in an echo canceller.

Assuming the increase of time $\Delta(iT)$ in Equation 2.66a is simply equal to the sampling period T, we get

$$\hat{e}(kT) \cong \sum_{i=0}^{M-1} (\tilde{c}(iT) \cdot T)a((k-i)T) \tag{2.66b}$$

And comparison of Equation 2.66b with 2.65a reveals that

$$c(i) \cong \tilde{c}(i) \cdot T \tag{2.66c}$$

holds. That is, the samples of the filter impulse response from Equation 2.65a incorporate the sampling rate.

Consider again expression 2.65e for the echo replica samples at the corresponding phases $\frac{0}{R}, \frac{1}{R}, \ldots, \frac{R-1}{R}$, and consider, in more detail, the calculation of the sample for the phase $\frac{0}{R}$, illustrated in Figure 2.29 for $R = 4$. First, note from Figure 2.29 that the value of $\hat{e}(k'T)$ is a sum of values of four convolution-summations realized with the use of the transversal filters with the coefficients taken from the impulse response $c(k)$ sampled at the rate R/T. See, however, that the coefficients of the above four transversal filters are so chosen that only each fourth is taken and a shift of T/R occurs between the neighboring sets of coefficients. Because the coefficients come from the impulse response $c(k)$ sampled at the rate R/T, they incorporate the sampling interval T/R, That is, according to Equation 2.66c, we have $c\left(\left(i + \frac{l}{4}\right)T\right) \cong \tilde{c}\left(\left(i + \frac{l}{4}\right)T\right) \cdot \frac{T}{R}$. Furthermore, observe that the values of the transmitted data $a\left(\left((k'-i) - \frac{0}{4}\right)T\right)$, $i = 0, \ldots, M-1$, involved in the convolution-summations in Figure 2.29 differ from the data $a\left(\left((k'-i) - \frac{1}{4}\right)T\right)$, $a\left(\left((k'-i) - \frac{2}{4}\right)T\right)$, and $a\left(\left((k'-i) - \frac{3}{4}\right)T\right)$ for the corresponding values of i. This is so because the values of the transmitted data $a(nT)$, n any integer, do not change only in the interval from nT to $(n + 1)T$. And because we want to use only one unique set of the transmitted data in the expression for $\hat{e}(k'T)$, we choose for our further approximation only one phase from all the four phases presented in Figure 2.29. It is convenient to choose the phase $\frac{l}{4} = \frac{0}{4}$ identical with the phase $\frac{n}{4} = \frac{0}{4}$ at the argument of $\hat{e}(k'T) = \hat{e}\left(\left(k' + \frac{0}{4}\right)T\right)$; see expressions 2.65d and 2.65e, which, in other words, means choosing $l = n = 0$.

Making the choice described above, we must, however, incorporate the sampling period T instead of $T/4$ because we decided to apply only one transversal filter. To express this mathematically, we use the following approximation:

$$\begin{aligned}\hat{e}(k'T) &\cong \sum_{i=0}^{M-1}\left(\tilde{c}\left(\left(i+\frac{0}{4}\right)T\right)\cdot T\right)a\left(\left((k'-i)+\frac{0}{4}\right)T\right) \\ &= 4\sum_{i=0}^{M-1}\left(\tilde{c}\left(\left(i+\frac{0}{4}\right)T\right)\cdot \frac{T}{4}\right)a\left(\left((k'-i)+\frac{0}{4}\right)T\right) \\ &= \sum_{i=0}^{M-1} c^{\#}\left(\left(i+\frac{0}{4}\right)T\right)a\left(\left((k'-i)+\frac{0}{4}\right)T\right)\end{aligned} \tag{2.67a}$$

for the case shown in Figure 2.29, or, more generally,

$$\hat{e}(k'T) \cong \sum_{i=0}^{M-1} c^{\#}\left(\left(i+\frac{0}{R}\right)T\right)a\left(\left((k'-i)+\frac{0}{R}\right)T\right) \tag{2.67b}$$

for the case of R phases. In Equations 2.67a and 2.67b, the coefficients $c^{\#}$ follow from a general expression

$$c^{\#}\left(\left(i+\frac{l}{R}\right)T\right)=\tilde{c}\left(\left(i+\frac{l}{R}\right)T\right)\cdot T=\tilde{c}\left(\left(i+\frac{l}{R}\right)T\right)\frac{T}{R}\cdot R\cong Rc\left(\left(i+\frac{l}{R}\right)T\right), 0\le l\le R-1 \tag{2.67c}$$

by substituting $l = 0$ in it.

Proceeding similarly with the remaining phases $\frac{1}{R}, \ldots, \frac{R-1}{R}$, as in the case of the phase $0/R$, assuming $l = n$ in all the remaining Equations 2.65e, and applying a similar approximation as in Equations 2.67a, we can write the counterparts of Equation 2.67b for other phases as

$$\begin{aligned}\hat{e}\left(\left(k'+\frac{1}{R}\right)T\right) &\cong \sum_{i=0}^{M-1} c^{\#}\left(\left(i+\frac{1}{R}\right)T\right)a\left(\left((k'-i)+\frac{1}{R}\right)T\right) \\ &\cdots \\ \hat{e}\left(\left(k'+\frac{R-1}{R}\right)T\right) &\cong \sum_{i=0}^{M-1} c^{\#}\left(\left(i+\frac{R-1}{R}\right)T\right)a\left(\left((k'-i)+\frac{R-1}{R}\right)T\right)\end{aligned} \tag{2.67d}$$

Finally, putting Equations 2.67b and 2.67d into one equation, we get

$$\hat{e}\left(\left(k' + \frac{l}{R}\right)T\right) \cong \sum_{i=0}^{M-1} c^{\#}\left(\left(i + \frac{l}{R}\right)T\right)a\left(\left((k'-i) + \frac{l}{R}\right)T\right), 0 \le l \le R-1 \quad (2.68a)$$

Furthermore, note that the values of the transmitted data symbols $a\left(((k' + i) + l/R)T\right)$ in Equation 2.68a do not change when the parameter l changes from $l = 0$ to $l = R - 1$ and $(k' - i)$ is fixed. That is, we can write

$$a\left(\left((k'-i) + \frac{l}{R}\right)T\right) = a((k'-i)T) \quad (2.68b)$$

for a fixed $(k' - i)$ and $0 \le l \le R - 1$. Further, substituting Equation 2.68b into 2.68a gives

$$\hat{e}\left(\left(k' + \frac{l}{R}\right)T\right) \cong \sum_{i=0}^{M-1} c^{\#}\left(\left(i + \frac{l}{R}\right)T\right)a((k'-i)T), \quad 0 \le l \le R-1 \quad (2.68c)$$

Moreover, assume for the purpose of this derivation the following notations:

$$c_l(iT) = c^{\#}\left(\left(i + \frac{l}{R}\right)T\right) \quad (2.69a)$$

and

$$\hat{e}_l(k'T) = \hat{e}\left(\left(k' + \frac{l}{R}\right)T\right) \quad (2.69b)$$

Forgetting that Equation 2.68c represents only an approximation, that is writing the "equal" symbol instead of "approximately equal to" and using Equations 2.69a and 2.69b, we can rewrite Equation 2.68c in the form

$$\hat{e}_l(k'T) = \sum_{i=0}^{M-1} c_l(iT)a((k'-i)T) \quad (2.70a)$$

or, equivalently, by dropping the interval T, as

$$\hat{e}_l(k') = \sum_{i=0}^{M-1} c_l(i)a(k'-i) \tag{2.70b}$$

where k' refers to sampling with the rate $1/T$.

Figure 2.30 illustrates the realization of the response given by Equation 2.65b with the use of the so-called interleaved transversal filters described by Equation 2.70b.

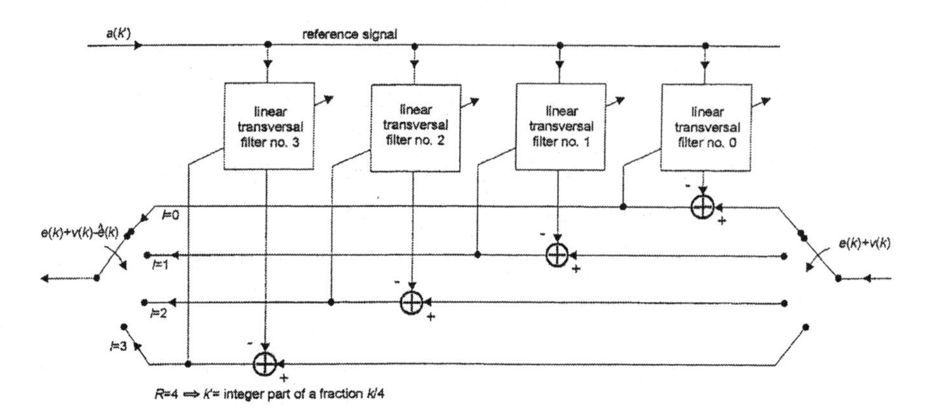

FIGURE 2.30 Interleaved adaptive linear transversal filters for achieving larger output sampling than that at the input (here four times larger, $R = 4$).

In Figure 2.30, the case is illustrated in which the output sampling rate of the canceller is four times larger than at its input. The linear transversal filters are given numbers corresponding to the values of the parameter $l = 0, 1, 2$, and 3, respectively. The structure of these filters is the same as that in Figure 2.11. Moreover, they are made adaptive, that is, the filter coefficients $c_l(i)$, $i = 0, 1, \ldots, M - 1$ are adapted in the adaptation process and, as it follows from the figure, they are adapted separately for each filter, no. 0, no. 1, no. 2, and no. 3. In other words, each of the transversal filters in Figure 2.30 adapts to the echo impulse response at the rate equal to the transmitted data rate, but with a different phase out of the number R possible phases $\frac{0}{R}, \frac{1}{R}, \ldots, \frac{R-1}{R}$. The filters converge and cancel independently. The filter output streams are recombined into a single data stream $e(k) - \hat{e}(k) + v(k)$ at the sampling rate $\frac{R}{T}$.

It follows from the above considerations that the solution of the problem with the use of the interleaved transversal filters, as illustrated in Figure 2.30 for the case of four phases, can be viewed as approximating the echo canceller response separately for each phase, and quite independently of the other phases; this fact is expressed by Equations 2.67b and 2.67d. Furthermore, the values of the echo replica at the corresponding phases are obtained using the sampling at the same rate as that at the canceller input.

Note that, in the case of nonlinear echo and its cancellation with the use of the nonlinear transversal filter, we can apply the same argument stated above of approximating independently for each phase. More precisely, when we use the notation of Equation 2.53e for expressing the nonlinear echo replica $\hat{e}(k)$, we arrive at the following counterparts of Equations 2.67b and 2.67d:

$$\hat{e}(k'T) \cong \mathbf{a}^T(k'T)\mathbf{c}^{\#}(k'T) + \mathbf{u}_1^T(k'T)\mathbf{c}_N^{\#}(k'T) \tag{2.71a}$$

$$\begin{aligned}\hat{e}\left(\left(k' + \frac{1}{R}\right)T\right) &\cong \mathbf{a}^T\left(\left(k' + \frac{1}{R}\right)T\right)\mathbf{c}^{\#}\left(\left(k' + \frac{1}{R}\right)T\right)\\ &+ \mathbf{u}_1^T\left(\left(k' + \frac{1}{R}\right)T\right)\mathbf{c}_N^{\#}\left(\left(k' + \frac{1}{R}\right)T\right)\end{aligned} \tag{2.71b}$$

$$\cdots$$

$$\begin{aligned}\hat{e}\left(\left(k' + \frac{R-1}{R}\right)T\right) &\cong \mathbf{a}^T\left(\left(k' + \frac{R-1}{R}\right)T\right)\mathbf{c}^{\#}\left(\left(k' + \frac{R-1}{R}\right)T\right)\\ &+ \mathbf{u}_1^T\left(\left(k' + \frac{R-1}{R}\right)T\right)\mathbf{c}_N^{\#}\left(\left(k' + \frac{R-1}{R}\right)T\right)\end{aligned} \tag{2.71c}$$

where the corresponding vectors $\mathbf{a}(\cdot)$, $\mathbf{c}^{\#}(\cdot)$, $\mathbf{u}_1(\cdot)$, and $\mathbf{c}^{\#}_N(\cdot)$ now assume the form:

$$\mathbf{a}\left(\left(k' + \frac{l}{R}\right)T\right) = \left[a\left(\left(k' + \frac{l}{R}\right)T\right), a\left(\left(k' - 1 + \frac{l}{R}\right)T\right), \ldots, a\left(\left(k' - M + 1 + \frac{l}{R}\right)T\right)\right]^{\mathrm{T}} \tag{2.71d}$$

$$\mathbf{c}^{\#}\left(\left(k' + \frac{l}{R}\right)T\right) = \left[c_0^{\#}\left(\left(k' + \frac{l}{R}\right)T\right), c_1^{\#}\left(\left(k' + \frac{l}{R}\right)T\right), \ldots, c_{M-1}^{\#}\left(\left(k' + \frac{l}{R}\right)T\right)\right]^{\mathrm{T}} \tag{2.71e}$$

$$\begin{aligned}\mathbf{u}_1\left(\left(k' + \frac{l}{R}\right)T\right) &= U_1\left(\mathbf{u}\left(\left(k' + \frac{l}{R}\right)T\right)\right)\\ = U_1\Bigg(\Bigg[1, a\left(\left(k' + \frac{l}{R}\right)T\right)&a\left(\left(k' - 1 + \frac{l}{R}\right)T\right), a\left(\left(k' + \frac{l}{R}\right)T\right)\left(\left(k' - 2 + \frac{l}{R}\right)T\right), \ldots,\\ a\left(\left(k' - M + 2 + \frac{l}{R}\right)T\right)a\left(\left(k' - M + 1 + \frac{l}{R}\right)T\right), &\ldots, a\left(\left(k' + \frac{l}{R}\right)T\right)a\left(\left(k' - 1 + \frac{l}{R}\right)T\right)\\ \cdots a\left(\left(k' - M + 1 + \frac{l}{R}\right)T\right)\Bigg]^{\mathrm{T}}\Bigg)\end{aligned} \tag{2.71f}$$

and

$$\mathbf{c}_N^{\#}\left(\left(k'+\frac{l}{R}\right)T\right) = C_N\left(\left[c_{00}^{\#}\left(\left(k'+\frac{l}{R}\right)T\right), c_{01}^{\#}\left(\left(k'+\frac{l}{R}\right)T\right), c_{02}^{\#}\left(\left(k'+\frac{l}{R}\right)T\right), \ldots, c_{0,1,\ldots,M-1}^{\#}\left(\left(k'+\frac{l}{R}\right)T\right)\right]^{\mathrm{T}}\right), 0 < l < R-1 \quad (2.71g)$$

As before, $l = 0, 1, \ldots, R-1$ in Equations 2.71d to g correspond to the phases $\frac{0}{R}, \frac{1}{R}, \ldots, \frac{R-1}{R}$, respectively. Moreover, k' and T refer to the sampling of the transmitted data symbols. The vector $\mathbf{a}\left(\left(k'+\frac{l}{R}\right)T\right)$ corresponds to the vector $\mathbf{a}(k)$ defined by expression 2.5a, and the vector $\mathbf{c}^{\#}\left(\left(k'+\frac{l}{R}\right)T\right)$ to the vector $\mathbf{c}(k)$ defined by expression 2.5e, with the # superscript to underline the fact of its occurrence in the approximations 2.71a to c. Also, note that the elements of the vector $\mathbf{c}^{\#}\left(\left(k'+\frac{l}{R}\right)T\right)$ depend upon the time; that is, they are put into the form suitable for performing an adaptation analysis. The operations U_1 and C_N in Equations 2.71f and 2.71g, respectively, are operations of filtering those elements of given vectors, which a particular nonlinear echo canceller structure takes into account. The resulting vectors, obtained through performing the filtering operations U_1 and C_N contain only elements involved in the description of a nonlinear canceller constructed. Furthermore, note that the vectors $\mathbf{u}_1\left(\left(k'+\frac{l}{R}\right)T\right)$ and $\mathbf{c}_N^{\#}\left(\left(k'+\frac{l}{R}\right)T\right)$ correspond to the vectors $\mathbf{u}_1(k)$ and $\mathbf{c}_N(k)$, used for the first time in Equation 2.53b and Equation 2.53e, respectively. These vectors applied in Equations 2.71a to c are defined separately for each of the phases $\frac{0}{R}, \frac{1}{R}, \ldots, \frac{R-1}{R}$.

Putting Equations 2.71a to c into one equation, taking into account relation 2.68b, and for simplicity dropping the sampling interval T, we get

$$\hat{e}\left(k'+\frac{l}{R}\right) \cong \mathbf{a}^T\left(k'+\frac{l}{R}\right)\mathbf{c}^{\#}\left(k'+\frac{l}{R}\right) + \mathbf{u}_1^T\left(k'+\frac{l}{R}\right)\mathbf{c}_N^{\#}\left(k'+\frac{l}{R}\right), \; 0 \le l \le R-1 \quad (2.72a)$$

Using then a similar notation as that used in Equations 2.69a and 2.69b to the vectors occurring in Equation 2.72a, that is, denoting

$$\hat{e}_l(k') = \hat{e}\left(k'+\frac{l}{R}\right) \quad (2.72b)$$

$$\mathbf{a}_l(k') = \mathbf{a}\left(k' + \frac{l}{R}\right) \tag{2.72c}$$

$$\mathbf{c}_l(k') = \mathbf{c}^{\#}\left(k' + \frac{l}{R}\right) \tag{2.72d}$$

$$\mathbf{u}_{1,l}(k') = \mathbf{u}_1\left(k' + \frac{l}{R}\right) \tag{2.72e}$$

$$\mathbf{c}_{N,l}(k') = \mathbf{c}^{\#}\left(k' + \frac{l}{R}\right) \tag{2.72f}$$

we can rewrite Equation 2.72a as

$$\hat{e}_l(k') = \mathbf{a}_l^T(k') \cdot \mathbf{c}_l(k') + \mathbf{u}_{1,l}^T(k') \cdot \mathbf{c}_{N,l}(k'), 0 \le l \le R-1 \tag{2.72g}$$

In Equation 2.72g, the symbol of "equality" is used in place of the symbol of "approximately equal to." We thereby ignore the fact that Equation 2.72g represents only an approximation.

Equation 2.72g is illustrated in Figure 2.31; in this figure, the nonlinear echo canceller with the use of interleaved nonlinear transversal filters is presented for the case of four ($R = 4$) times higher sampling rate at the canceller output than at its input.

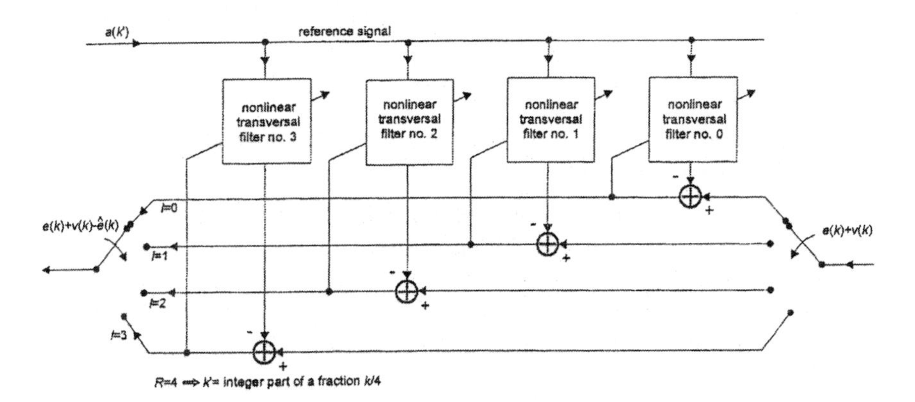

FIGURE 2.31 Nonlinear echo canceller consisting of interleaved adaptive nonlinear transversal filters for achieving higher output sampling than that at the input (here four times higher, $R = 4$).

Note from Figure 2.31 that the structure of the nonlinear echo canceller using nonlinear transversal filters is, in principle, the same as that in Figure 2.30, exploiting linear transversal filters. The only difference lies in the fact that we have here nonlinear transversal filters in place of linear ones.

Example 2.7

Let us assume that the interleaved adaptive nonlinear transversal filters of Figure 2.31 have the structure presented in Figure 2.24. For this structure, we illustrate here the form of the vectors $\mathbf{a}_l(k')$, $\mathbf{c}_l(k')$, $\mathbf{u}_{1,l}(k')$, and $\mathbf{c}_{N,1}(k')$. Keeping in mind Equations 2.71d to g and Equations 2.72c to f, we get

$$\mathbf{a}_l(k') = \left[a\left(k' + \frac{l}{4}\right), a\left(k' - 1 + \frac{l}{4}\right), a\left(k' - 2 + \frac{l}{4}\right)\right]^{\mathrm{T}}$$

$$= [a(k'), a(k' - 1), a(k' - 2)]^{\mathrm{T}}$$

$$\mathbf{c}_l(k') = \left[c_0^{\#}\left(k' + \frac{l}{4}\right), c_1^{\#}\left(k' + \frac{l}{4}\right), c_2^{\#}\left(k' + \frac{l}{4}\right)\right]^{\mathrm{T}}$$

$$\mathbf{u}_{1,l}(k') = \left[1, a\left(k' + \frac{l}{4}\right)a\left(k' - 1 + \frac{l}{4}\right), a\left(k' + \frac{l}{4}\right)a\left(k' - 2 + \frac{l}{4}\right)\right]^{\mathrm{T}}$$

$$= [1, a(k')a(k' - 1), a(k')a(k' - 2)]^{\mathrm{T}}$$

and

$$\mathbf{c}_{N,l}(k') = \left[c_{00}^{\#}\left(k' + \frac{l}{4}\right), c_{01}^{\#}\left(k' + \frac{l}{4}\right), c_{02}^{\#}\left(k' + \frac{l}{4}\right)\right]^{\mathrm{T}}, 0 \le l \le 3$$

Note that the form of the vectors $\mathbf{a}_l(k')$ and $\mathbf{u}_{1,l}(k')$ given above is simplified further using the relation 2.68b.

To start with passband echo cancellers, the second topic of this section, we first must introduce some new notions. We begin with the notion of an analytic signal, defining it as such a signal that has only positive frequency components in its spectrum. To get such a signal from a real-valued one, we need to use the so-called Hilbert transform filter,[40] or Hilbert transformer,[69] a linear filter with transfer function in the form

$$H(j\omega) = -j\mathrm{sgn}(\omega) \tag{2.73a}$$

where the "sign of" function, sgn(ω), is given by

$$\text{sgn}(\omega) = \begin{cases} 1 & \text{for} \quad \omega > 0 \\ 0 & \text{for} \quad \omega = 0 \\ -1 & \text{for} \quad \omega < 0 \end{cases} \tag{2.73b}$$

It follows from expressions 2.73a and 2.73b that the magnitude of $H(j\omega)$, $|H(j\omega)| = 1$, and the phase of $H(j\omega)$, $\varphi(j\omega) = -\frac{\pi}{2}$, radians for positive frequencies ($\omega > 0$) and $\varphi(j\omega) = \frac{\pi}{2}$ radians for negative frequencies ($\omega < 0$). In other words, we can say that the Hilbert transformer is in principle, a 90° phase shifter; this is illustrated in Figure 2.32.

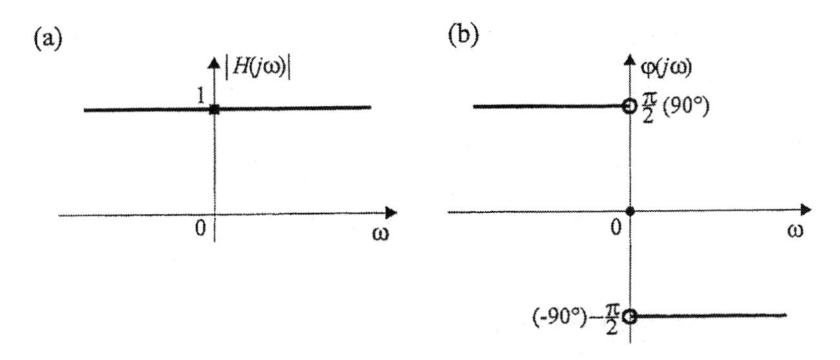

FIGURE 2.32 Sketch of the transfer function magnitude (a), and the transfer function phase (b) for the Hilbert transform filter.

Now let us take into account a real-valued continuous-time signal $x(t)$, and let the Fourier transform (see the defining Equation 1.31a) of this signal be $X(j\omega)$. Moreover, take into account a signal obtained by passing the signal $x(t)$ through the Hilbert transform filter; the Fourier transform of this signal will be $X(j\omega)H(j\omega) = -j\text{sgn}(\omega)X(j\omega)$. Furthermore, note that the resulting signal will be represented in the time-domain by a real-valued signal because the impulse response of the Hilbert transform filter is real-valued and given by

$$h(t) = \frac{1}{\pi t}, \quad -\infty \leq t \leq \infty \tag{2.74}$$

where t stands for the continuous time. With regard to the form of $h(t)$ see, for example, Reference 69 for more details. Thus, the product $X(j\omega)H(j\omega)$ corresponding to the convolution-integral in the time-domain will give as a result a real-valued signal in this domain. We will denote this signal by $x_H(t)$.

Further, having defined the signal $x_H(t)$ as the resulting signal of passing a real-valued signal $x(t)$ through the Hilbert transform filter, let us construct a complex-valued signal $x_a(t)$ as follows

$$x_a(t) = x(t) + jx_H(t) \tag{2.75a}$$

where the signals $x(t)$ and $x_H(t)$, according to the previous considerations, have the Fourier transforms denoted by $X(j\omega)$ and $X(j\omega)H(j\omega)$, respectively. The Fourier transform of the signal $x_a(t)$ can be expressed as

$$X_a(j\omega) = X(j\omega) + jX_H(j\omega) = X(j\omega) + jX(j\omega)H(j\omega) \tag{2.75b}$$

Using expression 2.73a in 2.75b gives

$$X_a(j\omega) = X(j\omega) + \mathrm{sgn}(\omega)X(j\omega) \tag{2.75c}$$

$$= \begin{cases} 2X(j\omega) & \text{for} \quad \omega > 0 \\ 0 & \text{for} \quad \omega < 0 \end{cases}$$

The resulting expressions, 2.75c show that the signal $x_a(t)$ given by Equation 2.75a is an analytic signal. According to our definition, it possesses only positive frequency components in its spectrum. Moreover, it follows from Equation 2.75a that the analytic signal is a complex-valued signal. We see also that this signal has an asymmetrical spectrum, in contrast to real-valued signals obeying the relation $X(-j\omega) = X^*(j\omega)$, where the asterisk means a complex-conjugate. That is, we have $|X(-j\omega)| = |X^*(j\omega))| = |X(j\omega)|$, $\omega > 0$, in the latter case, as opposed to the previous case, in which $|X_a(j\omega)| \neq |X_a(-j\omega)| = 0$, $\omega > 0$ holds. The realization of an analytic signal is through the use of the Hilbert transform filter providing the imaginary part of this signal. Knowing that the product $X(j\omega)H(j\omega)$ corresponds to the convolution-integral in the time-domain and using expression 2.74, we can write

$$x_H(t) = \frac{1}{\pi}\int_{-\infty}^{\infty}\frac{x(\tau)}{t-\tau}d\tau \tag{2.76}$$

Relation 2.75b can be also interpreted as getting an analytic signal (being a complex-valued signal) by passing a real-valued signal through a filter with the complex-valued impulse response. Observe from relation 2.75b that the Fourier transform of this complex-valued impulse response is equal to $1 + jH(j\omega)$; so it follows that the impulse response itself, using expression 2.74, is given by $\delta(t) + j\frac{1}{\pi t}$.

Assume now that the real-valued signal $x(t)$ taken into account in our considerations regarding the notion of the analytic signal is a bandpass signal. That is, $x(t)$ is a signal of which spectrum (the magnitude of its Fourier transform) is concentrated around some angular frequency ω_c, as illustrated in Figure 2.33. Then consider the analytic signal $x_a(t)$ resulting from a real-valued bandpass signal $x(t)$, of which spectrum is shown in Figure 2.33. It follows from expression 2.75c that this signal is a bandpass signal as well. Its spectrum is shown in Figure 2.34. This spectrum is asymmetrical (with respect to the "magnitude" axis), as previously mentioned. Moreover, the spectrum values at the corresponding frequencies are two times greater than those in Figure 2.33, according to relation 2.75c.

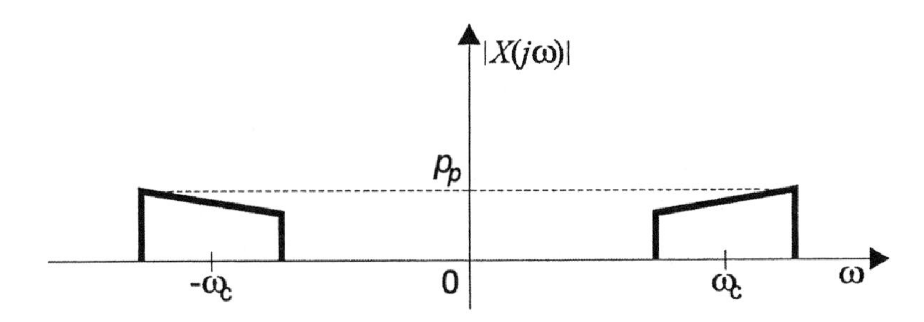

FIGURE 2.33 Spectrum of a real-valued bandpass signal.

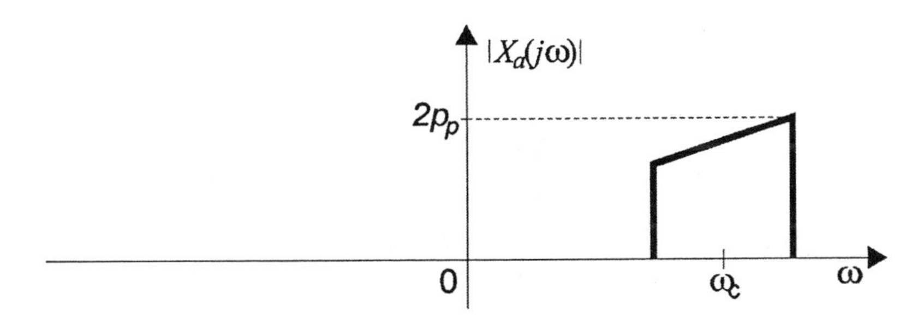

FIGURE 2.34 Spectrum of the analytic bandpass signal related with that in Figure 2.33.

In telecommunications,[24,69] a very popular means of representing bandpass signals is through their lowpass equivalents. Such lowpass equivalents are constructed by frequency shift to the zero frequency performed on the Fourier transform of the related analytic signal. Thus, considering a real-valued bandpass signal $x(t)$, which has a representation in the form of an analytic signal $x_a(t)$ with the Fourier transform $X_a(j\omega)$, we define its lowpass equivalent as

$$X_l(j\omega) \stackrel{df}{=} X_a(j(\omega + \omega_c)) \tag{2.77a}$$

The relation 2.77a is illustrated in Figure 2.35, showing the frequency shift of the signal spectrum from the angular frequency ω_c to the zero frequency.

The shift to the left in the frequency-domain is related with the multiplication by $\exp(-j\omega_c t)$ in the time-domain.[24] Hence, relation 2.77a can be rewritten equivalently in the time-domain as

$$x_l(t) \stackrel{df}{=} x_a(t)e^{-j\omega_c t} \tag{2.77b}$$

Further, substituting $x_a(t)$ given by Equation 2.75a into 2.77b, we get

$$x_l(t) = (x(t) + jx_H(t))e^{-j\omega_c t} \tag{2.77c}$$

which multiplied by $\exp(j\omega_c t)$, gives the following expression

$$x_a(t) = x(t) + jx_H(t) = x_l(t)e^{j\omega_c t} \tag{2.77d}$$

for the analytic signal versus the equivalent lowpass signal.

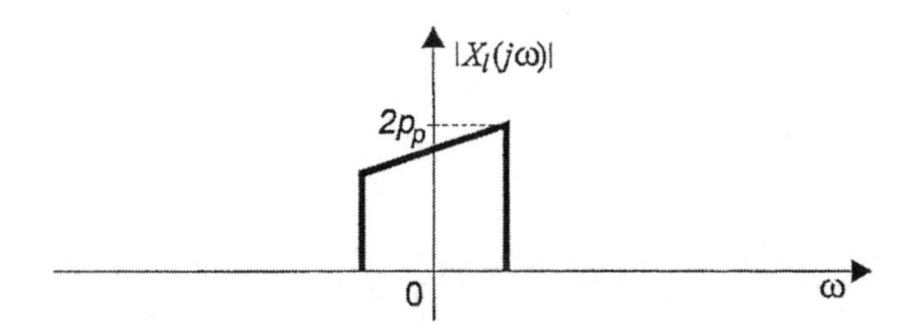

FIGURE 2.35 Spectrum of the equivalent lowpass signal.

In general, the equivalent lowpass signal $x_l(t)$ is not a real-valued signal because relation $X_l(-j\omega) = X_l^*(j\omega)$ or, equivalently, $X_a(j(-\omega + \omega_c)) = X_a^*(j(\omega + \omega_c))$, $\omega > 0$, does not hold. This fact is reflected in Figure 2.35, where the spectrum shown is asymmetrical with respect to the "absolute value of" axis (more precisely, a less restricted case of $|X_l(-j\omega)| \neq |X_l(j\omega)|$, $\omega > 0$, is shown in the figure). Considering in general the equivalent lowpass signal $x_l(t)$ as a complex-valued signal, we write down

$$x_l(t) = x_R(t) + jx_I(t) \tag{2.78}$$

for it, where $x_R(t)$ and $x_I(t)$ mean its real and imaginary part, respectively.

Substituting $x_l(t)$ given by Equation 2.78 into 2.77d, and equating then the real and imaginary parts on both sides of the resulting equation, we get

$$x(t) = x_R(t)\cos(\omega_c t) - x_I(t)\sin(\omega_c t) \tag{2.79a}$$

and

$$x_H(t) = x_R(t)\sin(\omega_c t) + x_I(t)\cos(\omega_c t) \tag{2.79b}$$

Note that Equation 2.79a provides a representation of a real-valued bandpass signal $x(t)$ through two real-valued lowpass signals $x_R(t)$ and $x_I(t)$ being respectively the real and imaginary parts of the complex-valued lowpass signal $x_l(t)$ (see Equation 2.78). The signal $x(t)$ can be viewed as a sum of two amplitude modulated signals $x_R(t)\cos(\omega_c t)$ and $-x_I(t)\sin(\omega_c t)$, that is, as the carrier $\cos(\omega_c t)$ modulated by the signal $x_R(t)$ plus the carrier $-\sin(\omega_c t) = \cos\left(\omega_c t + \frac{\pi}{2}\right)$ modulated by the signal $x_I(t)$. The phase difference between the above carriers is equal to $\frac{\pi}{2}$ radians, meaning that they are in phase quadrature. According to this, the signal $x_R(t)$ modulating the carrier $\cos(\omega_c t)$ is called the in-phase component, and the signal $x_I(t)$ modulating the carrier $\sin(\omega_c t)$ is called the quadrature component. Moreover, the modulated signal $x_R(t)\cos(\omega_c t)$ is called the in-phase signal, and the modulated signal $x_I(t)\sin(\omega_c t)$ is called the quadrature signal.

In the voiceband data transmission, a technique of modulation called the quadrature amplitude modulation (QAM) is used. It is convenient to describe this technique mathematically using the notions just introduced, such as the analytic signal, Hilbert transform filter, and equivalent lowpass representation. To this end, consider the form of the transmitted signal in a transmitter using the QAM technique. As in the pulse amplitude modulation technique in the baseband, applied in the digital subscriber loop, the pulse shape, that is, the impulse response of the transmit filter, is here a real-valued function, too. Hence, let us denote this impulse response by $p(t)$. Furthermore, a popular way[24] of describing the QAM technique is through consideration of the data symbols as being complex. Using this description here we can express the outgoing signal of the transmit filter in the following form as well,

$$\begin{aligned} y(t) &= \int_{-\infty}^{\infty} p(t-\tau)\cdot\left(\sum_{i=-\infty}^{\infty}(a(i)\delta(\tau-iT)+jb(i)\delta(\tau-iT))\right)d\tau \\ &= \sum_{i=-\infty}^{\infty} p(t(-iT))(a(i)+jb(i)) \end{aligned} \tag{2.80a}$$

which can be written equivalently as

$$y(t) = \sum_{i=-\infty}^{\infty} p(t - iT)\tilde{a}(i) \tag{2.80b}$$

where $\tilde{a}(i)$ means the ith complex-valued data symbol,

$$\tilde{a}(i) = a(i) + jb(i) \tag{2.80c}$$

In Equations 2.80b and 2.80c, T denotes the sampling period.

Equation 2.80a is illustrated in Figure 2.36.

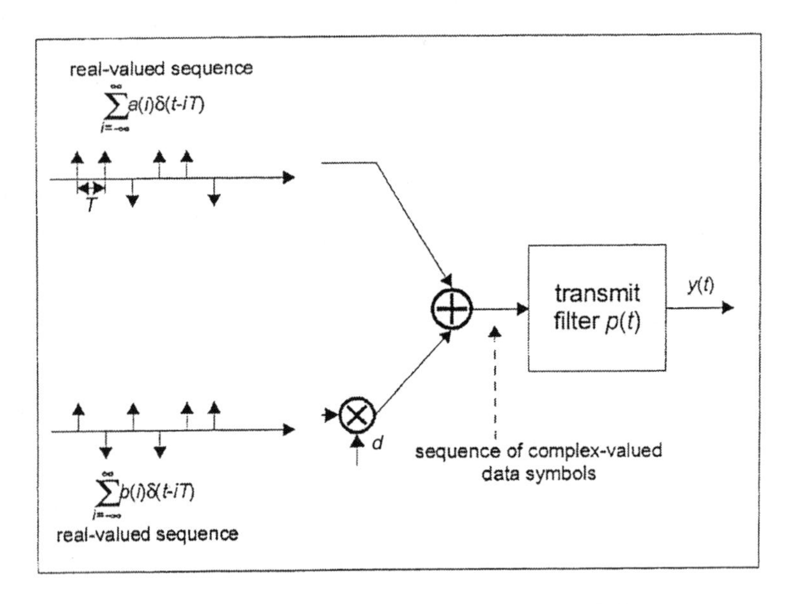

FIGURE 2.36 Illustration of applying a complex-valued data symbol sequence to a transmit filter having the real-valued impulse response.

To proceed further, let us simplify our considerations at this moment by assuming that a single complex-valued data symbol is transmitted, for example, that occurring at the time instant $iT = 0$. Then, $y(t)$ given by 2.80b simplifies to $p(t)(a(0) + jb(0)) = p(t)(a + jb)$, which is evidently a complex-valued signal for $b \neq 0$. The Fourier transform of this signal is equal to $P(j\omega)(a + jb)$, which does not obey the complex-conjugate equality that is $P(-j\omega)(a + jb) \neq (P(j\omega)(a + jb))^* = P^*(j\omega)(a - jb)$, $\omega > 0$. Note in this case, however, that the equality of the absolute values of the Fourier transforms $|P(-j\omega)(a + jb)| = |P^*(j\omega)(a - jb)| = |P(j\omega)(a + jb)|$ takes place because $|P^*(j\omega)| = |P(j\omega)|$ for real-valued impulse responses $p(t)$.

The pulse shapes $p(t)$ used in digital telecommunications are lowpass band-limited pulses. An ideal one is the pulse having the frequency characteristic (the Fourier transform)

$$P(j\omega) = \begin{cases} \frac{\pi}{B} & \text{for} \quad |\omega| < B \\ 0 & \text{for} \quad |\omega| \geq B \end{cases} \tag{2.81a}$$

or, equivalently, in the time-domain, the impulse response of the form

$$p(t) = \frac{\sin(Bt)}{Bt} \tag{2.81b}$$

where B means the pulse bandwidth. For more details regarding the pulse shape characterized by Equation 2.81 and other pulse shapes allowing the avoidance of the so-called intersymbol interference, see Reference 24.

For illustration, the pulse shape given by Equation 2.81b is sketched in Figure 2.37.

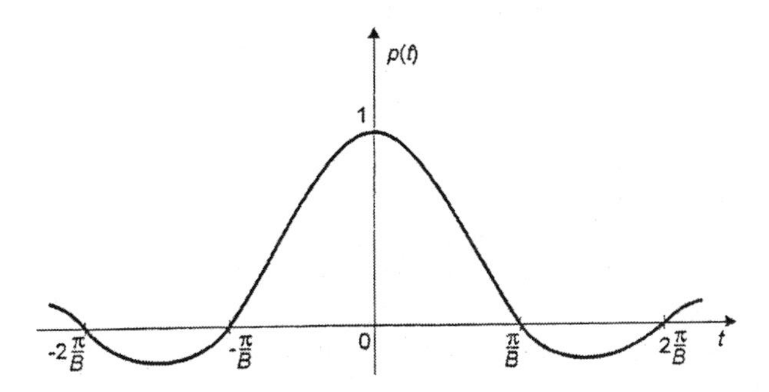

FIGURE 2.37 Sketch of the pulse shape $\frac{\sin(Bt)}{Bt}$.

Returning now to consideration of the signal $p(t)(a + jb)$, we see that, using in it the pulse shape given by Equation 2.81b or a related lowpass band-limited one, we get the resulting complex-valued signal possessing the spectrum concentrated at the zero frequency, such as the signal of which spectrum is presented in Figure 2.35. Then, if we multiply this signal by a complex exponential $\exp(j\omega_c t)$ with the angular frequency $\omega_c > B$, we get, evidently, according to Equation 2.77d, an analytic signal. This signal,

$$p(t)(a + jb)e^{j\omega_c t} \tag{2.82a}$$

will of course be a complex-valued bandpass signal with the spectrum located similarly as that shown in Figure 2.34. But what will be put into a bandpass channel for transmission, will be a real-valued bandpass signal, that is, a real part of Equation 2.82a. So this real-valued bandpass signal will have the form

$$\text{Re}(p(t)(a+jb)e^{j\omega_c t}) = ap(t)\cos(\omega_c t) - bp(t)\sin(\omega_c t) \qquad (2.82b)$$

where Re(·) means the operation of "taking a real part of a complex number." Furthermore, observe that the form of the signal given by Equation 2.82b is the same as that given by Equation 2.79a. Comparison of the above expressions shows that $x_R(t) = \text{Re}(p(t)(a + jb)) = ap(t)$ and that $x_I(t) = \text{Im}(p(t)(a + jb)) = bp(t)$, where Im(·) means the operation of "taking an imaginary part of a complex number." The spectrum of the signal 2.82b, a real-valued bandpass signal, is similar to that shown in Figure 2.33. That is, this spectrum has two parts symmetrically located about the zero frequency: one part around the carrier frequency ω_c and the second around the negative frequency $-\omega_c$. According to the terminology introduced, the lowpass signal $ap(t)$ modulating the carrier $\cos(\omega_c t)$ is the in-phase component, and the lowpass signal $bp(t)$ modulating the carrier $\sin(\omega_c t)$ is the quadrature component, respectively.

With the above explanation in mind, as regards the transmit filter output signal $y(t)$ in a simple form of the single pulse $(a + jb)\,p(t)$, consider its general form given by Equation 2.80a. The signal $y(t)$ given by 2.80a is a sum of lowpass signals $(a(i) + jb(i))\,p(t - iT)$ shifted in time by iT, $i = 0, \pm 1, \pm 2, \ldots$. To prove that the character of the frequency characteristic of this signal remains a lowpass one, a more advanced analysis is needed. In such an analysis,[24] one assumes that the transmitted data symbols $a(i) + jb(i)$ are random, so $y(t)$ given by Equation 2.80a represents outcomes of a complex-valued random process. Moreover, the notion of the power spectrum characterizing random processes, instead of the magnitude of the Fourier transform, is then used. With the assumption that the successive data symbols are uncorrelated, it is possible to show that the frequency characteristic of the power spectrum of a random process, of which outcomes are described by Equation 2.80a, is exclusively determined by the frequency characteristic of $|P(j\omega)|$. For more details, see, for example, Reference 24.

Multiplying the signal $y(t)$, given by Equation 2.80a by $\exp(j\omega_c t)$, we get

$$y_a(t) = \sum_{i=-\infty}^{\infty} p(t-iT)(a(i)+jb(i))e^{j\omega_c t} \qquad (2.83)$$

which is a sum of analytic signals. The expression $y_a(t)$ is also an analytic signal in the sense that the power spectrum related to it, as the representative of a random process, is nonzero only for the positive frequencies. This can easily be shown using the argument discussed above. Treating each $p(t - iT)$ together with $\exp(j\omega_c t)$ in Equation 2.83 as a deterministic multiplier of the random part $(a(i) + jb(i))$, we get for it, the Fourier transform in the form $P(j(\omega - \omega_c))\exp(-j\omega iT)$. The magnitude of this Fourier transform is equal to $|P(j(\omega - \omega_c))|$, and it corresponds to the spectrum $|P(j\omega)|$ shifted on the frequency axis to the point $\omega = \omega_c$. Because the power spectrum of the random process considered depends (as a function of frequency) exclusively upon the frequency characteristic of a pulse shape, here the frequency characteristic of the complex-valued pulse shape $p(t)\exp(j\omega_c t)$, this implies that the power spectrum frequency characteristic of $y_a(t)$ given by Equation 2.83 is fully determined by $|P(j(\omega - \omega_c))|$.

Taking a real part of $y_a(t)$ given by Equation 2.83, we obtain

$$\begin{aligned}\mathrm{Re}(y_a(t)) &= \sum_{i=-\infty}^{\infty} \mathrm{Re}(\tilde{a}(i))p(t-iT)\cos(\omega_c t) - \sum_{i=-\infty}^{\infty} \mathrm{Im}(\tilde{a}(i))p(t-iT)\sin(\omega_c t) \\ &= \sum_{i=-\infty}^{\infty} a(i)p(t-iT)\cos(\omega_c t) - \sum_{i=-\infty}^{\infty} b(i)p(t-iT)\sin(\omega_c t)\end{aligned} \tag{2.84}$$

Note that the real-valued bandpass signal $\mathrm{Re}(y_a(t))$ is presented in Equation 2.84 in two equivalent forms (one with the use of the complex-valued notation for the transmitted data symbols, $\tilde{a}(i)$).

A usual means of illustrating Equation 2.84 describing the quadrature amplitude modulation (QAM) used in the transmitter for transmitting in the voiceband channel is presented in Figure 2.38.

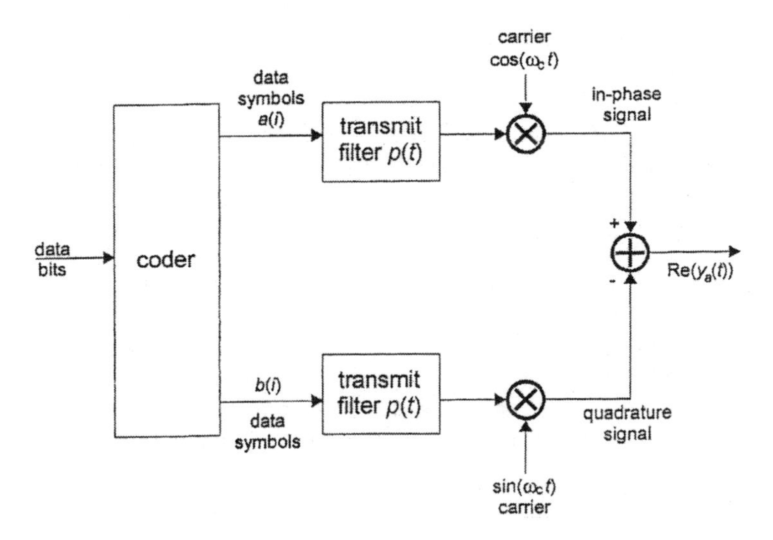

FIGURE 2.38 Illustration of structure of the QAM transmitter.

At this point, let us also illustrate a plot called a signal constellation, used in digital telecommunications for visualization of the signal alphabet. We know what the data symbols are. The set of all the data symbols available for transmission is called the signal alphabet. A baseband signal possesses the signal alphabet, which consists of real numbers, such as $\{-1,1\}$; note that this alphabet, $\{-1,1\}$, was used in the analysis presented in the previous section. The data symbols can also be complex numbers. To illustrate, consider the so-called 16-QAM that has the signal alphabet consisting of 16 complex numbers. Let this be the following set: $\{-3 - 3j, -3 - j, -3 + j, -3 + 3j, -1 - 3j, -1 - j, -1 + j, -1 + 3j, 1 - 3j, 1 - j, 1 + j, 1 + 3j, 3 - 3j, 3 - j, 3 + j, 3 + 3j\}$. Both the signal alphabets are illustrated in Figure 2.39; the plots presented in the figure are called signal constellations.

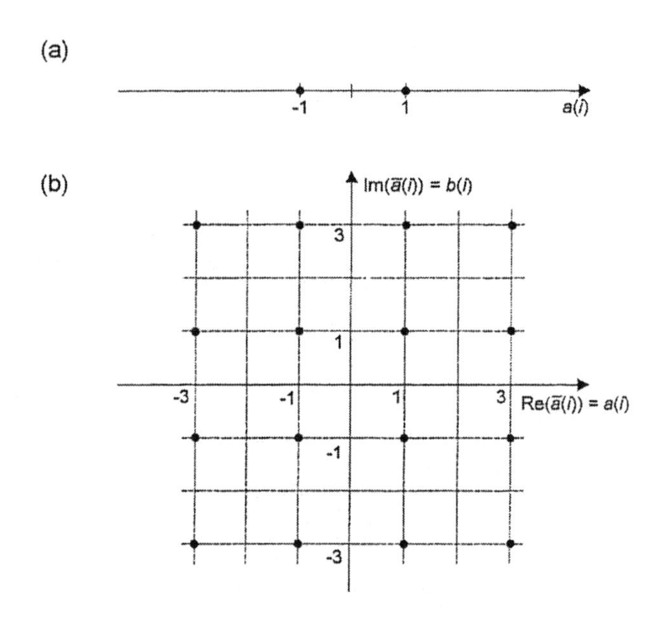

FIGURE 2.39 Signal constellations regarding the signal alphabets given in the text, (a) for baseband signal, and (b) for 16-QAM.

Observe from Figure 2.39a that the data symbols of a basisband signal lie on the real axis $a(i)$. This is in contrast to the case of the 16-QAM shown in Figure 2.39b, where the sixteen possible complex-valued data symbols are represented by the sixteen points on the complex plane. The encoder shown in Figure 2.38 performs in this case the mapping of four input bits, representing one of the $2^4 = 16$ different four-bit blocks, into one of the complex-valued data symbols shown in Fig.2.39b. Two of the input bits are mapped into the real part and two into the imaginary part of $\tilde{a}(i)$.

The first operation which must be performed in the QAM receiver is the recovery of the lowpass signal from the received bandpass signal. To do this, we can proceed as shown schematically in Figure 2.40; the correctness of the structure of Figure 2.40 will follow from the analysis presented below.

Observe that the received bandpass signal $z_b(t)$ in the QAM receiver of Figure 2.40 is a signal, obtained through performing the convolution operation of the transmitter output signal $\mathrm{Re}\left(y_a(t) \right)$ with a composite bandpass impulse response $r(t)$ of the communication channel and receive filter. That is we have

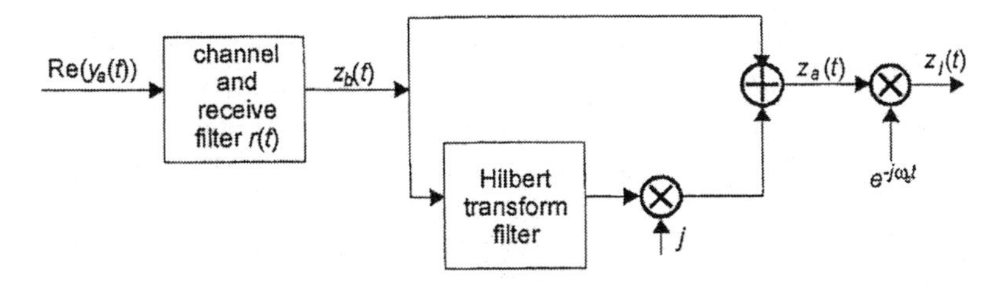

FIGURE 2.40 One of the possible schemes for performing signal demodulation in the QAM receiver.

$$z_b(t) = \int_{-\infty}^{\infty} r(\tau) \cdot \mathrm{Re}(y_a(t-\tau))d\tau \tag{2.85a}$$

where the subscript at $z_b(t)$ stands for emphasizing the bandpass character of this signal. Assuming that the communication channel is linear, the composite impulse response $r(t)$ in Equation 2.85a is determined by another convolution integral as

$$r(t) = \int_{-\infty}^{\infty} r_r(\tau)r_c(t-\tau)d\tau = \int_{-\infty}^{\infty} r_c(\tau)r_r(t-\tau)d\tau \tag{2.85b}$$

where $r_c(\cdot)$ stands for the channel impulse response, and $r_r(\cdot)$ for the receive filter impulse response, respectively. To get the composite characteristics of the communication channel and receive filter in the time and frequency domains, see Figure 2.41.

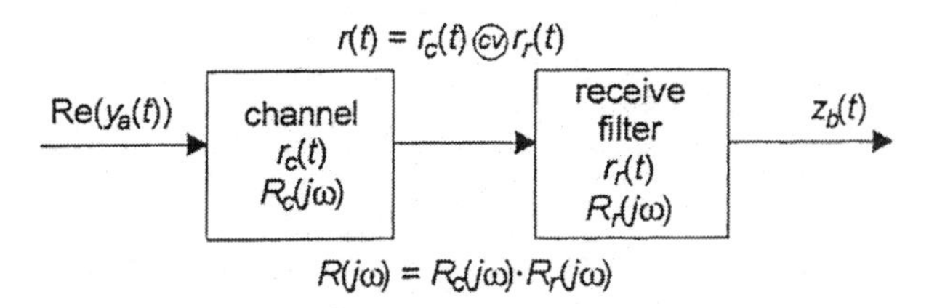

FIGURE 2.41 Getting the composite impulse response or composite transfer function for the communication channel together with the receive filter.

In our consideration, we assume that both the impulse responses $r_c(\cdot)$ and $r_r(\cdot)$ in Equation 2.85b are real-valued, and that they have bandpass character about the carrier frequency. Then, of course, the composite impulse response $r(t)$ is also real-valued, having the bandpass character as well.

Observe from Figure 2.40 that the real-valued bandpass signal $z_b(t)$ is filtered through the Hilbert transform filter to get the imaginary part of the analytic signal $z_a(t)$ (see the general relation given by expression 2.75a). Both the $\mathrm{Re}(z_a(t)) = z_b(t)$ and $j\mathrm{Im}(z_a(t))$ are added to each other, resulting in $z_a(t)$, which is then multiplied by $\exp(-j\omega_c t)$. Consequently, we get the lowpass signal $z_l(t)$ at the output. To see this, note that because $r(t)$ being real-valued, we can rewrite Equation 2.85a as

$$z_b(t) = \mathrm{Re}\left(\int_{-\infty}^{\infty} r(\tau)y_a(t-\tau)d\tau\right) \tag{2.85c}$$

Substituting then $y_a(\cdot)$ given by Equation 2.83 into 2.85c, we get from the latter equation

$$\begin{aligned} z_b(t) &= \mathrm{Re}\left(\int_{-\infty}^{\infty} r(\tau)\sum_{i=-\infty}^{\infty} p(t-\tau-iT)(a(i)+jb(i))e^{j\omega_c(t-\tau)}d\tau\right) \\ &= \mathrm{Re}\left(\sum_{i=-\infty}^{\infty}(a(i)+jb(i))\int_{-\infty}^{\infty} r(\tau)p(t-\tau-iT)e^{j\omega_c(t-\tau)}d\tau\right) \end{aligned} \tag{2.85d}$$

Furthermore, by introducing in Equation 2.85d a new equivalent complex-valued impulse response

$$m(t) = \int_{-\infty}^{\infty} r(\tau)p(t-\tau)e^{-j\omega_c z}d\tau \tag{2.85e}$$

we can rewrite Equation 2.85d in the following form:

$$z_b(t) = \mathrm{Re}\left(\sum_{i=-\infty}^{\infty} m(t-iT)(a(i)+jb(i))e^{j\omega_c t}\right) \tag{2.85f}$$

To proceed further, consider $r(\tau)$ together with $\exp(-j\omega_c z)$ in Equation 2.85e, and recall from earlier discussions in this section that the shift to the left in the frequency-domain is related with the multiplication by $\exp(-j\omega_c t)$ in the time-domain. Then, taking the Fourier transforms on both sides of Equation 2.85e gives

$$M(j\omega) = R(j(\omega+\omega_c))\cdot P(j\omega) \tag{2.85g}$$

From Equation 2.85g, it is evident that $M(j\omega)$ is a lowpass transfer function because the bandpass transfer function $R(j\omega)$ is shifted by ω_c to the dc frequency, giving the lowpass transfer function $R(j(\omega + \omega_c))$. And, finally, a result of multiplication of the lowpass $R(j(\omega + \omega_c))$ by lowpass $P(j\omega)$ remains lowpass.

Compare now the expression under the operation Re(·) in Equation 2.85f with the expression for $y_a(t)$ given by Equation 2.83. Note that the form of the expressions is the same, and $m(t - iT)$ in Equation 2.85f, corresponding to $p(t - iT)$ in Equation 2.83, is lowpass, similarly as the latter. By virtue of the discussion underlying the expression 2.83, the sum under the Re(·) operation in Equation 2.85f is a sum of analytic signals. Moreover, this sum is itself an analytic signal, and is given by

$$z_a(t) = \sum_{i=-\infty}^{\infty} m(t-iT)(a(i)+jb(i))e^{j\omega_c t} \tag{2.85h}$$

To get the lowpass signal $z_l(t)$ from the analytic signal $z_a(t)$, one multiplies the latter by $\exp(-j\omega_c t)$. Note that this operation is mathematically expressed by Equation 2.77b. As a result, one obtains

$$z_l(t) = \sum_{i=-\infty}^{\infty} m(t-iT)(a(i)+jb(i)) \tag{2.85i}$$

The expression $z_l(t)$ given by Equation 2.85i is a complex-valued lowpass signal. Furthermore, because as it follows from Equation 2.85e the impulse response $m(t)$ is, in general, complex-valued, Equation 2.85i represents a convolution (in the discrete time) of the complex-valued signal $a(i)+jb(i)$ with the complex-valued samples of the impulse response $m(t)$; more precisely, with $m(t - iT)$, $i = 0, \pm1, \pm2, \ldots$, where t is not the discrete but continuous time. Furthermore, each of the components of the sum in Equation 2.85i is obtained as illustrated in Figure 2.42.

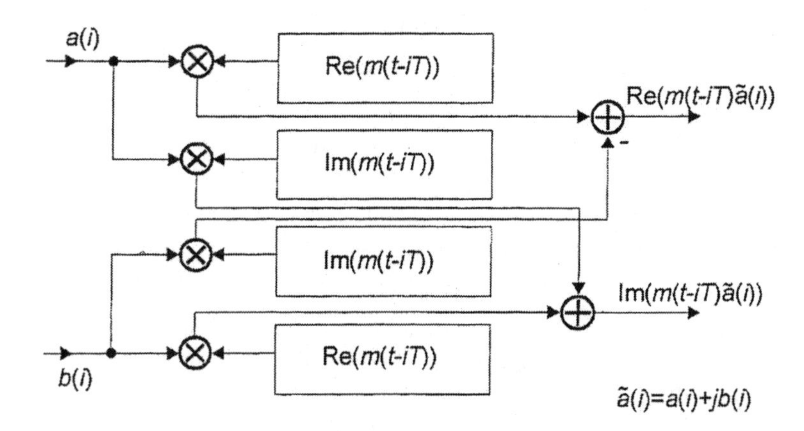

FIGURE 2.42 Illustration of the means of calculation of components in sum 2.85i.

Note that expression 2.85i for the signal $z_l(t)$ can be also rewritten in the following form:

$$z_l(t) = \mathrm{Re}(z_l(t)) + j\mathrm{Im}(z_l(t)) \tag{2.85j}$$

$$= \left(\sum_{i=-\infty}^{\infty} \mathrm{Re}(m(t-iT))a(i) - \sum_{i=-\infty}^{\infty} \mathrm{Im}(m(t-iT))b(i)\right)$$

$$+ j\left(\sum_{i=-\infty}^{\infty} \mathrm{Re}(m(t-iT))b(i) + \sum_{i=-\infty}^{\infty} \mathrm{Im}(m(t-iT))a(i)\right)$$

It follows from Equation 2.85j that the complex-valued signal $z_l(t)$ can be viewed as resulting from performing four real-valued convolutions, having the corresponding shares in its real and imaginary parts. This is shown in Figure 2.43, where the notation of Figure 2.44 is used for the convolution sum of the type occurring in Equation 2.85j.

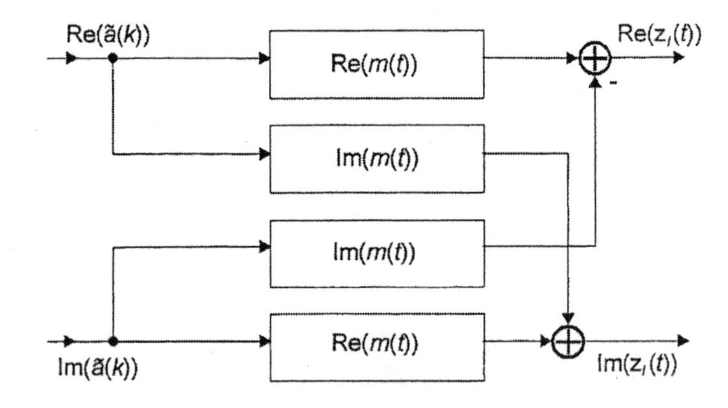

FIGURE 2.43 Getting the signal $z_l(t)$ by performing four real-valued convolutions.

x(k) → h(t) → z(t) ⇔ $z(t) = \sum_{i=-\infty}^{\infty} h(t-iT)x(i)$

x(k), h(t), and z(t) real-valued

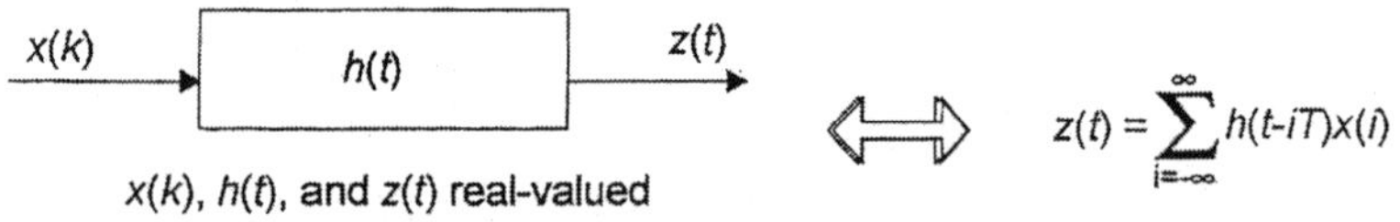

FIGURE 2.44 Graphical notation used for convolution sums occurring in Equation 2.85j.

After the discussion of the structure of Figure 2.40, note that a similar scheme can be used to describe the passband echo path. The only difference between both the schemes will lie in the interpretation of the word "channel" in the first block on the left-hand side of Figure 2.40. In the case of considering the echo path, it will simply mean an "echo channel." Furthermore, referring, for example, to References 24 and 40, we assume in our consideration that a signal taken for performing the echo cancellation is the one denoted by $z_a(t)$ in Figure 2.40.

Using notational convention introduced previously, let us now rewrite the signal $z_a(t)$, expressed by Equation 2.85h, in the form

$$e_a(t) = \sum_{i=-\infty}^{\infty} g(t-iT)(a(i)+jb(i))e^{j\omega_c t} \tag{2.86}$$

for the case of the passband echo channel. In Equation 2.86, $e_a(t)$ means the analytic signal of the echo channel, and $g(t)$ denotes its equivalent lowpass impulse response ($g(t)$ corresponds to $m(t)$, and the latter is an equivalent lowpass impulse response (see the explanation regarding Equation 2.85g).

Note that Equation 2.86 can be rewritten for the discrete-time $t = kT$ as

$$e_a(k) = \sum_{i=-\infty}^{\infty} g(k-i)(a(i)+jb(i))e^{j\omega_c kT} \tag{2.87}$$

where the sampling period T is dropped at $e_a(k) = e_a(kT)$ and $g(k-i) = g(kT-iT)$. Furthermore, substituting a new variable $i' = k - i$ into Equation 2.87, rearranging, and dropping finally the prime in i', we get from Equation 2.87

$$e_a(k) = \sum_{i=-\infty}^{\infty} g(i)(a(k-i)+jb(k-i))e^{j\omega_c kT} \tag{2.88}$$

The kind of characteristic represented by $g(i)$ allows[40] us to approximate the expression

$$\sum_{i=-\infty}^{\infty} g(i)(a(k-i)+jb(k-i)) \tag{2.89a}$$

in Equation 2.88 by a complex-valued transversal filter representation with the complex-valued coefficients and complex-valued data symbols. That is, we write

$$\sum_{i=-\infty}^{\infty} g(i)(a(k-i)+jb(k-i)) \cong \sum_{i=0}^{M_e-1} g(i)(a(k-i)+jb(k-i)) \tag{2.89b}$$

Using then Equation 2.89b for approximation of Equation 2.88, we arrive at

$$e_a(k) \cong \sum_{i=0}^{M_e-1} g(i)(a(k-i)+jb(k-i))e^{j\omega_c kT} \tag{2.90}$$

A good choice for a means of cancelling the echo signal described by Equation 2.90 is the use of a canceller having the description in a similar form. Let us write the corresponding relation for this canceller as

$$\hat{e}_a(k) = \sum_{i=0}^{M_{\hat{e}}-1} c(i)(a(k-i)+jb(k-i))e^{j\omega_c kT} \tag{2.91}$$

where $c(i)$, $i = 0, 1, 2, \ldots, M_{\hat{e}} - 1$ represent the complex-valued filter coefficients.

Having the expressions describing the bandpass echo channel and echo canceller, we can illustrate a means of cancellation in the case of bandpass voiceband channel. This is shown in Figure 2.45. In Figure 2.45a, $\mathrm{Re}(v_a(t))$ means the received real-valued signal, except for the echo signal; the latter signal is denoted by $\mathrm{Re}(e_a(t))$ in the figure. We assume that the effect of filtering through the received filter is already included in the signals mentioned above. The sum of these signals, that is, $\mathrm{Re}(e_a(t) + v_a(t))$, provides an input signal to the so-called phase splitter,[24] whose behavior is explained in Figure 2.45b. As shown in this figure, the task of the phase splitter is to provide an analytic signal from a real-valued one; its implementation is through the use of the Hilbert transform filter. For more explanation, compare the scheme of Figure 2.45a with that presented in Figure 2.40. The complex-valued transversal filter in Figure 2.45a works with the input signal as a complex-valued data symbol stream provided by the encoder shown in Figure 2.38.

The output signal of the complex-valued transversal filter in Figure 2.45a is given by

$$\hat{e}(k) = \sum_{i=0}^{M_{\hat{e}}-1} (\mathrm{Re}(c(i)) + j\mathrm{Im}(c(i)))(a(k-i)+jb(k-i)) \tag{2.92}$$

It is clearly seen from Equation 2.92 that the coefficients $c(i)$ of the transversal filter considered are complex-valued. The implementation of the complex-valued transversal filter having description in the form given by Equation 2.92 will look similar to that presented in Figure 2.43. For more detailed illustration, we redraw the scheme of the above figure for the transversal filter; it is shown in Figure 2.46.

Multiplying the output signal of the complex-valued transversal filter of Figure 2.45, $\hat{e}(k)$, by $\exp(j\omega_c kT)$ gives as a result the sampled analytic signal 2.91, which is subtracted from the sampled received analytic signal $v_a(k) + e_a(k)$. The resulting signal, $v_a(k) + e_a(k) - \hat{e}_a(k)$, is processed further in the QAM receiver.

The frequency characteristic of the complex-valued transversal filter in Figure 2.45 is lowpass because the complex-valued impulse response described by the coefficients $c(i)$, transformed into the frequency-domain, has such a character. Similarly, the data symbol stream $\{\tilde{a}(k)\}$ has a lowpass frequency characteristic. However,

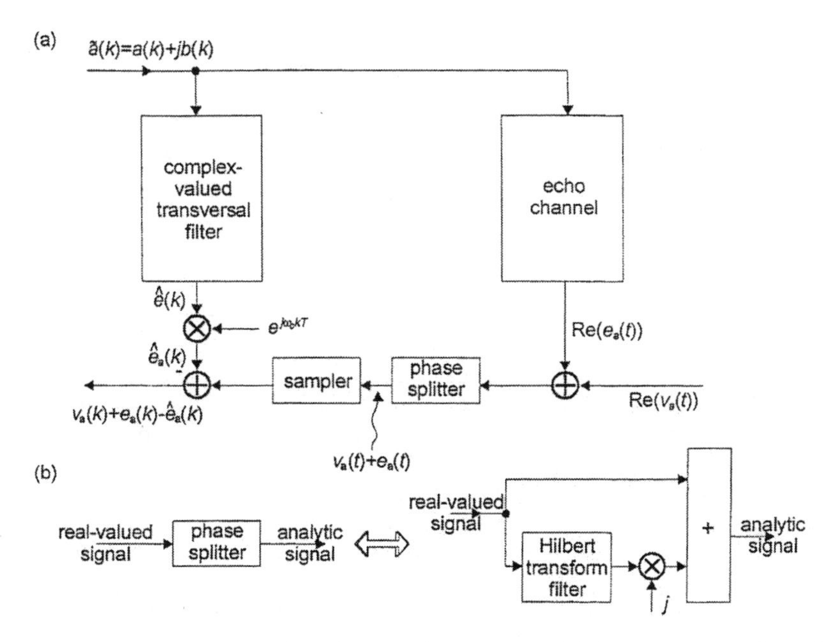

FIGURE 2.45 (a) Cancelling the voiceband channel echo with the use of a complex-valued transversal filter; (b) explanation of the function of the phase splitter introduced in the scheme above.

it is possible to interpret Equation 2.90 describing the echo channel in another way. The new interpretation leads to applying a complex-valued passband transversal filter, the point of which will be explained in more detail in what follows.

Consider again Equation 2.90, and rewrite it in the following form:

$$e_a(k) \cong \sum_{i=0}^{M_e-1} g(i)(a(k-i)+jb(k-i))e^{j\omega_c(k-i+i)T} \tag{2.93a}$$

Rearranging in Equation 2.93a gives

$$e_a(k) \cong \sum_{i=0}^{M_e-1} g(i)e^{j\omega_c iT}(a(k-i)+jb(k-i))e^{j\omega_c(k-i)T} \tag{2.93b}$$

Then introducing in Equation 2.93b an equivalent bandpass impulse response of the echo channel given by the coefficients

$$\breve{g}(k) = g(k)e^{j\omega_c kT}, k = 0, 1, \ldots, M_e - 1 \tag{2.93c}$$

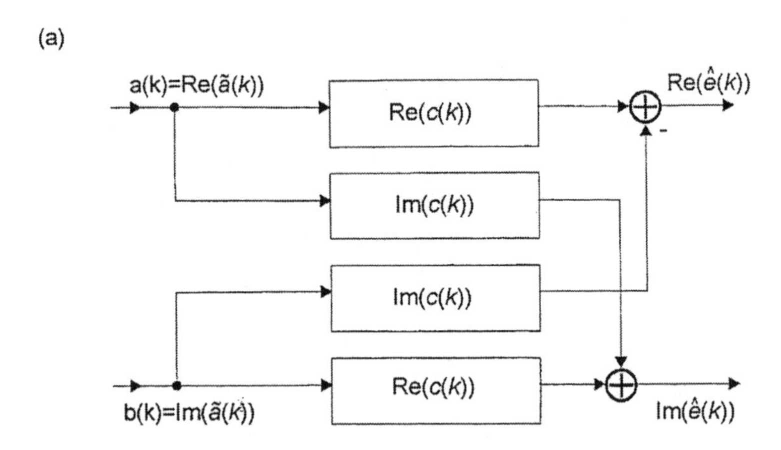

FIGURE 2.46 (a) Implementation of the complex-valued transversal filter of Figure 2.45 through four real-valued transversal filters; (b) graphical notation used for convolution sums occurring in Equation 2.92.

and the so-called rotated data symbols[24] given by

$$\breve{a}(k) = \tilde{a}(k)e^{j\omega_c kT} = (a(k) + jb(k))e^{j\omega_c kT} \tag{2.93d}$$

we arrive at

$$e_a(k) \cong \sum_{i=0}^{M_e - 1} \breve{g}(i)\breve{a}(k-i) \tag{2.93e}$$

Hence, the corresponding equation for an equivalent complex-valued passband transversal filter will have the form

$$\hat{e}_a(k) \cong \sum_{i=0}^{M_{\hat{e}} - 1} \breve{c}(i)\breve{a}(k-i) \tag{2.94a}$$

where the equivalent passband complex-valued filter coefficients will be given by

$$\breve{c}(k) = c(k)e^{j\omega_c kT}, k = 0, 1, \ldots, M_{\hat{e}} - 1 \tag{2.94b}$$

At this point, a notational remark: the equivalent complex-valued passband transversal filter can be also given a name of an equivalent complex-valued bandpass transversal filter, because its impulse response is a bandpass one, and it produces a bandpass signal. Furthermore, note that Equation 2.94a, describing the passband transversal filter, has the same form as Equation 2.65a for the baseband transversal filter, except that $\hat{e}_a(k)$ is a sampled analytic signal.

A cancelling scheme equivalent to that in Figure 2.45a, with the use of the passband transversal filter, is presented in Figure 2.47.

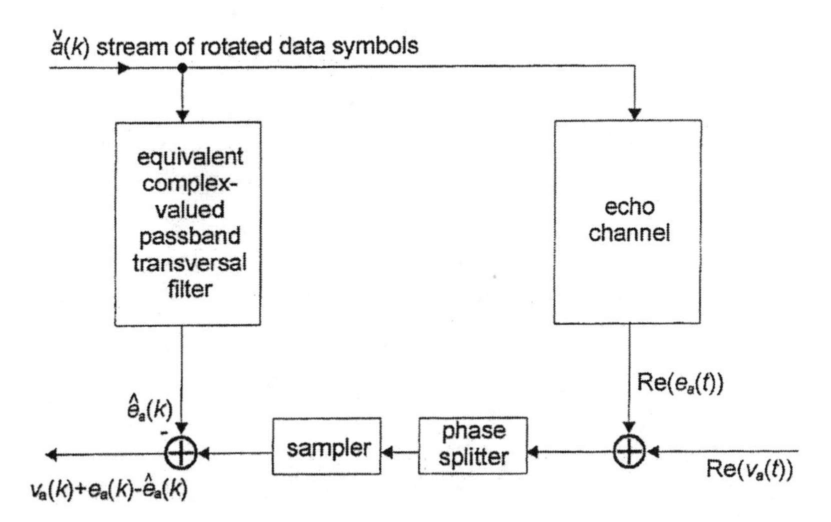

FIGURE 2.47 Cancelling scheme with the use of a passband transversal filter.

Similarly as in the baseband case, the timing recovery in the passband environment needs to implement an echo canceller that has the output sampling rate higher than that at its input. Assume that the canceller output sampling rate is R times higher than its input sampling rate. When the canceller input sampling period equals, say, T, we can distinguish R phases, $0\frac{T}{R}, 1\frac{T}{R}, \ldots, (R-1)\frac{T}{R}$ in this period. At these phases, the sampling is performed at the canceller output. Furthermore, we use the same argument as before of approximating the canceller output response independently at each of the phases. Exploiting the relation 2.68b and using a similar notation to that in expressions 2.69a and 2.69b, that is

$$c_l(k) = c^{\#}\left(k + \frac{l}{R}\right), 0 \le l \le R - 1 \tag{2.95a}$$

and

$$\hat{e}_{a,l}(k) = \hat{e}\left(k + \frac{l}{R}\right) \tag{2.95b}$$

we can write an equivalent of Equation 2.91 for the corresponding phases as

$$\hat{e}_{a,l}(k) = \sum_{i=0}^{M_{\hat{e}}-1} c_l(i)\tilde{a}(k-i)e^{j\omega_c\left(k+\frac{l}{R}\right)T}, 0 \le l \le R-1 \tag{2.95c}$$

In Equations 2.95, k refers to sampling with the rate $\frac{1}{T}$. Moreover, the superscript # at $c\left(k + \frac{l}{R}\right)$ in Equation 2.95a refers to a similar type of approximation as that shown in Equation 2.67c.

A similar relation to that given by Equation 2.95c can be also written for the echo canceller using an equivalent structure with the passband transversal filter. To this end, rewrite Equation 2.95c in the following form:

$$\hat{e}_{a,l}(k) = \sum_{i=0}^{M_{\hat{e}}-1} c_l(i)\tilde{a}(k-i)e^{j\omega_c\left(k+\frac{l}{R}-i+i\right)T} \tag{2.96a}$$

Rearranging then the expressions under the sum symbol in Equation 2.96a gives

$$\hat{e}_{a,l}(k) = \sum_{i=0}^{M_{\hat{e}}-1} c_l(i)e^{j\omega_c\left(i+\frac{l}{R}\right)T}\tilde{a}(k-i)e^{j\omega_c(k-i)T} \tag{2.96b}$$

And introducing in Equation 2.96b a notation similar to that in expression 2.94b for the coefficients of the equivalent passband filters for the corresponding phases $0\frac{T}{R}, 1\frac{T}{R}, \ldots, (R-1)\frac{T}{R}$, that is, denoting them by

$$\breve{c}_l(k) = c_l(k)e^{j\omega_c\left(k+\frac{l}{R}\right)T}, k = 0, 1, \ldots, M_{\hat{e}} - 1, 0 \le l \le R-1 \tag{2.96c}$$

and using also the notation of Equation 2.93d for the rotated data symbols, we arrive at

$$\hat{e}_{a,l}(k) = \sum_{i=0}^{M_{\hat{e}}-1} \breve{c}_l(i)\breve{a}(k-i), 0 \le l \le R-1 \tag{2.96d}$$

Both Equations 2.95c and 2.96d are illustrated through canceller implementations with interleaved transversal filters for four phases, $R = 4$, in Figure 2.48 and Figure 2.49, respectively. Observe that the form of Equation 2.96d is identical with that in Equation 2.70b for the baseband case, except that $\hat{e}_{a,l}(k)$ in Equation 2.96d is a sampled analytic signal and $\breve{c}_l(i), i = 0, 1, \ldots, M_{\hat{e}}-1$, represent coefficients of a passband transversal filter. The same form of the expressions mentioned is the reason the structures presented in Figures 2.30 and 2.49 look similar. The difference lies in the type of transversal filter used (lowpass or passband) and the form of the input data stream (baseband real-valued or passband complex-valued rotated data symbols).

Quite another form has the structure presented in Figure 2.48. Here, the interleaved lowpass transversal filters are used, of which the outputs are modulated by the carrier expressions $\exp(j\omega_c(k+\frac{l}{R})T)$, corresponding to the phases $0\frac{T}{R}, 1\frac{T}{R}, \ldots, (R-1)\frac{T}{R}$. Input data stream to the transversal filters are the complex-valued data symbols $\tilde{a}(k) = a(k) + jb(k)$.

Finally, we draw the reader's attention to the fact that k and k' used in Figures 2.30, and 2.31 have the opposite meaning of k and k' in Figures 2.48 and 2.49.

To consider cancellation of nonlinear passband echo, we need to use a continuous-time version of the Volterra series[1] for the underlying analysis. Such a series for nonlinear systems independent of time has, in analogy to Equations 1.1 and 1.3, the following form:

$$y(t) = h^{(0)} + \int_{-\infty}^{\infty} h^{(1)}(\tau)x(t-\tau)d\tau + \int_{-\infty}^{\infty}\int_{-\infty}^{\infty} h^{(2)}(\tau_1,\tau_2)x(t-\tau_1)x(t-\tau_2)d\tau_1 d\tau_2 \tag{2.97}$$
$$+ \int_{-\infty}^{\infty}\int_{-\infty}^{\infty}\int_{-\infty}^{\infty} h^{(3)}(\tau_1,\tau_2,\tau_3)x(t-\tau_1)x(t-\tau_2)x(t-\tau_3)d\tau_1 d\tau_2 d\tau_3 + \ldots$$

where t means a continuous time. Moreover, $y(t)$ and $x(t)$ in Equation 2.97 stand for the system output and system input signal, respectively.

With regard to the passband channel shown in Figure 2.40, assume that this channel behaves as a nonlinear system, which can be described by the time-independent, that is, stationary Volterra series (2.97). Furthermore, assume that the dc component $h^{(0)}$ does not occur here. Then, Equation 2.85a can be extended to the nonlinear case as

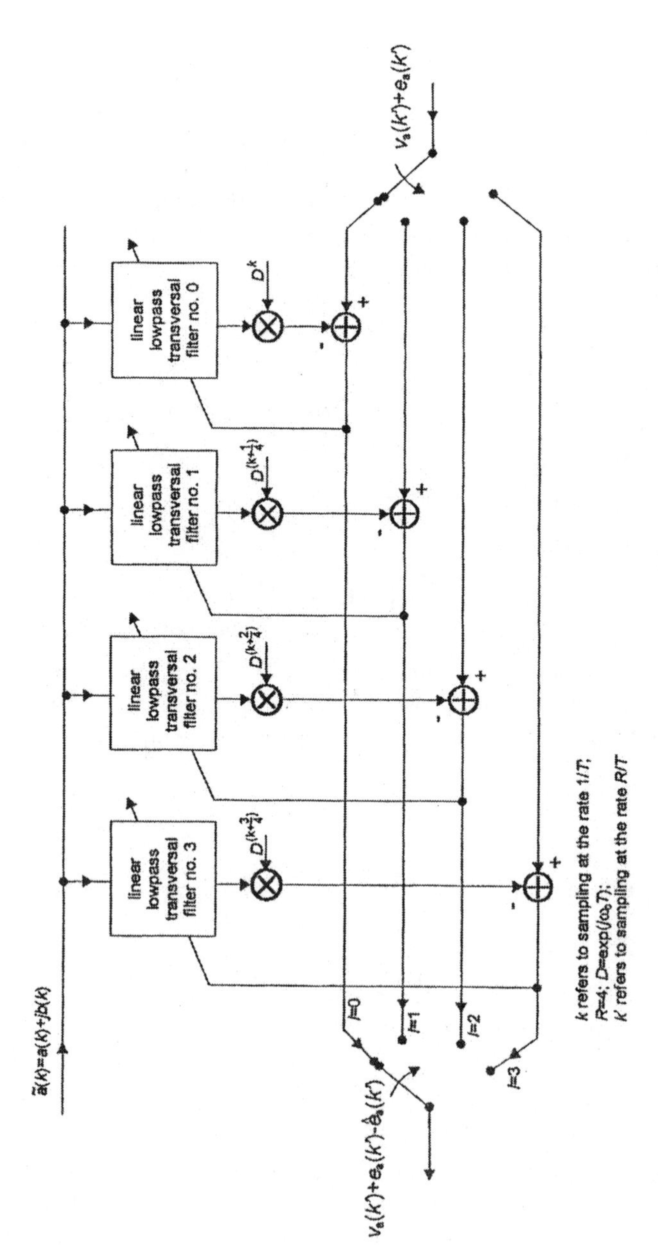

FIGURE 2.48 Interleaved adaptive linear lowpass transversal filters for achieving higher canceller output sampling rate in cancellation of passband echo.

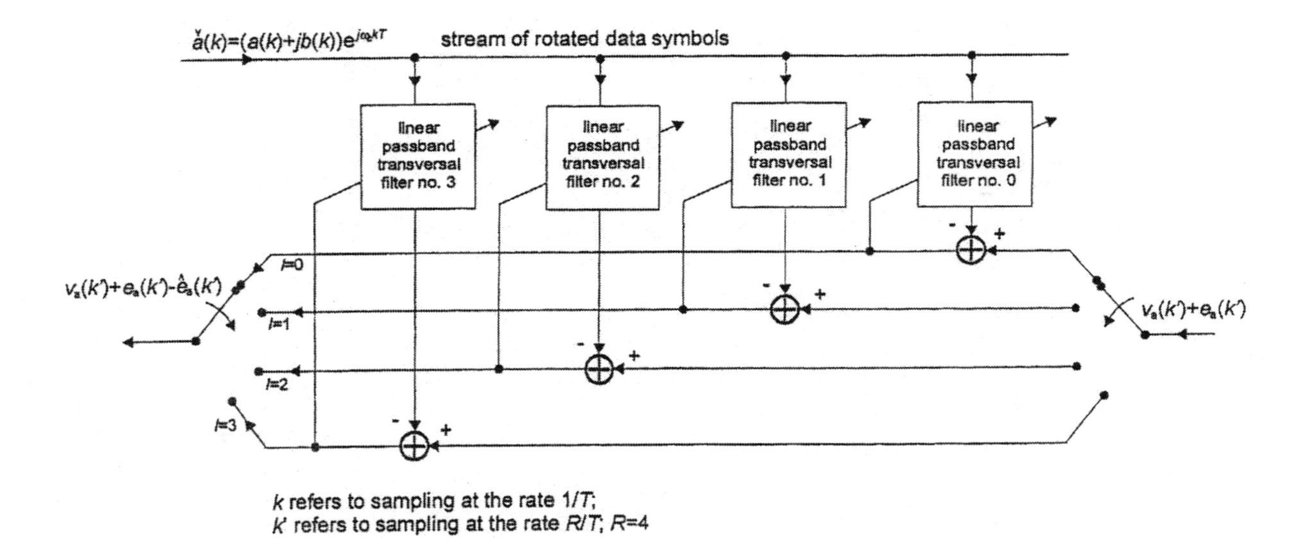

FIGURE 2.49 Interleaved adaptive linear passband transversal filters for achieving higher canceller output sampling rate in cancellation of passband echo.

$$z_b(t) = \int_{-\infty}^{\infty} r^{(1)}(\tau)\mathrm{Re}(y_a(t-\tau))d\tau$$

$$+ \int_{-\infty}^{\infty}\int_{-\infty}^{\infty} r^{(2)}(\tau_1, \tau_2)\mathrm{Re}(y_a(t-\tau_1))\mathrm{Re}(y_a(t-\tau_2))d\tau_1 d\tau_2$$

$$+ \int_{-\infty}^{\infty}\int_{-\infty}^{\infty}\int_{-\infty}^{\infty} r^{(3)}(\tau_1, \tau_2, \tau_3)\mathrm{Re}(y_a(t-\tau_1))\mathrm{Re}(y_a(t-\tau_2))\mathrm{Re}(y_a(t-\tau_3))d\tau_1 d\tau_2 d\tau_3 + \ldots \tag{2.98a}$$

where $r^{(1)}(\tau)$, $r^{(2)}(\tau_1, \tau_2)$, $r^{(3)}(\tau_1, \tau_2, \tau_3)$, mean the nonlinear impulse responses of the first, second, and third order, respectively, of the communication channel and receive filter connected with each other in series, as shown in Figure 2.41. Furthermore, assume that the communication channel in Figure 2.41 now has a nonlinear bandpass characteristic, and the receive filter, as before, is a linear bandpass filter. For this case, using the theory of the continuous-time version of the Volterra series, presented, for example, in Reference 29, one can show that the relations between the impulse responses $r^{(1)}(\cdot)$, $r^{(2)}(\cdot,\cdot)$, $r^{(3)}(\cdot,\cdot,\cdot)$, and the impulse responses of the nonlinear communication channel and linear receive filter are

$$r^{(1)}(\tau) = \int_{-\infty}^{\infty} r_r(\varphi)r_c^{(1)}(\tau-\varphi)d\varphi \tag{2.98b}$$

$$r^{(2)}(\tau_1, \tau_2) = \int_{-\infty}^{\infty} r_r(\varphi)r_c^{(2)}(\tau_1-\varphi, \tau_2-\varphi)d\varphi \tag{2.98c}$$

$$r^{(3)}(\tau_1, \tau_2, \tau_3) = \int_{-\infty}^{\infty} r_r(\varphi)r_c^{(3)}(\tau_1-\varphi, \tau_2-\varphi, \tau_3-\varphi)d\varphi \tag{2.98d}$$

and so on. In Equations 2.98b, c, and d, $r_r(\cdot)$ is the (linear) impulse response of the (linear) bandpass receive filter, and $r_c^{(1)}(\cdot)$, $r_c^{(1)}(\cdot,\cdot)$, and $r_c^{(3)}(\cdot,\cdot,\cdot)$ are the first, second, and third order, respectively, nonlinear impulse responses of the nonlinear communication channel. Moreover, note that Equation 2.98b corresponds to Equation 2.85b for the linear case. The next equations are the result of nonlinear behavior of the communication channel.

To proceed further, consider now the analytic signal $y_a(t)$ given by Equation 2.83, and note that the real part of this signal can be written as

$$\mathrm{Re}(y_a(t)) = \frac{y_a(t) + y_a^*(t)}{2} = \frac{1}{2}\sum p(t-iT)(\tilde{a}(i)e^{j\omega_c t} + \tilde{a}^*(i)e^{-j\omega_c t}) \tag{2.98e}$$

Applying then Equation 2.98e in Equation 2.98a leads to

$$z_b(t) = \frac{1}{2}\int_{-\infty}^{\infty} r^{(1)}(\tau) \sum_{i=-\infty}^{\infty} p(t-\tau-iT)(\tilde{a}(i)e^{j\omega_c(t-\tau)} + \tilde{a}^*(i)e^{-j\omega_c(t-\tau)})d\tau$$

$$+\frac{1}{4}\int_{-\infty}^{\infty}\int_{-\infty}^{\infty} r^{(2)}(\tau_1,\tau_2) \sum_{i_1=-\infty}^{\infty}\sum_{i_2=-\infty}^{\infty} p(t-\tau_1-i_1T)p(t-\tau_2-i_2T)\cdot(\tilde{a}(i_1)\tilde{a}(i_2)e^{j\omega_c(2t-\tau_1-\tau_2)}$$

$$+\tilde{a}(i_1)\tilde{a}^*(i_2)e^{j\omega_c(\tau_2-\tau_1)} + \tilde{a}^*(i_1)\tilde{a}(i_2)e^{j\omega_c(\tau_1-\tau_2)} + \tilde{a}^*(i_1)\tilde{a}^*(i_2)e^{-j\omega_c(2t-\tau_1-\tau_2)})d\tau_1 d\tau_2$$

$$+\frac{1}{8}\int_{-\infty}^{\infty}\int_{-\infty}^{\infty}\int_{-\infty}^{\infty} r^{(3)}(\tau_1,\tau_2,\tau_3) \sum_{i_1=-\infty}^{\infty}\sum_{i_2=-\infty}^{\infty}\sum_{i_3=-\infty}^{\infty} p(t-\tau_1-i_1T)p(t-\tau_2-i_2T)p(t-\tau_3-i_3T)$$

$$\cdot(\tilde{a}(i_1)\tilde{a}(i_2)\tilde{a}(i_3)e^{j\omega_c(3t-\tau_1-\tau_2-\tau_3)} + \tilde{a}(i_1)\tilde{a}^*(i_2)\tilde{a}(i_3)e^{j\omega_c(t-\tau_1+\tau_2-\tau_3)}$$

$$+\tilde{a}^*(i_1)\tilde{a}(i_2)\tilde{a}(i_3)e^{j\omega_c(t+\tau_1-\tau_2-\tau_3)} + \tilde{a}^*(i_1)\tilde{a}^*(i_2)\tilde{a}(i_3)e^{-j\omega_c(t-\tau_1-\tau_2+\tau_3)}$$

$$+\tilde{a}(i_1)\tilde{a}(i_2)\tilde{a}^*(i_3)e^{j\omega_c(t-\tau_1-\tau_2+\tau_3)} + \tilde{a}(i_1)\tilde{a}^*(i_2)\tilde{a}^*(i_3)e^{-j\omega_c(t+\tau_1-\tau_2-\tau_3)}$$

$$+\tilde{a}^*(i_1)\tilde{a}(i_2)\tilde{a}^*(i_3)e^{-j\omega_c(t-\tau_1+\tau_2-\tau_3)} + \tilde{a}^*(i_1)\tilde{a}^*(i_2)\tilde{a}^*(i_3)e^{-j\omega_c(3t-\tau_1-\tau_2-\tau_3)})d\tau_1 d\tau_2 d\tau_3 + \ldots \quad (2.98f)$$

Furthermore, observe that rearranging the terms in Equation 2.98f, we can rewrite the latter equation in the following form:

$$z_b(t) = \frac{1}{2}\sum_{i=-\infty}^{\infty}\left(e^{j\omega_c t}\tilde{a}(i)\int_{-\infty}^{\infty} r^{(1)}(\tau)p(t-\tau-iT)e^{-j\omega\tau}d\tau\right.$$

$$\left.+e^{-j\omega_c t}\tilde{a}^*(i)\int_{-\infty}^{\infty} r^{(1)}(\tau)p(t-\tau-iT)e^{j\omega_c\tau}d\tau\right)$$

$$+\frac{1}{4}\sum_{i_1=-\infty}^{\infty}\sum_{i_2=-\infty}^{\infty}\left(e^{j2\omega_c t}\tilde{a}(i_2)\tilde{a}(i_2)\int_{-\infty}^{\infty}\int_{-\infty}^{\infty} r^{(2)}(\tau_1,\tau_2)\cdot p(t-\tau_1-i_1T)p(t-\tau_2-i_2T)e^{-j\omega_c\tau_1}e^{-j\omega_c\tau_2}d\tau_1 d\tau_2\right.$$

$$+e^{j0\omega_c t}\tilde{a}(i_1)\tilde{a}^*(i_2)\int_{-\infty}^{\infty}\int_{-\infty}^{\infty} r^{(2)}(\tau_1,\tau_2)\cdot p(t-\tau_1-i_1T)p(t-\tau_2-i_2T)e^{-j\omega_c\tau_1}e^{j\omega_c\tau_2}d\tau_1 d\tau_2$$

$$+e^{j0\omega_c t}\tilde{a}^*(i_1)\tilde{a}(i_2)\int_{-\infty}^{\infty}\int_{-\infty}^{\infty} r^{(2)}(\tau_1,\tau_2)$$

$$\cdot p(t-\tau_1-i_1T)p(t-\tau_2-i_2T)e^{j\omega_c\tau_1}e^{-j\omega_c\tau_2}d\tau_1 d\tau_2$$

$$+e^{-j2\omega_c t}\tilde{a}^*(i_1)\tilde{a}^*(i_2)\int_{-\infty}^{\infty}\int_{-\infty}^{\infty} r^{(2)}(\tau_1,\tau_2)$$

$$\cdot\, p(t-\tau_1-i_1T)p(t-\tau_2-i_2T)e^{j\omega_c\tau_1}e^{j\omega_c\tau_2}d\tau_1 d\tau_2)$$

$$+\frac{1}{8}\sum_{i_1=-\infty}^{\infty}\sum_{i_2=-\infty}^{\infty}\sum_{i_3=-\infty}^{\infty}(e^{j3\omega_c t}\tilde{a}(i_1)\tilde{a}(i_2)\tilde{a}(i_3)$$

$$\cdot\int_{-\infty}^{\infty}\int_{-\infty}^{\infty}\int_{-\infty}^{\infty} r^{(3)}(\tau_1,\tau_2,\tau_3)p(t-\tau_1-i_1T)p(t-\tau_2-i_2T)$$

$$\cdot\, p(t-\tau_3-i_3T)e^{-j\omega_c\tau_1}e^{-j\omega_c\tau_2}e^{-j\omega_c\tau_3}d\tau_1 d\tau_2 d\tau_3$$

$$+e^{j\omega_c t}\tilde{a}(i_1)\tilde{a}^*(i_2)\tilde{a}(i_3)\int_{-\infty}^{\infty}\int_{-\infty}^{\infty}\int_{-\infty}^{\infty} r^{(3)}(\tau_1,\tau_2,\tau_3)p(t-\tau_1-i_1T)p(t-\tau_2-i_2T)$$

$$\cdot\, p(t-\tau_3-i_3T)e^{-j\omega_c\tau_1}e^{j\omega_c\tau_2}e^{-j\omega_c\tau_3}d\tau_1 d\tau_2 d\tau_3$$

$$+e^{j\omega_c t}\tilde{a}^*(i_1)\tilde{a}(i_2)\tilde{a}(i_3)\int_{-\infty}^{\infty}\int_{-\infty}^{\infty}\int_{-\infty}^{\infty} r^{(3)}(\tau_1,\tau_2,\tau_3)p(t-\tau_1-i_1T)p(t-\tau_2-i_2T)$$

$$\cdot\, p(t-\tau_3-i_3T)e^{j\omega_c\tau_1}e^{-j\omega_c\tau_2}e^{-j\omega_c\tau_3}d\tau_1 d\tau_2 d\tau_3$$

$$+e^{j\omega_c t}\tilde{a}(i_1)\tilde{a}(i_2)\tilde{a}^*(i_3)\int_{-\infty}^{\infty}\int_{-\infty}^{\infty}\int_{-\infty}^{\infty} r^{(3)}(\tau_1,\tau_2,\tau_3)p(t-\tau_1-i_1T)p(t-\tau_2-i_2T)$$

$$\cdot\, p(t-\tau_3-i_3T)e^{-j\omega_c\tau_1}e^{-j\omega_c\tau_2}e^{j\omega_c\tau_3}d\tau_1 d\tau_2 d\tau_3$$

$$+e^{-j\omega_c t}\tilde{a}^*(i_1)\tilde{a}^*(i_2)\tilde{a}(i_3)\int_{-\infty}^{\infty}\int_{-\infty}^{\infty}\int_{-\infty}^{\infty} r^{(3)}(\tau_1,\tau_2,\tau_3)p(t-\tau_1-i_1T)p(t-\tau_2-i_2T)$$

$$\cdot\, p(t-\tau_3-i_3T)e^{j\omega_c\tau_1}e^{j\omega_c\tau_2}e^{-j\omega_c\tau_3}d\tau_1 d\tau_2 d\tau_3$$

$$+e^{-j\omega_c t}\tilde{a}(i_1)\tilde{a}^*(i_2)\tilde{a}^*(i_3)\int_{-\infty}^{\infty}\int_{-\infty}^{\infty}\int_{-\infty}^{\infty} r^{(3)}(\tau_1,\tau_2,\tau_3)p(t-\tau_1-i_1T)p(t-\tau_2-i_2T)$$

$$\cdot\, p(t-\tau_3-i_3T)e^{-j\omega_c\tau_1}e^{j\omega_c\tau_2}e^{j\omega_c\tau_3}d\tau_1 d\tau_2 d\tau_3$$

$$+e^{-j\omega_c t}\tilde{a}^*(i_1)\tilde{a}(i_2)\tilde{a}^*(i_3)\int_{-\infty}^{\infty}\int_{-\infty}^{\infty}\int_{-\infty}^{\infty} r^{(3)}(\tau_1,\tau_2,\tau_3)p(t-\tau_1-i_1T)p(t-\tau_2-i_2T)$$

$$\cdot\, p(t-\tau_3-i_3T)e^{j\omega_c\tau_1}e^{-j\omega_c\tau_2}e^{j\omega_c\tau_3}d\tau_1 d\tau_2 d\tau_3$$

$$+ e^{-j3\omega_c t}\tilde{a}^*(i_1)\tilde{a}^*(i_2)\tilde{a}^*(i_3) \int_{-\infty}^{\infty}\int_{-\infty}^{\infty}\int_{-\infty}^{\infty} r^{(3)}(\tau_1, \tau_2, \tau_3)p(t-\tau_1-i_1T)p(t-\tau_2-i_2T)$$

$$\cdot\, p(t-\tau_3-i_3T)e^{j\omega_c\tau_1}e^{j\omega_c\tau_2}e^{j\omega_c\tau_3}d\tau_1 d\tau_2 d\tau_3) + \ldots \tag{2.98g}$$

Let us now extend the notion of the equivalent lowpass complex-valued impulse response $m(t)$ given by Equation 2.85e to the nonlinear case. Then, we shall speak about the nonlinear first-order, second-order, third-order, and so on, equivalent low-pass complex-valued impulse responses, and we shall express these impulse responses as

$$m^{(1)}(\eta) = \int_{-\infty}^{\infty} r^{(1)}(\tau)p(\eta-\tau)e^{-j\omega_c\tau}d\tau \tag{2.98h}$$

$$m^{(2)}(\eta_1, \eta_2) = \int_{-\infty}^{\infty}\int_{-\infty}^{\infty} r^{(2)}(\tau_1, \tau_2)p(\eta_1-\tau_1)p(\eta_2-\tau_2)e^{-j\omega_c\tau_1}e^{-j\omega_c\tau_2}d\tau_1 d\tau_2 \tag{2.98i}$$

$$m^{(3)}(\eta_1, \eta_2, \eta_3) = \int_{-\infty}^{\infty}\int_{-\infty}^{\infty}\int_{-\infty}^{\infty} r^{(3)}(\tau_1, \tau_2, \tau_3)p(\eta_1-\tau_1) \tag{2.98j}$$

$$\cdot\, p(\eta_2-\tau_2)p(\eta_3-\tau_3)e^{-j\omega_c\tau_1}e^{-j\omega_c\tau_2}e^{-j\omega_c\tau_3}d\tau_1 d\tau_2 d\tau_3$$

and so on.

Furthermore, to simplify Equation 2.98g, we shall use the fact that the receive filter in Figure 2.41 is a bandpass filter with the passband about the carrier frequency ω_c. So this filter filters out all those new frequency components in the signal, which eventually appear in the nonlinear communication channel but are outside the receive filter passband. Without going into the mathematical details (for more theory, the interested reader is referred to Reference 70 and references cited there), we observe that the components related with exponentials: $\exp(j2\omega_c t)$, $\exp(j0\omega_c t)$, $\exp(-j2\omega_c t)$, $\exp(j3\omega_c t)$, and $\exp(-j3\omega_c t)$ in Equation 2.98g are those that are outside the filter passband. Hence, we can assume that the above components are equal to zero. Using this fact in Equation 2.98g, and definitions 2.98h, 2.98i, 2.98j, we can rewrite Equation 2.98g in the following form:

$$z_b(t) = e^{j\omega_c t}\left(\frac{1}{2}\sum_{i=-\infty}^{\infty} m^{(1)}(t-iT)\tilde{a}(i) + \frac{1}{8}\sum_{i_1=-\infty}^{\infty}\sum_{i_2=-\infty}^{\infty}\sum_{i_3=-\infty}^{\infty} m_*^{(3)}(t-i_1T, t-i_2T, t-i_3T)\right.$$
$$\left.\cdot(\tilde{a}(i_1)\tilde{a}^*(i_2)\tilde{a}(i_3) + \tilde{a}^*(i_1)\tilde{a}(i_2)\tilde{a}(i_3) + \tilde{a}(i_1)\tilde{a}(i_2)\tilde{a}^*(i_3)) + \ldots\right)$$
$$+ e^{-j\omega_c t}\left(\frac{1}{2}\sum_{i=-\infty}^{\infty} m^{(1)*}(t-iT)\tilde{a}^*(i) + \frac{1}{8}\sum_{i_1=-\infty}^{\infty}\sum_{i_2=-\infty}^{\infty}\sum_{i_3=-\infty}^{\infty} m_{**}^{(3)}(t-i_1T, t-i_2T, t-i_3T)\right.$$
$$\left.\cdot(\tilde{a}^*(i_1)\tilde{a}^*(i_2)\tilde{a}(i_3) + \tilde{a}(i_1)\tilde{a}^*(i_2)\tilde{a}^*(i_3) + \tilde{a}^*(i_1)\tilde{a}(i_2)\tilde{a}^*(i_3)) + \ldots\right) \tag{2.99}$$

Note that, to arrive at the form of $z_b(t)$ given by Equation 2.99, we also assumed in derivation of this equation that the nonlinear impulse responses involved are symmetric. That is, they are symmetrized according to definition 1.15, when applied to the continuous-time nonlinear impulse responses, and where the index "*sym*" is dropped after performing the symmetrization. More precisely, we assumed

$$r^{(3)}(\tau_1, \tau_2, \tau_3)e^{-j\omega_c\tau_1}e^{j\omega_c\tau_2}e^{-j\omega_c\tau_3} \doteq r^{(3)}(\tau_1, \tau_2, \tau_3)e^{j\omega_c\tau_1}e^{-j\omega_c\tau_2}e^{-j\omega_c\tau_3} \tag{2.100}$$
$$\doteq r^{(3)}(\tau_1, \tau_2, \tau_3)e^{-j\omega_c\tau_1}e^{-j\omega_c\tau_2}e^{j\omega_c\tau_3}$$

where the symbol $\doteq$ means "symmetric to." We treated all the other nonlinear impulse responses occurring in Equation 2.99 accordingly. Moreover, we used in Equation 2.99 a special notation for "partly conjugate" versions of expression defined by 2.98j; that is,

$$m_*^{(3)}(\eta_1, \eta_2, \eta_3) = \int_{-\infty}^{\infty}\int_{-\infty}^{\infty}\int_{-\infty}^{\infty} r^{(3)}(\tau_1, \tau_2, \tau_3)p(\eta_1-\tau_1)p(\eta_2-\tau_2)p(\eta_3-\tau_3) \tag{2.101a}$$
$$\cdot e^{j\omega_c\tau_1}e^{-j\omega_c\tau_2}e^{-j\omega_c\tau_3}d\tau_1 d\tau_2 d\tau_3$$

and

$$m_{**}^{(3)}(\eta_1, \eta_2, \eta_3) = \int_{-\infty}^{\infty}\int_{-\infty}^{\infty}\int_{-\infty}^{\infty} r^{(3)}(\tau_1, \tau_2, \tau_3)p(\eta_1-\tau_1)p(\eta_2-\tau_2)p(\eta_3-\tau_3) \tag{2.101b}$$
$$\cdot e^{j\omega_c\tau_1}e^{j\omega_c\tau_2}e^{-j\omega_c\tau_3}d\tau_1 d\tau_2 d\tau_3$$

The third-order impulse response $m_*^{(3)}$ given by Equation 2.101a is named $m^{(3)}$ partly conjugate "with one star," because one of the exponents, $\exp(j\omega_c\tau_1)$ in determining expression 2.101a is the complex-conjugate of $\exp(-j\omega_c\tau_1)$ occurring in $m^{(3)}$ given by Equation 2.98j. Similarly, the third-order impulse response $m_{**}^{(3)}$ given by Equation 2.101b is named $m^{(3)}$ partly conjugate "with two stars" because two of the exponents, $\exp(j\omega_c\tau_1)$ and $\exp(j\omega_c\tau_2)$, are the complex-conjugates of $\exp(-j\omega_c\tau_1)$ and $\exp(-j\omega_c\tau_2)$, respectively, occurring in $m^{(3)}$ given by Equation 2.98j.

At first glance, the nonlinear impulse responses written in Equation 2.100 seem to be not equal to each other. Therefore, a special notation ($\doteq$) for "approaches" rather than the symbol for "equal" is used for denoting that they are symmetric. On the other hand, however, because we assumed that the nonlinear impulse responses in Equation 2.100 are symmetric in the sense of definition 1.15, they must be equal to each other. In what follows below, we shall show that this is the case. To this end, let us begin with the assumption that the Volterra series representation (2.98a) involves the symmetric impulse responses. This means, according to definition 1.15, that we have

$$\begin{aligned} r^{(3)}(\tau_1,\tau_2,\tau_3) &= r^{(3)}(\tau_1,\tau_3,\tau_2) = r^{(3)}(\tau_2,\tau_1,\tau_3) \\ &= r^{(3)}(\tau_2,\tau_3,\tau_1) = r^{(3)}(\tau_3,\tau_1,\tau_2) = r^{(3)}(\tau_3,\tau_2,\tau_1) \end{aligned} \tag{2.102a}$$

where, at each of $r_c^{(3)}(\cdot,\cdot,\cdot)$'s, the subscript "*sym*" is omitted.

Now note that formally interchanging the variables τ_1 with τ_2 (i.e., renaming τ_1 as τ_2, and τ_2 as τ_1) in the first expression in Equation 2.100, and τ_1 with τ_3 (i.e., renaming τ_1 as τ_3, and τ_3 as τ_1) in the third expression in Equation 2.100, we can rewrite Equation 2.100 as

$$\begin{aligned} r^{(3)}(\tau_2,\tau_1,\tau_3)e^{j\omega_c\tau_1}e^{-j\omega_c\tau_2}e^{-j\omega_c\tau_3} &\doteq r^{(3)}(\tau_1,\tau_2,\tau_3)^{e^{j\omega_c\tau_1}e^{-j\omega_c\tau_2}e^{-j\omega_c\tau_3}} \\ &\doteq r^{(3)}(\tau_3,\tau_2,\tau_1)e^{j\omega_c\tau_1}e^{-j\omega_c\tau_2}e^{-j\omega_c\tau_3} \end{aligned} \tag{2.102b}$$

Because of the relation 2.102a, the expressions in Equation 2.102b are equal to each other. That is, the symbol for "approaches" in Equation 2.102b can be replaced by the "equal" symbol. Of course, similar equalities to those in 2.102b can also be written for all the other nonlinear impulse responses of a similar form to that represented by the impulse responses in Equation 2.100. For example, we can write for the impulse responses involved in the expressions standing by the exponential $\exp(-j\omega_c t)$ in Equation 2.98g,

$$\begin{aligned} r^{(3)}(\tau_1,\tau_2,\tau_3)e^{j\omega_c\tau_1}e^{j\omega_c\tau_2}e^{-j\omega_c\tau_3} &\doteq r^{(3)}(\tau_3,\tau_2,\tau_1)^{e^{j\omega_c\tau_1}e^{j\omega_c\tau_2}e^{-j\omega_c\tau_3}} \\ &\doteq r^{(3)}(\tau_1,\tau_3,\tau_2)e^{j\omega_c\tau_1}e^{j\omega_c\tau_2}e^{-j\omega_c\tau_3} \end{aligned} \tag{2.102c}$$

Application of relations such as Equations 2.102b and 2.102c in Equation 2.98g led to the simplification of the resulting form for $z_b(t)$ given by Equation 2.99.

Keeping in mind our purpose to get from Equation 2.99 an analog of the expression 2.85f, we check now whether the $m_{**}^{(3)}(\eta_1,\eta_2,\eta_3)$ given by Equation 2.101b is a complex-conjugate of the $m_*^{(3)}(\eta_1, \eta_2, \eta_3)$ given by Equation 2.101a. To this end, take into account the following nonlinear impulse response:

$$r^{(3)}(\tau_1, \tau_2, \tau_3)e^{-j\omega_c\tau_1}e^{j\omega_c\tau_2}e^{j\omega_c\tau_3} \tag{2.103a}$$

Using equivalences 2.102b and 2.102a, we easily show that the above impulse response can be rewritten as

$$\begin{aligned} r^{(3)}(\tau_1, \tau_2, \tau_3)e^{-j\omega_c\tau_1}e^{j\omega_c\tau_2}e^{j\omega_c\tau_3} &= r^{(3)}(\tau_3, \tau_2, \tau_1)e^{-j\omega_c\tau_3}e^{j\omega_c\tau_2}e^{j\omega_c\tau_1} \\ &= r^{(3)}(\tau_1, \tau_2, \tau_3)e^{j\omega_c\tau_1}e^{j\omega_c\tau_2}e^{-j\omega_c\tau_3} \end{aligned} \tag{2.103b}$$

Furthermore, note that relation

$$(e^{j\omega_c\tau_1}e^{-j\omega_c\tau_2}e^{-j\omega_c\tau_3})^* = e^{-j\omega_c\tau_1}e^{j\omega_c\tau_2}e^{j\omega_c\tau_3} \tag{2.103c}$$

holds. Then, using Equations 2.103b and 2.103c in Equation 2.101b gives

$$\begin{aligned} m_{**}^{(3)}(\eta_1, \eta_2, \eta_3) &= \int_{-\infty}^{\infty}\int_{-\infty}^{\infty}\int_{-\infty}^{\infty} r^{(3)}(\tau_1, \tau_2, \tau_3)p(\eta_1 - \tau_1)p(\eta_2 - \tau_2)p(\eta_3 - \tau_3) \\ &\quad \cdot e^{-j\omega_c\tau_1}e^{j\omega_c\tau_2}e^{j\omega_c\tau_3}d\tau_1 d\tau_2 d\tau_3 \\ &= \int_{-\infty}^{\infty}\int_{-\infty}^{\infty}\int_{-\infty}^{\infty} r^{(3)}(\tau_1, \tau_2, \tau_3)p(\eta_1 - \tau_1)p(\eta_2 - \tau_2)p(\eta_3 - \tau_3) \\ &\quad \cdot (e^{j\omega_c\tau_1}e^{-j\omega_c\tau_2}e^{-j\omega_c\tau_3})^* d\tau_1 d\tau_2 d\tau_3 = m_*^{(3)*}(\eta_1, \eta_2, \eta_3) \end{aligned} \tag{2.103d}$$

In another form, Equation 2.103d can be rewritten as

$$m_*^{(3)}(\eta_1, \eta_2, \eta_3) = m_{**}^{(3)*}(\eta_1, \eta_2, \eta_3) \tag{2.103e}$$

Equation 2.103d gives evidence to the fact that $m_{**}^{(3)}(\eta_1, \eta_2, \eta_3)$ is really a complex-conjugate of $m_*^{(3)}(\eta_1,\eta_2, \eta_3)$. Similar relations can also be found for the corresponding partially conjugate equivalent lowpass complex-valued impulse responses of higher odd orders, like the fifth, seventh, ninth order, and so on.

Using in Equation 2.99 the fact that

$$\begin{aligned} &(\tilde{a}^*(i_1)\tilde{a}^*(i_2)\tilde{a}(i_3) + \tilde{a}(i_1)\tilde{a}^*(i_2)\tilde{a}^*(i_3) + \tilde{a}^*(i_1)\tilde{a}(i_2)\tilde{a}^*(i_3))^* \\ &= \tilde{a}(i_1)\tilde{a}^*(i_2)\tilde{a}(i_3) + \tilde{a}^*(i_1)\tilde{a}(i_2)\tilde{a}(i_3) + \tilde{a}(i_1)\tilde{a}(i_2)\tilde{a}^*(i_3) \end{aligned} \tag{2.104a}$$

and the relation 2.103d, we can rewrite Equation 2.99 as

$$z_b(t) = e^{j\omega_c t}\left(\frac{1}{2}\sum_{i=-\infty}^{\infty} m^{(1)}(t-iT)\tilde{a}(i) + \frac{1}{8}\sum_{i_1=-\infty}^{\infty}\sum_{i_2=-\infty}^{\infty}\sum_{i_3=-\infty}^{\infty} m_*^{(3)}(t-i_1T, t-i_2T, t-i_3T)\right.$$
$$\left.\cdot\left(\tilde{a}(i_1)\tilde{a}^*(i_2)\tilde{a}(i_3) + \tilde{a}^*(i_1)\tilde{a}(i_2)\tilde{a}(i_3) + \tilde{a}(i_1)\tilde{a}(i_2)\tilde{a}^*(i_3)\right) + \ldots\right)$$
$$+ e^{-j\omega_c t}\left(\frac{1}{2}\sum_{i=-\infty}^{\infty} (m^{(1)}(t-iT)\tilde{a}(i))^* + \frac{1}{8}\sum_{i_1=-\infty}^{\infty}\sum_{i_2=-\infty}^{\infty}\sum_{i_3=-\infty}^{\infty} (m_*^{(3)}(t-i_1T, t-i_2T, t-i_3T)\right.$$
$$\left.\cdot\left(\tilde{a}(i_1)\tilde{a}^*(i_2)\tilde{a}(i_3) + \tilde{a}^*(i_1)\tilde{a}(i_2)\tilde{a}(i_3) + \tilde{a}(i_1)\tilde{a}(i_2)\tilde{a}^*(i_3)\right))^* + \ldots\right)$$

(2.104b)

Moreover, it can be shown that $z_b(t)$, being real-valued, has the form of the expression $\exp(j\omega_c t)(\cdot) + \exp(-j\omega_c t)(\cdot)^*$. Using this fact in Equation 2.104b, we can write the latter equation in the following form:

$$z_b(t) = \mathrm{Re}\left(\left(\sum_{i=-\infty}^{\infty} m^{(1)}(t-iT)\tilde{a}(i) + \frac{1}{4}\sum_{i_1=-\infty}^{\infty}\sum_{i_2=-\infty}^{\infty}\sum_{i_3=-\infty}^{\infty} m_*^{(3)}(t-i_1T, t-i_2T, t-i_3T)\right.\right.$$
$$\left.\left.\cdot\left(\tilde{a}(i_1)\tilde{a}^*(i_2)\tilde{a}(i_3) + \tilde{a}^*(i_1)\tilde{a}(i_2)\tilde{a}(i_3) + \tilde{a}(i_1)\tilde{a}(i_2)\tilde{a}^*(i_3)\right) + \ldots\right)e^{j\omega_c t}\right)$$

(2.104c)

From the analysis just presented for the linear communication channel, it follows that the form of the expression 2.104c allows us to write the analytic signal $z_a(t)$, after performing the Hilbert transform filtering in the QAM receiver of Figure 2.40 as

$$z_a(t) = \left(\sum_{i=-\infty}^{\infty} m^{(1)}(t-iT)\tilde{a}(i) + \frac{1}{4}\sum_{i_1=-\infty}^{\infty}\sum_{i_2=-\infty}^{\infty}\sum_{i_3=-\infty}^{\infty} m_*^{(3)}(t-i_1T, t-i_2T, t-i_3T)\right.$$
$$\left.\cdot\left(\tilde{a}(i_1)\tilde{a}^*(i_2)\tilde{a}(i_3) + \tilde{a}^*(i_1)\tilde{a}(i_2)\tilde{a}(i_3) + \tilde{a}(i_1)\tilde{a}(i_2)\tilde{a}^*(i_3)\right) + \ldots\right)e^{j\omega_c t}$$

(2.104d)

The expression 2.104d regards the case in which the communication channel shown in Figure 2.40 is a channel that has nonlinear characteristics.

Similar to the case of the linear echo passband channel discussed previously, we model the nonlinear echo passband channel using the description derived just for the nonlinear communication channel. That is, we use the same form of description of an analytic signal related with the nonlinear echo channel as that given by

Equation 2.104d for the nonlinear communication channel. Hence, denoting the analytic signal related with the nonlinear echo channel by $e_a(t)$, we write for it

$$e_a(t) = \left(\sum_{i=-\infty}^{\infty} g^{(1)}(t-iT)\tilde{a}(i) + \frac{1}{4} \sum_{i_1=-\infty}^{\infty} \sum_{i_2=-\infty}^{\infty} \sum_{i_3=-\infty}^{\infty} g_*^{(3)}(t-i_1T, t-i_2T, t-i_3T) \right.$$
$$\left. \cdot (\tilde{a}(i_1)\tilde{a}^*(i_2)\tilde{a}(i_3) + \tilde{a}^*(i_1)\tilde{a}(i_2)\tilde{a}(i_3) + \tilde{a}(i_1)\tilde{a}(i_2)\tilde{a}^*(i_3)) + \ldots \right) e^{j\omega_c t} \tag{2.105}$$

In Equation 2.105, $g^{(1)}(\cdot)$ and $g_*^{(3)}(\cdot,\cdot,\cdot)$ for the nonlinear echo passband channel correspond to $m^{(1)}(\cdot)$ and $m_*^{(3)}(\cdot,\cdot,\cdot)$, respectively, which were introduced previously. The $g^{(1)}(\cdot)$ and $g_*^{(3)}(\cdot,\cdot,\cdot)$ represent the nonlinear first-order and third-order, respectively, equivalent lowpass complex-valued impulse responses. Additionally, $g_*^{(3)}(\cdot,\cdot,\cdot)$ is partly-conjugate compared to $g^{(3)}(\cdot,\cdot,\cdot)$, similarly as was $m_*^{(3)}(\cdot,\cdot,\cdot)$, compared to $m^{(3)}(\cdot,\cdot,\cdot)$.

For the discrete-time $t = kT$ and with the substitution of new variables, as in Equation 2.87, the expression for $e_a(t)$ given by Equation 2.105 can be rewritten as

$$e_a(k) = \left(\sum_{i=-\infty}^{\infty} g^{(1)}(i)\tilde{a}(k-i) + \frac{1}{4} \sum_{i_1=-\infty}^{\infty} \sum_{i_2=-\infty}^{\infty} \sum_{i_3=-\infty}^{\infty} g_*^{(3)}(i_1, i_2, i_3) \right. \tag{2.106a}$$
$$\cdot (\tilde{a}(k-i_1)\tilde{a}^*(k-i_2)\tilde{a}(k-i_3) + \tilde{a}^*(k-i_1)\tilde{a}(k-i_2)\tilde{a}(k-i_3)$$
$$\left. + \tilde{a}(k-i_1)\tilde{a}(k-i_2)\tilde{a}^*(k-i_3)) + \ldots \right) e^{j\omega_c kT}$$

where $\tilde{a}(k - i)$ stands for

$$\tilde{a}(k-i) = a(k-i) + jb(k-i) \tag{2.106b}$$

and similarly,

$$\tilde{a}(k-i_i) = a(k-i_i) + jb(k-i_i),\ i_i = i_1, i_2, i_3, \ldots \tag{2.106c}$$

Furthermore, let us assume the same kind of approximation of $e_a(k)$ given by Equation 2.106a as that used in Equation 2.89b, that is, by a complex-valued transversal filter representation with the complex-valued coefficients and complex-valued data symbols; however, now being a nonlinear transversal filter description. Using this approximation in Equation 2.106a gives

$$e_a(k) \cong \left(\sum_{i=0}^{M_e-1} g^{(1)}(i)\tilde{a}(k-i) + \frac{1}{4} \sum_{i_1=0}^{M_e-1} \sum_{i_2=0}^{M_e-1} \sum_{i_3=0}^{M_e-1} g_*^{(3)}(i_1, i_2, i_3) \right. \tag{2.107}$$

$$\cdot (\tilde{a}(k-i_1)\tilde{a}^*(k-i_2)\tilde{a}(k-i_3) + \tilde{a}^*(k-i_1)\tilde{a}(k-i_2)\tilde{a}(k-i_3)$$

$$\left. + \tilde{a}(k-i_1)\tilde{a}(k-i_2)\tilde{a}^*(k-i_3)) + \ldots \right) e^{j\omega_c kT}$$

A good choice for cancellation of the echo signal described by Equation 2.107 is the use of a canceller possessing the description in a similar form to that represented by Equation 2.107. If we restrict ourselves to the components, for example, up to the third order in the description of a canceller, we shall have the relation:

$$\hat{e}_a(k) = \left(\sum_{i=0}^{M_{\hat{e}}-1} c(i)\tilde{a}(k-i) + \frac{1}{4} \sum_{i_1=0}^{M_{\hat{e}}-1} \sum_{i_2=0}^{M_{\hat{e}}-1} \sum_{i_3=0}^{M_{\hat{e}}-1} c(i_1, i_2, i_3) \right. \tag{2.108}$$

$$\cdot \tilde{a}(k-i_1)\tilde{a}^*(k-i_2)\tilde{a}(k-i_3) + \tilde{a}^*(k-i_1)\tilde{a}(k-i_2)\tilde{a}(k-i_3)$$

$$\left. + \tilde{a}(k-i_1)\tilde{a}(k-i_2)\tilde{a}^*(k-i_3) \right) e^{j\omega_c kT}$$

for this nonlinear passband canceller. In Equation 2.108, $c(i)$, $i = 0, 1, \ldots, M_{\hat{e}} - 1$, correspond to $g^{(1)}(i)$, $i = 0, 1, \ldots, M_{\hat{e}} - 1$, and $c(i_1, i_2, i_3)$, $i_1, i_2, i_3 = 0, 1, \ldots, M_{\hat{e}} - 1$, correspond to $g_*^{(3)}(i_1, i_2, i_3)$, $i_1, i_2, i_3 = 0, 1, \ldots, M_e - 1$, respectively. Furthermore, when $M_e = M_{\hat{e}}$, the number of elements $c(i)$ will be identical with that of $g^{(1)}(i)$; and similarly, the numbers of elements $c(i_1, i_2, i_3)$ and $g_*^{(3)}(i_1, i_2, i_3)$ will be identical as well. In the case of $M_e \neq M_{\hat{e}}$, some elements in one set will not be represented by the corresponding elements in the second set.

A general scheme of nonlinear echo cancellation with the use of a nonlinear complex-valued transversal filter, having the description in the form of $\hat{e}_a(k)$ given by Equation 2.108, is presented in Figure 2.50.

Comparing the structure of Figure 2.50 with that of Figure 2.45a, we see that the only difference between them lies in the type of transversal filter applied. In Figure 2.45a, we have a linear complex-valued transversal filter, in contrast to the structure of Figure 2.50, where a nonlinear complex-valued transversal filter is applied. Furthermore, comparing the structure of Figure 2.50 with the corresponding description given by Equation 2.108, observe that the nonlinear transversal filter output signal $\hat{e}(k)$ is given by

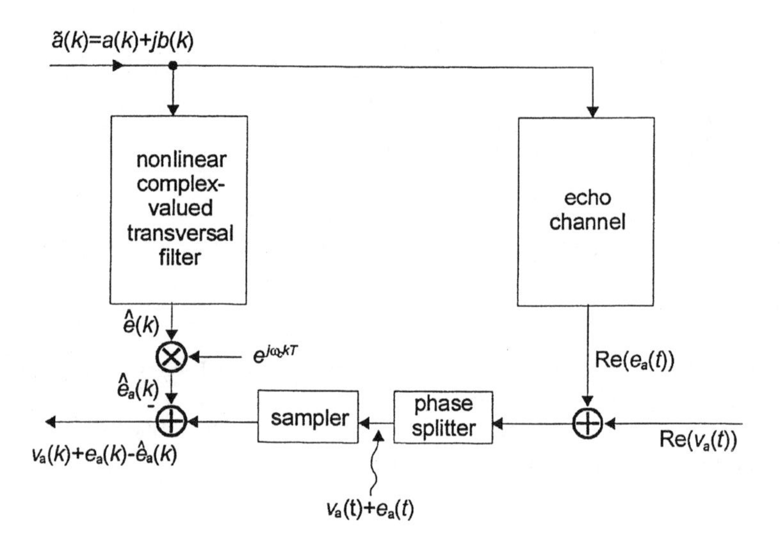

FIGURE 2.50 Cancelling the voiceband nonlinear channel echo with the use of a nonlinear complex-valued transversal filter.

$$\hat{e}(k) = \hat{e}_a(k)e^{-j\omega_c kT} = \sum_{i=0}^{M_{\hat{e}}-1} c(i)\tilde{a}(k-i) + \frac{1}{4}\sum_{i_1=0}^{M_{\hat{e}}-1}\sum_{i_2=0}^{M_{\hat{e}}-1}\sum_{i_3=0}^{M_{\hat{e}}-1} c(i_1, i_2, i_3) \tag{2.109}$$

$$\cdot (\tilde{a}(k-i_1)\tilde{a}^*(k-i_2)\tilde{a}(k-i_3) + \tilde{a}^*(k-i_1)\tilde{a}(k-i_2)\tilde{a}(k-i_3)$$

$$+ \tilde{a}(k-i_1)\tilde{a}(k-i_2)\tilde{a}^*(k-i_3))$$

Note that the form of the expression in Equation 2.109 resembles more the general form of the discrete-time truncated Volterra series than the form of the Volterra series for binary signals used previously in the discussion of canceller structures for baseband applications. In fact, expression 2.109 represents nothing other than a truncated Volterra series that works on the complex-valued data symbols and has complex-valued impulse responses. We also have to work here with multiplications of the data symbols by their complex-conjugates (in the second component of Equation 2.109).

To show that the coefficients (samples of the impulse responses) $c(i)$ and $c(i_1, i_2, i_3)$ change in the adaptation process, we rewrite Equation 2.109 in another form:

$$\hat{e}(k) = \sum_{i=0}^{M_{\hat{e}}-1} c_i(k)\tilde{a}(k-i) + \frac{1}{4}\sum_{i_1=0}^{M_{\hat{e}}-1}\sum_{i_2=0}^{M_{\hat{e}}-1}\sum_{i_3=0}^{M_{\hat{e}}-1} c_{i_1, i_2, i_3}(k) \tag{2.110}$$

$$\cdot (\tilde{a}(k-i_1)\tilde{a}^*(k-i_2)\tilde{a}(k-i_3) + \tilde{a}^*(k-i_1)\tilde{a}(k-i_2)\tilde{a}(k-i_3)$$

$$+ \tilde{a}(k-i_1)\tilde{a}(k-i_2)\tilde{a}^*(k-i_3))$$

Note that, in Equation 2.110 the notational convention of Equation 2.1c is used. That is, the arguments i, and i_1, i_2, i_3 at $c(i)$ and $c(i_1, i_2, i_3)$, respectively, are "shifted" to form the corresponding subscripts: as an argument occurs the discrete time k, which changes in an adaptation process.

Keeping in mind the form of $\hat{e}(k)$ given by Equation 2.110, we redraw the corresponding fragment of Figure 2.50 to underline the fact that the nonlinear complex-valued transversal filter must be an adaptive transversal filter in echo cancellation applications. This is shown in Figure 2.51.

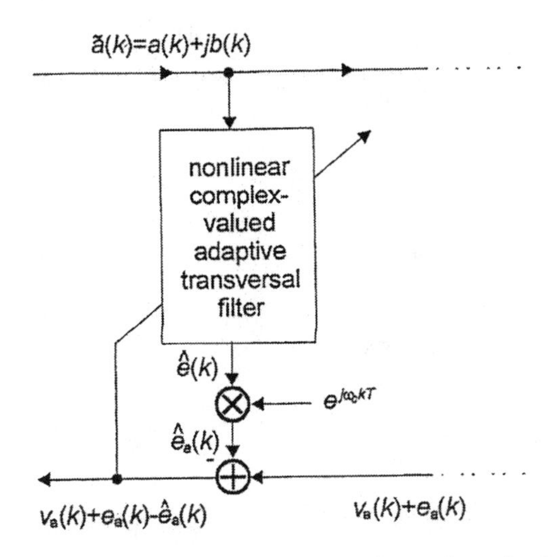

FIGURE 2.51 Adaptive cancelling the voiceband nonlinear channel echo.

To give some view into the internal structure of the nonlinear complex-valued adaptive transversal filter of Figure 2.51, consider the following simple example.

Example 2.8

Assume for simplicity of presentation that $M_{\hat{e}} = 2$. Then, to realize the linear part of Equation 2.110, we need two taps, and, to realize the strictly nonlinear part of Equation 2.110, we need $2^3 = 8$ taps. For this choice, the structure of a nonlinear complex-valued adaptive transversal filter is shown in Figure 2.52.

Looking at the structure of Figure 2.52, we see that it presents "a true cobweb" even, as here, for such a small $M_{\hat{e}} = 2$ (in practice, of course, $M_{\hat{e}}$ is much greater than 2), and when taking into account only nonlinearities up to the third order. This poses true difficulty in practical implementation of a nonlinear complex-valued transversal filter.

The structure of the nonlinear echo canceller of Figure 2.50 can be also presented in another equivalent form using an equivalent nonlinear complex-valued passband transversal filter, in analogy to the linear one shown in Figure 2.47. To get such a structure, one needs to repeat in Equation 2.108 the steps performed successively in Equations 2.93a to e. As a result, one obtains a similar structure to that presented

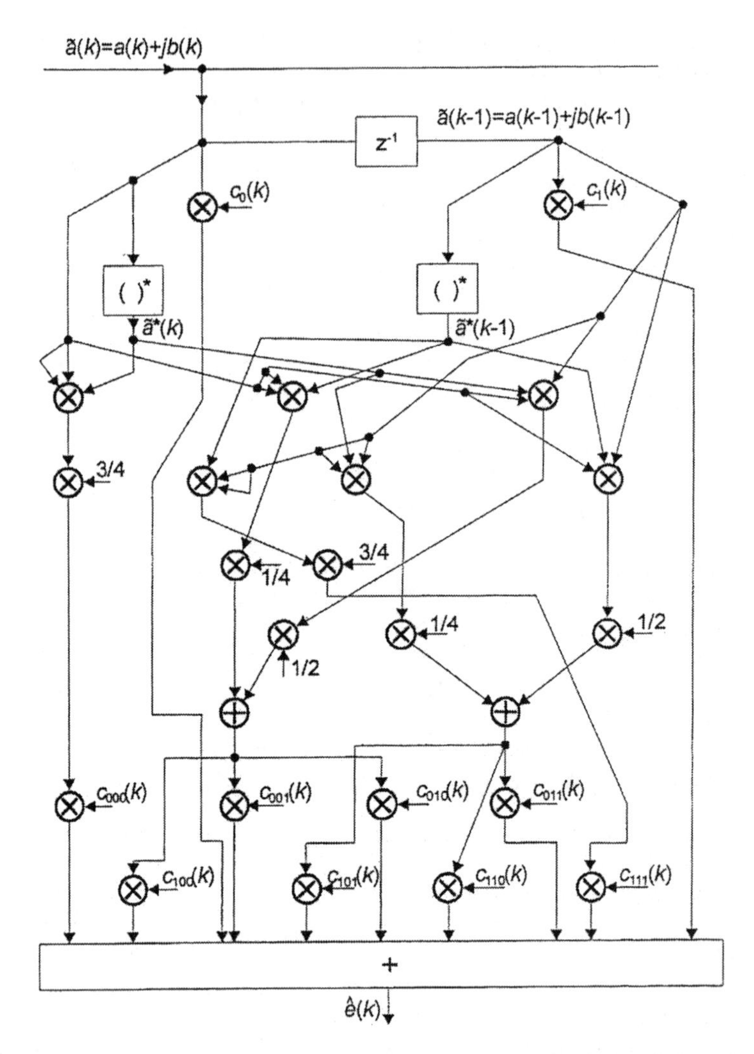

FIGURE 2.52 Nonlinear complex-valued adaptive transversal filter with memory length $M_{\hat{e}} = 2$.

in Figure 2.47, with the rotated data symbols at the input and an equivalent nonlinear complex-valued passband transversal filter in place of the linear one.

Structures for nonlinear interleaved complex-valued transversal filters can be obtained easily applying the argument used before: of approximating the echo replica separately in each of the phases, here with the use of Equation 2.108. Then we get structures similar to those presented in Figures 2.48 and 2.49, with nonlinear transversal filters in place of linear ones.

References

1. Boyd, S. and Chua, L. O., Fading memory and the problem of approximating nonlinear operators with Volterra series, *IEEE Trans. Circuits and Syst.*, CAS-32, 1150-1161, Nov. 1985.
2. Sandberg, I. W., Approximation theorems for discrete-time systems, *IEEE Trans. Circuits and Syst.*, CAS-38, 564-566, May 1991.
3. Sandberg, I. W., Uniform approximation with doubly finite Volterra series, *IEEE Trans. Signal Process.*, 40, 1438-1441, June 1992.
4. Park, J. and Sandberg, I. W., Criteria for the approximation of nonlinear systems, *IEEE Trans. on Circuits and Syst.*, I: *Fundamental Theory and Applications*, CAS-39, 673-676, August 1992.
5. Rugh, W. J., *Nonlinear System Theory: The Volterra/Wiener Approach*, Johns Hopkins University Press, Baltimore, 1981.
6. Alper P., A consideration of the discrete Volterra series, *IEEE Trans. Autom. Control*, AC-10, 322-327, July 1965.
7. Barker, H. A. and Ambati, S., Nonlinear sampled-data system analysis by multidimensional Z transforms, *Proc. IEE*, 119, 1407-1413, Sept. 1972.
8. Jagan, N. C. and Reddy, D.C., Evaluation of response of nonlinear discrete systems using multidimensional Z transforms, *Proc. IEE*, 119, 1521-1525, Oct. 1972.
9. Bronstein, I. N. and Semendjajew, K. A., *Taschenbuch der Mathematik*, Teubner, Leipzig, 1962.
10. Fliege, N. J., *Systemtheorie*, Teubner, Stuttgart, 1991.
11. Sandberg, I. W., Discrete-time p-powers and stability, *Circuits, Systems, and Signal Processing*, 9, 435–448, 1990.
12. Saleh, A. A. M., Matrix analysis of mildly nonlinear, multiple-input, multiple-output systems with memory, *Bell Syst. Tech.* J., 61, 2221-2243, Nov. 1982.
13. Rudin, W., *Functional Analysis*, McGraw-Hill, New York, 1974.
14. Kantorovich, L. V. and Akilov G. P., *Functional Analysis*, Pergamon, Oxford, 1982.
15. Matthews, M. B., Approximating nonlinear fading-memory operators using neural network models, *Circuits, Syst., Signal Process.*, 12, 279-307, 1993.
16. Borys A., On fading memory and asymptotic properties of steady state solutions for digital systems, *Int. J. Circuit Theor. Appl.*, 24, 593-596, 1996.
17. Sandberg, I. W., The mathematical foundations of associated expansions for mildly nonlinear systems, *IEEE Trans. Circuits Syst.*, CAS-30, 441-455, July 1983.
18. Sandberg, I. W., Nonlinear input-output maps and approximate representations, *AT&T Tech. J.*, 64, 1967-1983, Oct. 1985.
19. Sandberg, I. W., Approximately-finite memory and the theory of representations, Archiv für Elektronik und Übertragungstechnik, 46, 191-199, July 1992.
20. Sandberg, I. W., Approximately-finite memory and input-output maps, I: *Fundamental Theory and Applications*, *IEEE Trans. Circuits* Syst., CAS-39, 549-556, July 1992.
21. Dieudonne, J., *Foundations of Modern Analysis*, Academic Press, New York, 1969.
22. Jenkins, G. M. and Watts, D. G., *Spectral Analysis and Its Applications*, Holden-Day, Oakland, 1968.
23. Borys, A., Rupprecht, W., and Trick, U., Influence of nonlinearities on echo cancellation in two-wire full-duplex data transmission, *ntzArchiv*, 8, 185-190, Aug. 1986.

24. Lee, E. A. and Messerschmitt, D. G., *Digital Communication*, 2nd ed., Kluwer Academic Publishers, Boston/Dordrecht, 1994.
25. Volterra, V., Sopra le Funzioni che Dipendono da altre Funzioni, Nota 1, *Rend. Lincei Ser.*, 4, 97-105, 1887.
26. Volterra, V., *The Theory of Functionals and of Integral and Integro-differential Equations*, Dover, New York, 1959.
27. Schetzen, M., Nonlinear system modeling based on the Wiener theory, *Proc. IEEE*, 69, 1557-1573, Dec. 1981.
28. Chua, L. O. and Ng, C. Y., Frequency domain analysis of nonlinear systems: general theory, *IEE J. Electron. Circuits Syst.*, 3, 165-185, July 1979.
29. Kuo, Y. L., Frequency-domain analysis of weakly nonlinear networks, "canned" Volterra analysis, parts 1 and 2, *Circuits and Syst.*, 11, 2-8, Aug. 1977 (Part 1), and 2-6, Oct. 1977 (Part 2).
30. Bedrosian, E. and Rice, S. O., The output properties of Volterra systems (nonlinear systems with memory) driven by harmonic and Gaussian inputs, *Proc. IEEE*, 59, 1688-1707, Dec. 1971.
31. Boyd, S., Chua, L. O., and Desoer, C. A., Analytical foundations of Volterra series, *IMA* J. *Math. Control and Inf.*, 1, 243-282, 1984.
32. Sandberg, I. W., A perspective on system theory, *IEEE Trans. Circuits Syst.*, CAS-31, 88-103, Jan. 1984.
33. Sandberg, I. W., Uniform approximation and the circle criterion, *IEEE Trans. Autom. Control*, 38, 1450-1458, Oct. 1993.
34. Chua, L. O., *Introduction to Nonlinear Circuit Theory*, McGraw-Hill, New York, 1969.
35. Matthews, M. B. and Moschytz, G. S., The identification of nonlinear discrete-time fading-memory systems using neural network models, *II: Analog and Digital Signal Processing, IEEE Trans. Circuits Syst.*, CAS-41, 740-751, Nov. 1994.
36. Powell, M. J. D., Radial basis functions for multi-variable interpolation: a review, IMA Conference on Algorithms for the Approximation of Functions and Data, RMCS, Shrivenham, UK, 1985.
37. Park, J. and Sandberg, I. W., Universal approximation using radial-basis-function networks, *Neural Computation*, 3, 246-257, 1991.
38. Rudin, W., *Real and Complex Analysis*, 2nd ed., McGraw-Hill, New York, 1974.
39. Trick, U., Ebenig, H., and Rupprecht, W., Kombinierte analog/digitale Echolöschung unter Berücksichtigung von Nichtlinearitäten, *ntzArchiv*, 10, 59-68, March 1988.
40. Messerschmitt, D. G., Echo cancellation in speech and data transmission, *IEEE J. Selected Areas Commun.*, SAC-2, 283-297, March 1984.
41. Bingham, J. A. C., *The Theory and Practice of Modem Design*, John Wiley & Sons, New York, 1988.
42. Stapleton, J. C. and Bass, S. C., Adaptive noise cancellation for a class of nonlinear, dynamic reference channels, *IEEE Trans. Circuits Syst.*, CAS-32, 143-150, Feb. 1985.
43. Gritton, C. W. K. and Lin, D. W., Echo cancellation algorithms, *IEEE ASSP Mag.*, 30-37, April 1984.
44. Thomas, E. J., An adaptive echo canceller in a nonideal environment (nonlinear or time variant), *Bell Syst. Tech. J.*, 50, 2779-2794, Oct. 1971.
45. Ulrich, F., Ein ISDN – Echokompensator für Übertragungstechnik und Ortung von Leitungsfehlern, Mikroelektronik für die Informationstechnik – Vermittlung, Übertragung und Verarbeitung, Vorträge der NTG – Fachtagung, NTG, Fachberichte 96, 25-30, 1986.

46. Verhoeckx, N. A. M., van den Elzen, H. C., Snijders, F. A. M., and van Gerwen, P. J., Digital echo cancellation for baseband data transmission, *IEEE Trans. Acoust., Speech, Signal Process.*, ASSP-27, 768-781, Dec. 1979.
47. Kanbach, A. and Körber, A., *ISDN – Die Technik*, Hüthig Buch Verlag, Heidelberg, 1991.
48. Hauk, W., Entwurf und Realisierung der Übertragungseinheit für den ISDN – Basisanschluss, Mikroelektronik für die Informationstechnik – Vermittlung, Übertragung und Verarbeitung, Vorträge der NTG – Fachtagung, NTG, Fachberichte 96, 14-24, 1986.
49. Trick, U. and Rupprecht, W., Einfluß des Leitungscodes auf die Echokompensation bei Vollduplex - Basisband – Datenübertragung, *ntzArchiv*, 9, 219-229, Sept. 1987.
50. Mueller, K.H., Combining echo cancellation and decision feedback equalization, *Bell Syst. Tech. J.*, 58, 491-500, Feb. 1979.
51. Holte, N. and Stueflotten, S., A new digital echo canceller for two-wire subscriber lines, *IEEE Trans. Communi.*, COM-29, 1573-1581, Nov. 1981.
52. Falconer, D. D., Adaptive reference echo cancellation, *IEEE Trans. on Communi.*, Vol. COM-30, pp. 2083-2094, Sept. 1982.
53. Agazzi, O., Hodges, D. A., and Messerschmitt, D. G., Large-scale integration of hybrid-method digital subscriber loops, *IEEE Trans. Communi.*, COM-30, 2095-2108, Sept. 1982.
54. Agazzi, O., Messerschmitt, D. G., and Hodges, D. A., Nonlinear echo cancellation of data signals, *IEEE Trans. Communi.*, COM-30, 2421-2433, Nov. 1982.
55. Smith, M. J., Cowan, C. F. N., and Adams, P. F., Nonlinear echo cancellers based on transpose distributed arithmetic, *IEEE Trans. Circuits Syst.*, CAS-35, 6-18, Jan. 1988.
56. Borys, A., Some considerations on nonlinear echo cancellation and discrete Volterra series for binary signals, *ntzArchiv*, 11, 73-75, Feb. 1989.
57. Naylor, J. R., Testing digital/analog and analog/digital converters, *IEEE Trans. Circuits Syst.*, CAS-25, 526-538, July 1978.
58. Fatouhi, B., and Hodges, D. A., High-resolution A/D conversion in MOS/LSI, *IEEE J. Solid-State Circuits*, SC-14, 920-925, Dec. 1979.
59. Freeman, D. M., Slewing distortion in digital-to-analog conversion, *J. Audio Eng. Soc.*, 4, 178-183, April 1977.
60. Bossche Vanden M., Schoukens, J., and Renneboog, J., Dynamic testing and diagnostics of A/D converters, *IEEE Trans. Circuits Syst.*, CAS-33, 775-785, Aug. 1986.
61. Thomas, E. J., Some considerations on the application of the Volterra representation of nonlinear networks to adaptive echo cancellers, *Bell Syst. Tech. J.*, 50, 2797-2805, Oct. 1971.
62. Coker, M. J. and Simkins, D. N., A nonlinear adaptive noise canceller, Proceedings of the Conference: Int. Conf. Acoust., Speech, Signal Proc., 470-473, April 1980.
63. Peled, A. and Liu, B., A new hardware realization of digital filters, *IEEE Trans. Acoust., Speech, and Signal Proc.*, ASSP-22, 456-462, Dec. 1974.
64. Burrus, C. S., Digital filter structures described by distributed arithmetic, *IEEE Trans. Circuits Syst.*, CAS-24, 674-680, Dec. 1977.
65. Weruaga-Prieto, L. and Figueiras-Vidal, A. R., Nonlinear echo cancelling using look-up tables and Volterra systems, *IEE Proc. Visual Image Signal Proc.*, 141, 357-364, Dec. 1994.
66. Lin, J.-N. and Unbehauen, R., Adaptive nonlinear digital filter with canonical piece-wise–linear structure, *IEEE Trans.* Circuits *Syst.*, CAS-37, 347-353, March 1990.

67. Borys, A., Rupprecht, W., and Trick, U., Nonlinear echo cancellation and multi-input discrete Volterra series, Proc. Third Euro. Signal Process. Conf., 1129-1132, Sept. 1986.
68. Papoulis, A., *Probability, Random Variables, and Stochastic Processes*, McGraw-Hill, New York, 1991.
69. Proakis, J. G., *Digital Communications*, 3rd ed., McGraw-Hill, New York, 1995.
70. Benedetto, S., Optimization and performance evaluation of digital satellite transmission systems, Proc. Sixth Summer Symp. Circuit Theor., 185-205, July 1982.

Index

F

G

H

I

S

W

Z